A UNITED KINGDOM?

A UNITED KINGDOM?

Economic, Social and Political Geographies

Edited by
JOHN MOHAN
Reader in Geography, University of Portsmouth

HODDER EDUCATION
PART OF HACHETTE LIVRE UK

First published in Great Britain in 1999
This impression reprinted in 2004 by
Hodder Education, part of Hachette Livre UK,
338 Euston Road, London NW1 3BH

http://www.hoddereducation.co.uk

British Library Cataloguing in Publication Data
A catalogue entry for this book is available from the British Library

Library of Congress Cataloging-in-Publication Data
A catalog record for this book is available from the Library of Congress

ISBN 978 0 340 67752 0

Typeset in 9.5pt Palatino by J&L Composition Ltd, Filey, North Yorkshire

If you have any comments to make about this, or any of our
titles, please send them to educationenquiries@hodder.co.uk

CONTENTS

FIGURES

TABLES

PREFACE

It is barely fifteen years since Stamp and Beaver *The British Isles: A geographic and economic survey* finally went out of print, half a century and several editions after it was first published. Developments in human geography and British politics have largely rendered redundant the kind of approach taken by Stamp and Beaver, as I argue in chapter 1, but I wrote this book because of a dissatisfaction with several more recent accounts of the UK's human geography which had their own limitations. We have had some none-too-subtle cartographies of distress which have said a great deal about spatial inequality at various scales, but have been rather less satisfactory in analytical terms. *A United Kingdom?* aims to steer a path between the detailed representations of Dorling's *A new social atlas of Britain*, and the macro-scale arguments of works such as Hutton's *The state we're in*. Rather than becoming fixated with patterns, I wanted to focus on the processes reshaping the human geographies of the UK, and on the lively academic debates surrounding the interpretation of these processes.

There are several consequences of this approach. First, the book does not offer the kind of detailed statistical portrait available elsewhere, for example in Dorling (1995) or Champion *et al.* (1996). Given the ready availability of statistics over the Web, such mappings would almost certainly be out of date before the book came off the press. Second, there is less detailed discussion of nuts-and-bolts issues (e.g. chronologies of legislation) so as to make space for consideration of underlying principles. There is also a smattering of millennial futurology; the temptation was more difficult to resist as the year 2000 approached. Third, constraints on length necessitated hard decisions on content. Readers may therefore detect several sins: omissions certainly; eclecticism probably; maybe apostasy. These may be inevitable consequences of attempts to synthesise, in an integrated way, a large volume of material, but I hope the book adequately represents the vigour of contemporary debates on what I believe to be the most important processes reshaping the human geographies of the UK.

Although the bibliography indicates my intellectual debts to many authors, I would like to thank certain individuals more formally, since without them this book would not have been produced. I would like to begin by gratefully acknowledging the prodigious speed and accuracy with which Margaret Fairhead typed this book and then edited what survived from the first draft. I am also grateful to Carol Derrick and Sharon Jakobek for assistance with the typing, and to Bill Johnson for meeting unreasonable deadlines for illustrations.

My academic colleagues at Portsmouth have contributed directly and indirectly to the development of this book. My thanks, then, to John Bradbeer and Carol Ekinsmyth, for comments on draft chapters, and to Rob Atkinson, Kelvyn Jones, Giles Mohan, Pam Shurmer-Smith, Liz Twigg and Mike Taylor for helpful suggestions.

Outside Portsmouth, Roger Lee and Steven Pinch generously read several chapters and I am grateful for their valuable comments. Others who kindly sent copies of recent offprints or pre-publication material include: Tony Champion, Danny Dorling, Nick Fyfe, Alex Gibson, Brian Graham, Chris

Hamnett, John Lovering, Jim McCormick, John McKendrick, Ron Martin, Colin Mason, Jamie Peck, Peter Phillimore, Deborah Phillips, Robert Rogerson, Martyn Senior, Humphrey Southall and Ivan Turok. If I have left anybody out, I apologise; if I had incorporated everything they sent me this would have been an even longer book, and I hope that none of the above objects if I have not done justice to their work.

Some of this book was test-driven on students at Portsmouth and Cambridge Universities. I would like to thank the students in both places for their tolerance; I am also grateful to Ron Martin and St. Catharine's College for their hospitality.

Once again my family – Ellie, Jennifer and Clare – contributed enormously, by keeping out of the way, strategically obstructing it, and (in Ellie's case) copy-editing the manuscript. They deserve far more than the prospect of another book after this.

John Mohan

ACKNOWLEDGEMENTS

The author and publishers would like to thank the following for permission to use copyright material in this book:

Macmillan Press Ltd for Table 2.1, from A. Gamble (1994) *Britain in Decline*; the Royal Scottish Geographical Society for Figure 2.1; Danny Dorling for Figures 1.1 and 8.1; Humphrey Southall and the Great Britain Historical GIS, Queen Mary and Westfield College, for Figures 1.2 and 1.3; *Regional Studies* and Carfax Publishing for Figures 5.2, 5.4 and 12.2 and Tables 4.5 and 8.2; *Oxford Review of Education* and Carfax Publishing for Figure 9.2; Kluwer Academic Publishers for Figure 10.2; Pion Limited, London, for Figure 9.1 and Table 10.1; Cambridge University Press for Table 8.1; the editor, *Sociological Research Online* for Table 9.3; the editor, *New Economy*, for Table 4.4; *Parliamentary Affairs* and Oxford University Press for Figure 10.1; the Child Poverty Action Group for Figure 7.2; the Policy Studies Institute for Table 6.2; Social and Community Planning Research for Table 9.1; *Town Planning Review* for Figure 13.1.

Figures 4.1, 5.1 and 5.3 and Table 4.2 draw on data provided by the National On-Line Manpower Information System (NOMIS), University of Durham; various employment statistics quoted in chapters 4–6 also draw on this source.

CARTOGRAPHIES AND GEOGRAPHIES

1.1 INTRODUCTION

It is a salutary experience to begin by looking back at several texts widely used in courses on the UK's human geography as recently as the early 1970s. Dury (1968) and Stamp and Beaver (1971) bore the imprint of geography's determinist legacy in their inventories of physical resources and descriptions of the physical characteristics of places. Particularly noteworthy was their coverage of primary production and industrial patterns within a conventional framework of location theory. House (1973) and Chisholm and Manners (1971) took a rather different approach. Reflecting the spirit of the times, they displayed considerable faith in the capacity of various forms of planning. House (1973), for example, spoke optimistically of the scope for 'more comprehensive regional planning'. He concluded with the statement that 'the necessary further management of the UK space ... will not be feasible without ... greater and more decisive public intervention to channel market forces in the national interest' (p.359). If the quotation ceased at 'intervention' we might regard this as 'old Labour' thinking, but the emphasis on working with the grain of market forces has a very contemporary ring.

These works were products of their times and should probably not be read through twenty-first-century lenses. Stamp and Beaver was in its seventh edition, having

originally been published in the mid-1930s. Nevertheless several important lacunae should be noted. Almost totally absent was a consideration of political geographies: even that subdiscipline's traditional concerns (electoral geography; subnational government) hardly feature. Nor is there explicit consideration of the role of the state except in a rather technocratic way rather than as a political entity. Social polarisation and poverty are hardly mentioned, notwithstanding the rediscovery of poverty in other disciplines during the early 1960s. Other planes of social division do not figure. There are some references to the uneven landscapes of welfare provision albeit at a regional scale (Coates and Rawstron, 1971). Apart from discussion of the imminent accession to the EC (in House, 1973) there is little sense of the UK's place in a wider scheme of things while, despite concern about limits to growth, discussion of environmental issues receives little attention beyond resource inventories and consideration of how to maximise the supply of them.

These silences and exclusions reflect a discipline which was only beginning to engage with social issues and break out of its regionalist and determinist legacy. They also reflect what, with hindsight, appears as an era of steady growth, social harmony and rising prosperity, and political consensus. Neither of these sets of circumstances still apply; nor is there much industry to write about. Perhaps three contingent influences precipitated a dramatic change in the character of work on the UK's human geography. First, there was the rising tide of radical ideas, drawn primarily from Marxist economic and political theory. Second, recessionary conditions in the early 1970s posed a challenge to conventional accounts of regional development. It became clear that explanations of the impacts of restructuring needed to go beyond narrow 'location factors' to consider the role places played in spatial divisions of labour. Third, the 1979 election result prompted a re-evaluation of the role of the state and, for some, symbolised a decisive break from an era of 'one nation' politics and its replacement by a divisive politics of inequality, with profound geographical consequences. Consequently, the widening of regional divisions in the UK during the 1980s, popularised in the notion of the 'North–South divide', was followed by a rash of texts on the human geography of the UK (Dickens, 1988; Ball et al. 1989; Hudson and Williams, 1995; Lewis and Townsend, 1989; Mohan, 1989; Balchin, 1990; Champion and Townsend, 1990; Cloke, 1992).

It is, of course, undeniable that the Conservative governments not only rewrote history but also reshaped geography, selectively reconstructing the map of regional inequality as part of a spatially selective 'two nations' project (see chapter 3). It was therefore an essential political and academic task that new academic geographies were developed to engage with the geopolitical bases and geographical consequences of the Thatcher administrations. But ultimately several stones were left unturned. Put crudely, some could be described as 'cartographies of distress' or as the 'geographical consequences of Mrs Thatcher'. There were several demonstrations of the extent and character of regional inequality in the UK, often through rather broad-brush regional contrasts on a range of socio-economic indicators, with some disaggregation to the subregional level. There was a tendency to treat topics (deindustrialisation, reindustrialisation, housing, education, electoral geography, etc.) as discrete entities, rather than demonstrating the links between them. Such a cartography also requires a discussion of why, precisely, geography matters to the operation of society: of course there are

variations in economic activity and levels of human welfare – there always have been – but are they anything other than cartographic curiosities? If not, then the authoritative evidence of inquiries such as the Rowntree Report (1995) has demonstrated the facts of socio-spatial inequality so convincingly that it is debatable what more can be contributed by geography.

Analysing spatial inequality clearly requires more than mapping a 'two nations' project. What this book attempts to do is shift attention from spatial patterns *per se* to analysing academic and political debates about these patterns. For example chapters 4 and 5 focus on how contemporary trends in the organisation of production are to be understood while chapter 6 addresses debates about flexibility in the labour market. Chapter 7 moves away from simply mapping incomes to considering the institutional influences on the distribution of money and finance. Other chapters are organised in similar ways.

The book also addresses the fact that regional and inter-regional divisions have not been abolished but have rather emerged in new forms. Questions of spatial separation and segregation have received increasing attention; for example, social policy is increasingly about the management of a spatially concentrated and marginalised segment of the population, which some would pejoratively describe as an 'underclass'. The relationships between diverse social groups in an increasingly fragmented socio-economic landscape raise the question of whether support for a comprehensive welfare state can be maintained. Questions of territorial management have been thrown into sharp relief by the referendum campaigns in Scotland and Wales, the very different results of which reveal much about the differing bases of nationalist movements in the two countries (chapters 10 and 11). The pressures of globalisation have necessitated a reappraisal of the state's conventional role in regional policy which appears set to re-emphasise the institutional capacities and resources of particular places (chapters 3 and 12). These are all grounds for saying that there is a need for a fresh discussion of the

UK's human geography. To frame the chapters which follow, then, there is first a discussion of the ways in which regional inequalities have been mapped. A consideration of myths of regional inequality indicates why these matters are of more than just cartographic significance, and a section on 'realities' attempts to put regional inequality into a comparative context. Discussion then moves to the *geographies* – the academic accounts of and explanations for the UK's human geography, focusing principally on the changing economic landscape.

1.2 CARTOGRAPHIES: REPRESENTING REGIONAL INEQUALITY

Myths . . .

Regional inequalities have in a sense entered the realms of popular and literary mythology, in the form of stereotypical accounts of the character of places, and of their residents. Jewell (1994) demonstrates the longevity of such myths, showing that stereotyping of 'Northerners' as 'ferocious, obstinate and unyielding' is evident in the twelfth century. For this period there is much more evidence of what 'Southerners' thought than there is for their northern counterparts; Southerners, 'from the more confident, dominant culture, were quick to categorise what they saw' (p. 209) – an account which could clearly be given an Orientalist inflection. She attributes what were recognisable differences prior to industrialisation to different forms of agricultural organisation.

Much of the image of the 'North' derives from novels from the nineteenth-century onwards describing the sharp social and economic contrasts between London and the emerging industrial heartlands of the West Midlands and Lancashire. The novels and writings of Dickens, Disraeli, Orwell, Bennett and Sillitoe have forged an enduring image. Shields (1991, 210–11) shows how the 'North' is constructed as alien, if utilitarian: Dickens, in *Hard times*, for example, referred to 'Coketown', widely believed to be based on Preston, Lancashire, as being a 'town of

unnatural red and black, like the painted face of a savage', with chimneys belching 'interminable serpents of smoke'. There was 'nothing in this town but what was severely workful' (quoted in Shields, 1991, 210). Such writings helped establish contrasts not just between places but between their inhabitants. Northerners came to be seen as warmhearted, generous, hardworking, dogged, collectivist, brusque and aggressive; Southerners as snobbish, soft and lazy, individualistic and living by brain rather than brawn. The North's image of itself was as a place where wealth was made, while the South was where it was squandered (Pocock, 1979). Such literature has created influential images which have been difficult to break down.

Thus, following the re-emergence of concern about the 'North–South' divide in the 1980s, media coverage established stereotypes of 'two nations'. Shields (1991) summarises much of the writing on this topic. The 'North' is never a precisely defined and mapped-out region with clear borders, but despite this, certain defining features can be discerned. Consider the following:

> Britain is split by a North–South Divide running from Bristol to the Wash. The victims of decaying smokestack industry live in the North, the beneficiaries of new high-tech finance, scientific and service industries, plus London's cultural and political elite, are in the South. Cross the Divide, going north, and visibly the cars get fewer, the clothes shabbier, the people chattier.
> (*The Economist*, 7 February 1987; capitalisation of 'Divide' in original!) (quoted in Shields, 1991, 232)

This reiterates the demarcation of the divide (the Severn to the Wash) and the package of images which make up the myth of the 'North'. These include the industrial character of the region, the chattiness of its denizens, who live in the conditions of an earlier age untouched by technical progress and economic growth, and a general atmosphere of shabbiness. Although there were attempts to overcome such crude journalism, the effect was in general to reconfirm 'the position of the North in the national discourse on the British regions even while providing evidence of its nonconformity to the stereotypes of that discourse' (Shields, 1991, 240). Such evidence included demonstrations of affluence and depression (e.g. the presence of very wealthy areas in some northern cities) or reports which showed that not all talent and entrepreneurship was confined to the South, nor was the North devoid of places of interest. Yet the kind of anecdotes typically produced (successful football teams, regional shopping centres, museums or individual entrepreneurs) do not, individually or collectively, constitute a revival of the North.

As Taylor (1991a and 1993) demonstrates, such regional images also have their negative consequences. He describes the difficulties of persuading civil servants to move to the North East because of their perceptions about the culture of the region and of its remoteness; perceptions shaped, one presumes, largely by media and journalistic coverage. There is clearly much to do in order to shake off the image of the North as 'an abnormal place to which one travelled in order to be bitten by the surroundings' (Colls, 1992, 15). Conversely, such regional images and myths can form the basis of territorially based identities and, on occasion, political movements, while they can also be mobilised in defence of – or in order to promote the interests of – particular regions. Consider, for example, the role of regional images in promotional campaigns, such as the 'Great North' campaign mounted in the mid-1980s to help attract investment to the North East. The stress on the heritage of the region, its proud history of industrial innovation, and the salt-of-the-earth characteristics of its workforce – these characteristics were all used selectively to promote an image of a place amply deserving additional investment. Of course, such campaigns often rely in themselves on a highly partial history which typically erases class conflict and poverty (Sadler, 1993).

...And realities?

Is it possible to demonstrate, in any 'objective' or accurate way, the reality behind these myths? The endless procession of maps at the level of standard regions in academic texts certainly demonstrated the multifaceted nature of regional variation. Attempts at disaggregation of statistics naturally succeeded in showing the extent of within-region variation. These achieved a high level of sophistication in Dorling's (1995) *A new social atlas of Britain*. He devised a cartogram in which it was possible to incorporate all electoral wards in the UK (approximately 10 000 in total) while maintaining their relative location. The consequence was, of course, massive distortion of the physical outline of the UK, but the benefit was a level of detail unmatched elsewhere. Moreover, the cartogram is scaled so that a constant area of map represents a constant unit of population. This focuses attention on areas where population is greatest and discounts peripheral rural areas with small populations. Figure 1.1 uses this technique to portray child poverty for electoral wards in England and Wales for 1991. The proxy variable used is the proportion of children living in a household where no adult was working at the time of the 1991 Census. Three types of ward are identified. In the richest third less than 5 per cent of children lived in households without earners. By contrast, in the poorest third of wards between 10 per cent and 42 per cent of children lived in households without earners. Attention is drawn immediately to the principal conurbations and former coalfields, where the cartographic technique used highlights wards with large populations.

Despite its technical sophistication, reliance on such quantitative indicators had its limitations. One response has been attempts to measure the more subjective dimensions of 'quality of life' (e.g. Rogerson, 1997). There are several reasons for this interest. Quality of life issues are held to be significant in underpinning migration decisions and processes, and in influencing the location decisions of private (and some public) organisations. Concern about simplistic and unreflexive use of economic indicators as one-dimensional measures of social development has also led to attempts to devise indicators of environmental quality. These debates are all tied in to the growing import-

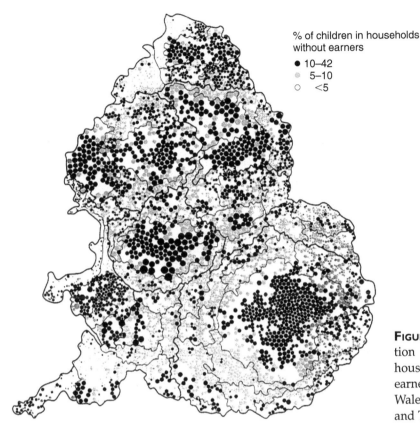

% of children in households
without earners

● 10–42
● 5–10
○ <5

FIGURE 1.1 Proportion of children in households without earners, England and Wales, 1991 (Dorling and Tomaney, 1995)

ance of environmentalism and to contemporary trends in social stratification – for example the 'service class' (see chapter 8) is said to attach considerable important to environmental quality. Drawing on national surveys which invited respondents to rank a number of attributes, the Quality of Life group produced weightings which were applied to measures drawn from published and unpublished sources (e.g. crime rates; house prices; leisure provision; scenic quality). The weighted measures were combined into one index of quality of life for 189 places. The contrasts between top and bottom on this index are revealing. The highest-ranking places are typically medium-sized towns in rural areas, with easy access to high-quality countryside and/or the sea, good educational and shopping facilities, and strong employment growth (e.g. Dumfries, Kendal, Hereford, Eastleigh) combined with historical attractions (York, Shrewsbury, Edinburgh). At the other end of the index, large cities are ranked badly (Hull, Nottingham, Bristol) as are a cluster of medium-sized towns in industrial parts of the Midlands and the North (Stockton, Wigan, Barnsley, Walsall, Mansfield). Interpreting these patterns is not easy, as the authors acknowledge; for example, there is some evidence of a preference for small towns but Edinburgh was ranked fourteenth despite being the fourth-largest city included in the study. These studies nevertheless provide endless ammunition for salvoes in place-marketing wars between local authorities competing for investment. More seriously they also reveal that, for those constrained to live in undesirable environments, a low quality of life can be a major source of stress and illness. Burrows and Rhodes' (1998) study of unpopular places is a good example. They estimate that up to 25 per cent of residents in certain wards (dominated by public-sector estates) in many northern cities expressed high levels of dissatisfaction with their neighbourhood. This compares with less than 5 per cent at the other end of the spectrum.

Thus far the spatial units used in the works referred to have all been based on geographical regionalisations – that is, each division of the country is self-contained and contiguous. Emphasising formal differences between places, based purely on geographical location, may conceal more fundamental functional differences. One way of conceptualising this is to examine a functional regionalisation of the UK. Champion and Townsend (1990) demonstrated substantial differences between different types of place, regardless of latitude or longitude: urban versus rural, small town versus large town, inner city versus suburbs. Consequently they grouped together places sharing these characteristics, regardless of location. The advantages of such analyses are clearly seen from an examination of manufacturing statistics. Relatively speaking, whether located in the 'North' or 'South', the key contrast for many years has been between small towns, regardless of location, and unsuccessful large cities. The former all either gained manufacturing jobs, or lost them at a rate less than the national average, while the latter experienced a veritable haemorrhage of manufacturing employment (see chapter 4; Turok, 1999). This moves away from a kind of locational determinism, but arguably has its own limitations: why should the position of a place on an arbitrary index – in this case population size – have any bearing on its economic success?

A valid response to the evidence of spatial inequality might be that both historically and contemporaneously, regional problems are nothing new. Here, for example, are the views of Lord David Young, a member of Mrs Thatcher's Cabinet, on the North–South divide:

There was more industrialisation in the North originally, therefore there now has to be more deindustrialisation. Until 70 years ago the North was always the richest part of the country. The two present growth industries – the City and tourism – are concentrated in the South. I try to encourage people to go North; that is

where all the great country houses are because that's where the wealth was. Now some of it is in the South. It's our turn, that's all.

(Quoted in Martin, 1988a)

Lord Young's rather simplistic use of indicators – number of country houses, for instance – should not obscure the fact that his view is not unlike a conventional wisdom on the origins of the regional problem. According to Southall (1988), a whole succession of authors regard the First World War as a turning point, marking a reversal in North–South disparities. The conventional wisdom is that during the nineteenth century and up to about 1914, the North was the most dynamic and prosperous part of Britain. This geographical structure was forged by the Industrial Revolution and was centred on cotton in Lancashire and reinforced by export-based industries such as shipbuilding, engineering, coalmining and textiles, associated with the expansion of empire and international dominance (see chapter 2). Considering regional shares of employment, there is evidence of a familiar split, with the North having a major share of manufacturing while the South is dominant in services. But, relatively speaking, the industrial periphery was in some sense in decline from the mid-nineteenth century. The share of manufacturing employment accounted for by the North, Wales and Scotland, was declining steadily from 1841, albeit in a context of overall steady growth. By contrast, manufacturing employment was not trivial in the South East – London was after all the largest centre of manufacturing – and while much of the South and East was predominantly agricultural, even here manufacturing employment had been growing faster than in the peripheral regions.

In this conventional wisdom, unemployment has been seen primarily as a problem of the South, with its difficulties of agricultural depression, and of London where the handicraft industries were in decline. Using trade union statistics, which allow the estimation of unemployment for specific occupational groups, Southall shows that unemployment was not, in fact, highest in the South of England; in several occupations it was actually higher in the industrial areas of Northern Britain. The lowest-wage areas were those where agricultural employment was dominant; in that sense there was more prosperity in the North than in the rural South. However, that prosperity depended to a large degree on the trade cycle – and the staple industries of the North were vulnerable to fluctuations in export demand. Consequently, unemployment was an ever-present risk to the working classes of Scotland, the North East, Lancashire and Yorkshire, and South Wales. These differentials can be traced to the early specialisation of the metropolitan economy in services. The financial, governmental and international roles of the South East were far more important and profitable than links with the industrial regions, producing per capita incomes at least twice the national average (Martin, 1988a, 394). We can see, therefore, that, making due allowance for urban poverty in London, the North–South divide is an enduring phenomenon.

Demonstrating long-run changes in regional inequality has been difficult because of changes in the spatial units for which data are available, and also because of changes in the definitions of variables. However, Southall et al. (1998) have used GIS (Geographical Information Systems) technology to overlay different sets of boundaries and then produce estimates of data values for one set of units from available data for the other. Figure 1.2 thus maps unemployment statistics for 1931 and 1991 onto the Poor Law Unions as they existed in 1898. It is evident that a North–South divide persists, although the 1990s' recession had a greater impact on London. A further example of such analysis is given in figure 1.3, which presents an index of inequality – the ratio between the 10 worst-off and 10 best-off areas throughout the twentieth century. All the indicators used have seen a decline over the century but the gap, as measured by the inequality ratio, is showing a long-run upward trend. The easy conclusion is that society has been becoming more unequal. A plausible alternative is that the *scale* of social segregation and exclusion has changed, with some locations pulling ahead while others fall behind in relative terms.

Finally, in a contemporary context, the scale of regional inequality in the UK is hardly exceptional. Variations in living standards within the UK bear no relation to variations between states on a global scale. Even within Europe such variations are generally more substantial than within Britain. In figure 1.4, GDP per head is standardised around the EC average so that a score of 100 indicates that a region has the same GDP per head as the EC. With the exception of London, all the UK regions fell between +/− 25 per cent of the EC average; this range is comparable with Spain or Belgium and is rather less than in Germany or Italy. France, however, shows much less disparity, with the exception of the Ile de France (Dunford and Perrons, 1994). The figures for Germany are affected by reunification but the former West Germany is notable because *all* its regions had a per capita GDP above the EC average whereas, by contrast, 30 per cent of the UK's population inhabited regions where GDP per head is under 90 per cent of the EC average.

There are, then, many different possible cartographies of economic and social change in the UK. These have been the basis of enduring and influential images – myths, even – about places. But mapping spatial distributions is only a beginning. Such mapping exercises can, in their more simplistic forms, reduce to a mere recording of the degree of presence or absence of particular phenomena; space is reduced to a surface, and places are characterised purely (and narrowly) by differences from each other. However, as Allen et al. (1998, 50) observe, space should be understood as the 'product of social relations . . . the product of the networks, interactions, juxtapositions and articulations of the myriad of connections through which social phenomena are lived out'. Allen et al. (1998) suggest, therefore, that what is important about spatial variation is not continuous-but-varied surfaces, but differentiated articulations of social relations/processes in different places. How far does work on the UK's human geography match up to this way of conceptualising spatial variation?

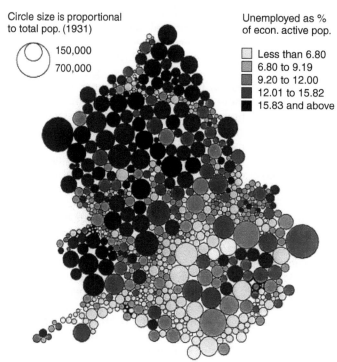

Circle size is proportional
to total pop. (1931)

150,000
700,000

Unemployed as %
of econ. active pop.

☐ Less than 6.80
☐ 6.80 to 9.19
☐ 9.20 to 12.00
☐ 12.01 to 15.82
■ 15.83 and above

FIGURE 1.2A 1931 Unemployment mapped onto 1898 Poor Law Unions (Great Britain Historical GIS, Queen Mary and Westfield College, London)

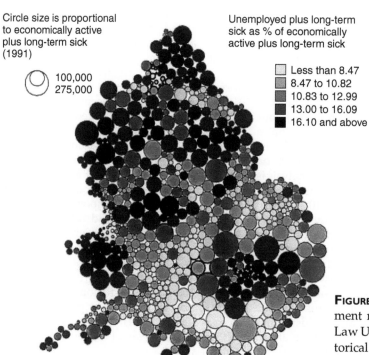

Circle size is proportional
to economically active
plus long-term sick
(1991)

100,000
275,000

Unemployed plus long-term
sick as % of economically
active plus long-term sick

☐ Less than 8.47
☐ 8.47 to 10.82
☐ 10.83 to 12.99
☐ 13.00 to 16.09
■ 16.10 and above

FIGURE 1.2B 1991 Unemployment mapped onto 1898 Poor Law Unions (Great Britain Historical GIS, Queen Mary and Westfield College, London)

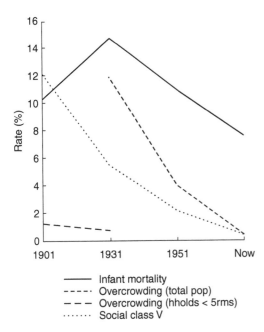

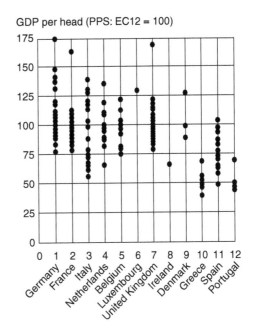

FIGURE 1.3A Median rates for infant mortality, overcrowded housing, and persons in social class V in 1898 Poor Law Unions, *c.* 1901–91 (Great Britain Historical GIS, Queen Mary and Westfield College, London)

FIGURE 1.4 Regional inequality in the EU, 1989 (GDP per capita at current market prices and purchasing power standards in 1989) (Dunford and Perrons, 1994)

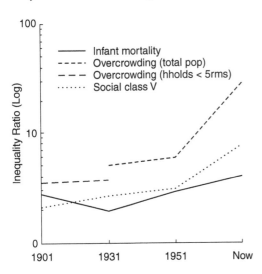

FIGURE 1.3B Inequality ratios for infant mortality, overcrowded housing, and persons in social class V in 1898 Poor Law Unions, *c.* 1901–91 (Great Britain Historical GIS, Queen Mary and Westfield College, London)

1.3 GEOGRAPHIES: EXPLAINING UNEVEN DEVELOPMENT

A useful place to start, which has affinities with the approach of Champion and Townsend (1990), is with Fothergill and Gudgin's (1982) quantitative analysis of regional employment changes from 1952–75. This analysis provoked an extensive debate over the conceptualisation and analysis of regional change. They identified associations between various 'location factors' and patterns of economic change. Critics argued that this was limited, and proposed instead the concept of a spatial division of labour, which would analyse the part played by different regions and places in successive rounds of capitalist investment. Drawing on this, there developed an awareness that broad-brush regional contrasts were of limited value, and that an examination was required of the ways processes of uneven development worked out in specific subregional units – localities (Cooke, 1989). This stimulated a vigorous debate on the theoretical status of the concept of locality and its political place. Finally, and perhaps most pervasively, attempts have been made to conceptualise regional change in terms of long-run theories of transitions in the mode of regulation governing the operation of capitalist societies. Here the key debate has focused on whether the era of 'Fordism' is being replaced by 'post-Fordism' and, if so, what the consequences will be for patterns of uneven development.

Shift–share analysis and the urban–rural shift

These analytical frameworks are most closely associated with the work of Fothergill and Gudgin (1982). They attempted to understand patterns of regional change in the UK (especially the apparent urban–rural shift of manufacturing) over the 1952–76 period by disaggregating the components of employment change into those related to the national and local structure of employment. Their basic proposition was that manufacturing employment change depended on particular characteristics inherited from previous rounds of industrial development – the industrial structure, the urban structure and the size structure of factories, for example. Because different industries experience different rates of growth (and decline) and because declining and growing industries are not evenly spread, areas could be regarded as having 'favourable' or 'unfavourable' employment structures, and they would be expected to grow or decline differentially. Moreover, focusing specifically on urban decline, they stressed the specific location 'factors' – particularly land availability, and the cost of labour – which rendered cities unattractive to manufacturing.

Their emphasis on such 'location factors' was criticised by authors arguing from a Marxist perspective that what really determines the fortunes of places is processes taking place at the level of the national economy or at the international level. Fothergill and Gudgin argued, in response, that it is one-sided simply to focus on either the national economy or the characteristics of localities in explaining economic change. All firms face external competitive pressures, but the extent to which each area is affected by these depends on the mix of firms therein, and on the costs, opportunities and constraints posed in each area. From this perspective, it is still defensible to identify specific national trends and the characteristics of cities which result in industrial decline. This provides a justification for their use of shift–share techniques to analyse regional trends in employment. Their discussion of the urban–rural shift differentiates between places according to whether they had a favourable or unfavourable industrial structure. Similarly, the precipitate decline of manufacturing in cities was because of the locational constraints which differentiate cities from other areas: if cost factors were the main or the only determinant of locational choice, then the urban–rural contrast would be weaker than it is and they stress the impacts of the availability of labour and land on firms' decisions.

This focus on location factors provoked the reaction that Fothergill and Gudgin were abstracting spatial form from social process. Certain spatial patterns were undoubtedly evident – the urban–rural shift, for instance, remains an enduring feature of the UK's economic geography according to Townsend (1993), for whom its salience in the late 1980s reflects some classic 'cost-push' factors (congestion and land prices) prompting dispersal (see also Turok, 1999). However, Fothergill and Gudgin pay less attention to influences such as the internal organisation of production, on one hand, and the form of social relations, on the other (Massey and Meegan, 1985, 90). Such influences attracted the attention of theorists working from an alternative point of departure in Marxist theory.

Spatial divisions of labour and the 'restructuring thesis'

The central argument of Marxist-inspired work, notably that of Massey (1997) was that the geography of production and uneven development could only be conceptualised adequately through an analysis of the social relations of production. Moreover the relations of production are organised spatially; they might involve the geographical separation of headquarters and branch plants, the separation of research and development from direct production, or the separation of plants within the same firm performing different tasks in the technical division of labour. Consequently, relations between areas 'are actually those relations of production stretched out between areas' (Massey, 1988, 252). Places may be specialised in the performance of a small number of functions, and these in turn are indicative of the distribution of economic power between regions. For this reason Massey refers to the *spatial division of labour*: the allocation of particular elements of the production process to specific places. Classically, for example, the core regions of national economies are characterised as possessing key command functions, while peripheral regions play a subordinate role, e.g. in being the location for assembly or branch plants. Massey points out the implications of these arguments for the geography of class structures and, by implication, of political power. Because these divisions of labour consist of mutually defining elements, then the functional and social characteristics of some areas must define the functional and social characteristics of others. It follows that any local area can only be understood when analysed in relation to the functions within a wider division of labour which are performed within it. The associated distinctive spatial structures are likely to have different implications for the nature and dynamics of growth: the nature and density of inter-firm linkages, and the extent of multiplier effects, are two of the most obvious examples. A crucial consequence is that uneven development varies not only in degree (the extent of disparities in unemployment rates) but also in character; there are qualitatively different kinds of regional problem.

The importance and force of Massey's arguments become clear when we consider her critique of other writers operating within a more conventional framework. Attributing the urban–rural shift of manufacturing to cost differentials (e.g. wages or property prices) went only so far; it neglected the external circumstances which made cost-cutting necessary. The necessity to cut costs is not an immutable constraint; for Massey these external pressures (and the responses to them) are precisely what is to be explained. If one wishes to understand why decentralisation of industry is taking place, one has to comprehend the circumstances which lead firms to consider it as an option at particular times.

Massey developed these ideas in her influential 'restructuring thesis'. This referred to: the ways enterprises respond to competition by altering their product mix and the ways production and distribution are organised; the ways these changes draw upon, and affect, the organisation of economic activity across space; and the social and political consequences of those changes, i.e. the linkages between spatial divisions of labour and the geographical pattern of social relations. She demonstrated both the diversity of repertoires used by capital to respond to external change and the ways in which these responses had quite different, place-specific impacts. What was distinctive, methodologically and politically, about Massey's approach was her attempt to analyse industrial location and regional development not in terms of the character of places or location factors affecting the decisions of individual firms, but in terms of the social relations of production. Rather than simply reading off spatial patterns from social structure, however, Massey sought to identify the mechanisms in capitalist society which generated specific and unique outcomes, by working with general theories of capitalist development and linking these, through successively lower levels of abstraction, to contingent empirical outcomes. She also emphasised that capital's responses to uneven development were a product of interactions between existing spatial patterns and the requirements of production. Thus uneven development was the outcome of successive 'rounds of investment' which would of necessity produce quite different outcomes from place to place.

Allen *et al.*'s (1998) recent work on South East England exemplifies and develops the use of these ideas. The authors emphasise that regions 'only take shape in particular contexts and from specific perspectives'. Consequently, 'there will always be multiple, coexisting characterisations' of places, and these characterisations will vary between social groups and over time. Second, drawing on the idea of a spatial division of labour, regions are constituted by 'their place within [the] overall constellation of forces which constitutes social space' (p. 50). South East England is defined as a region by, among other things, the national and international links of strategic sectors of its economy, the control functions exercised from within it, and a constellation of intersecting institutional geographies (chapter 2; see also Taylor, 1991a). This 'region' is not confined by administrative boundaries, therefore; it is defined by its external links. Moreover, not all places within it are articulated in the same way with national or international social and political structures. Finally, Allen *et al.* (1998, 60) insist that different spatial levels (global economy, nation-state, region, locality) should not be treated in sequence, as an ordered hierarchy, but should be conceptualised together from the start.

Critics of this concept of a spatial division of labour have questioned whether in some respects it is anything more than descriptive and taxonomic, particularly with respect to the restructuring strategies adopted by firms and the concept of a spatial division of labour. Identifying types of restructuring strategies or divisions of labour does not, of itself, provide explanations. Consequently, for Warde (1985) the concept of a spatial division of labour remained a heuristic device. Claims as to the status of restructuring theory have to be accorded a much more provisional status, and explanations of the characteristics of places cannot be achieved by relying on one cause or set of causes. However, Massey (1997) has insisted that she was not simply proposing a very passive view of social and economic change, in which successive 'rounds of investment' were merely 'deposited' on a pre-existing landscape. This misrepresented her argument in at least two ways. First, it simplified her analogy which had emphasised the *combination* of layers

arising from rounds of investment: the notion that geological layers could be combined was surely implausible. Thus the layers themselves (rounds of investment) were constituted out of *social relations*, implying, second, that places were not just the victims of capital (as some of her critics appeared to suggest). Instead, places interact with wider social structures through active human agency. A third criticism of *Spatial divisions of labour* was its apparent failure to specify 'rules of combination' between shifts in spatial divisions of labour, within labour markets, and between struggles over production and consumption issues. However, if this was meant as a call to clarify and classify the typical effects of different combinations of class, gender and popular forces, Massey argues, it is a requirement which it is impossible to satisfy, because of the complexity and indeterminacy of social life. Consequently Massey claimed (1997, 322–3) that her approach sought to identify 'key causal relations, including combinations, which may be replicated in numbers of situations' but this did *not* mean formulating rigid rules for understanding social change. Nor does it mean endless contemplation of uniqueness and difference, leading to 'endless micro studies from which . . . no larger lessons can be learned' (Massey, 1997, 324). Such studies must always be conducted within a wider theoretical understanding of the social relations of production. The main impacts of such ideas have been twofold: in discussions of the most appropriate scale at which research should be conducted, and in theorising change in capitalist society.

Scale: locality and region

The question of scale is fundamental to discussions of social change in contemporary Britain, but journalistic notions of the 'North–South divide' and the use of statistics for what are rather large regions, severely limit analysis. Massey's work had pointed out that in terms of understanding many dimensions of social and political life, the local labour market was crucial; for the majority of people, social life was spatially bounded in this way. Debates about social and political change, and their interrelationships with economic changes, were taking place on the basis of national- or regional-level statistics, but this was plainly untenable given the uneven impacts of component causal processes. Further, it followed that any conclusions about policy and politics would necessarily differ as between distinct parts of the country. Finally, recognising this diversity in no way implied abandoning a concern with wider processes, but it did mean avoiding explanations which reduced local change to the outcome of global processes. Building on these foundations there followed a major research programme into the dimensions and dynamics of economic and social change in a number of localities (Cooke, 1989; Harloe et al., 1990). The initial aims were to investigate the effects of international economic restructuring on different areas, but there were also examinations of the complex connections between economic change and other dimensions of political, social and cultural life: gender relations, political attitudes, local policy-making and so on.

Claims were made as to the status of the term 'locality' which some clearly found exaggerated, if not illegitimate. Cooke (1989, 296–305) suggested that localities could act as viable bases for social and political mobilisation. He was drawing on evidence from some case studies which appeared to suggest that certain localities (e.g. Swindon) had been swift to seize opportunities. In its most exaggerated form this resulted in the claim that localities could be pro-active and, in some sense, masters of their own destinies. This position, and indeed the whole research programme on 'locality studies', were challenged on conceptual and political grounds. Conceptually, there was no agreement over the meaning and causal status of 'locality', and the methodology used implied an implicit spatial and economic determinism. There were also wider concerns that locality studies were simply a licence for empiricism, that is, they encouraged researchers to excavate every conceivable piece of information about a locality without really thinking about how to integrate and analyse it (Smith, 1987).

Duncan (1989) argued that the term 'locality' referred to a 'mixture of local differences, locally-contextual action, and locally-referent consciousness' which had socially significant effects. As well as empirically observable variations in living standards, it meant that people's actions and beliefs were shaped locally. However, he felt that this was no more than the almost self-evident argument that space makes a difference, or that social processes combine to produce spatially specific outcomes, and that these observations did not make the case for localities as objects of study. Although social systems are spatially situated and bounded, this does not mean that social systems are created by local areas. Duncan therefore accused locality studies of reintroducing spatial determinism, albeit one in which the productive base of a locality was the primary determinant of social and political developments.

Others argued that such economism was unfortunate at a time when the social sciences were in the throes of the 'cultural turn' which emphasised that human action was far from being determined economically. Rose (1989), for instance, accused locality studies of concentrating on wage labour and the relations of production, to the neglect of other dimensions of social division and mobilisation such as gender and community. Likewise, Jackson (1991) argued that while studies of contemporary regional differentiation in Britain had not neglected cultural issues, they had done so in an unreflective manner redolent of an 'outmoded style of regional geography'. They had reproduced stereotypical images of places, failed successfully to connect the economic and the cultural (pp. 217–18) and read off socio-political change from economic change.

Massey (1991) made some broader political points about why, and how, localities might be studied. This was in part a response to one implication of Harvey's (1989a) *The condition of postmodernity*, which had suggested that a focus on place and the local is, by its very nature, antiprogressive in political terms, though this is not to argue that such a focus is necessarily reactionary. What Harvey seems to imply here is that a concern with place inevitably leads to an aestheticisation of place, and thence to a reactionary, defensive politics which slides into 'parochialism, myopia and self-referentiality' (p. 351; see also Byrne, 1995a). Elsewhere Harvey elaborates, suggesting that the problem with much Left politics of resistance is that it amounts to 'militant particularisms' produced under an 'oppressive and uncaring industrial order'; it is a politics which is potentially conservative inasmuch as it primarily aims to defend established patterns of production. Indeed, 'perpetuation of these political identities and loyalties requires the perpetuation of the conditions that gave rise to them' (1996, 40). He elaborates on this with an account of academic and political discussions over the running down of a car plant in Oxford: should union members aim to keep the oppressive jobs in the plant going at all costs, probably under increasingly harsh conditions, or should they aim at producing commodities which are more socially useful and less harmful to the environment? A consequence of the latter position might be the closure of the plant.

Massey (1991) points out that 'militant particularisms' are not the only or the inevitable outcome of place-based political movements, and elsewhere she gives examples of progressive senses of place which have produced place-*based* but not place-*bound* action. Put another way, place-bound action is simply about defending what exists in place, whereas place-based action seeks to use local bases to address wider issues. She cites progressive local authority economic initiatives in London; these forged links between workers in London and their counterparts abroad, as far afield as Latin America. Far from drawing on a static concept of place, such initiatives can promote dynamic and progressive political thinking.

The localities debate faded away after a few heated exchanges, but similar issues have resurfaced in debates about the alleged 'resurgence of regional economies' (Storper, 1995). There are links here with the question of whether localities could proactively secure control over their own destiny. The arguments about successful regional economies turn on the evidence of 'clustering' of and specialisation in certain forms of activity

(chapter 5), and these arguments have had some influence on policy-makers (chapter 12). However, critics suggest that over-optimistic conclusions have been drawn from a restricted set of case studies, and that it is by no means clear that the UK has (or is likely to obtain) 'new industrial spaces'. In particular the idea that the deindustrialised regions of peripheral Britain can emulate European success stories is viewed as wildly optimistic (Lovering, 1999). One reason for this is that although such regions have deindustrialised, the precise character of what follows deindustrialisation is a matter for debate.

Transitions in the organisation of production

The second principal way debates on Britain's human geography have evolved is by a recognition that theorising the changing geography of production requires an understanding of systemic shifts in the organisation of capitalism. Contemporary economic geographies must now confront the following: a 'new information-based techno-economic paradigm' which is transforming the organisation of production as a result of the automation of processes and improvements in information processing; the tertiarisation of the economy, a long-term trend but one which is of heightened relative importance given the recent pace of deindustrialisation; the growing importance (at least for those wealthy enough to participate in it) of consumption, stimulating new patterns and landscapes of consumption based on the commodification of the visual, aesthetic and symbolic; the impact of globalisation; and the changing character of state intervention (Martin, 1994, 23–5). In contrast to the relatively stable order depicted by many theories of industrial location and uneven development, the global economy is now characterised by disorder, rapid change, and uncertainty. How has geography risen to the challenges thus posed?

According to Martin (1994, 30–1) the primary focus has been on deriving a 'general, synoptic theorisation of the new realities'; there are two main strands to this. First, there are arguments that there is emerging a shift towards a new production system – for example, post-Fordism, flexible specialisation, or lean production. These are sometimes associated with long-wave theories of capitalist development, which associate upsurges in economic activity with major innovations in the economy (for example, Marshall, 1987). Thus, patterns of regional development are said to reflect the impress of Kondratieff cycles, so that the recent comparative good fortune of the South East is because the region was particularly well-placed to benefit from the latest of these cycles, in which growth in high-technology industry and information-processing activities led the way (P. Hall, 1985). Second, and ranging more widely in their focus on non-economic realms, are those accounts which postulate a more general shift in the structure and organisation of the accumulation process (Martin, 1988b). An example would be Lash and Urry's (1987) notion of 'disorganised capitalism', in which analysis goes beyond economic changes to incorporate related shifts in the welfare state, political behaviour, social structures and so on.

Of these various accounts, perhaps the most influential within geography have been discussions regarding the emergence of a 'post-Fordist' regime of accumulation, and analyses of flexible specialisation. Discussion of post-Fordism is usually (though not exclusively) associated with the French 'regulationist' school. The goal of the regulationists was to explain a paradoxical tendency in capitalism: on the one hand, the inherent tendency of capitalist development towards instability and crisis; on the other, the ability of capitalism to coalesce and stabilise around a set of institutions, norms and rules which serve to secure a relatively long period of economic stability (Amin, 1994, 7). For regulationists, two key concepts are the 'regime of accumulation' and the 'mode of regulation'. The *regime of accumulation* indicates a set of regularities at the level of the whole economy which enable a coherent accumulation process. These include: norms pertaining to the organisation of the labour process; relationships and forms of exchange between branches of the economy; common

rules of industrial and commercial management; norms of consumption and patterns of demand; and, *inter alia*, other aspects of the macroeconomy. The *mode of regulation* refers to institutions and conventions which serve to regulate and reproduce a given accumulation regime, including law, state policies, governance philosophies, rules of negotiation and bargaining, cultures of consumption, and social expectations (Amin, 1994, 8). Regulationists also make reference to other concepts and mechanisms, both to explain why capitalism is able to achieve relatively stable growth and to illuminate the reasons why regimes of accumulation break down. The key empirical focus here has been the alleged breakdown of the Fordist regime of intensive accumulation, and the specification of its successor (this discussion draws on Amin (1994) and Peck and Tickell (1994)).

Fordist intensive accumulation had four central features. As a *labour process* it was characterised by mass production based on assembly-line techniques operated with semi-skilled labour. As a *mode of macroeconomic growth* it was characterised by a virtuous circle linking mass production, rising productivity, rising incomes, growing demand, and increased profits and investment. As a *mode of (social and economic) regulation*, Fordism entailed: the separation of ownership and control in large corporations; union recognition and collective bargaining; wage indexation; and monetary policies geared towards (Keynesian) demand management. Fourth, Fordism may be viewed as a general pattern of social organisation (what Jessop terms *'societalisation'*). This involves the consumption of standardised, mass commodities in nuclear family households, and provision of standardised collective goods and services by a bureaucratic state.

This is very much a description of an 'ideal-type' of Fordism and while regulationists acknowledge this, critics argue that admitting national diversity has the effect of undermining the concept of a dominant Fordist logic. Others question what they regard as the functionalism and teleology inherent in regulation theory, as well as the limited space it allows for human agency and political struggle. However, regulation theory has informed – explicitly or implicitly – a substantial volume of geographical scholarship concerned to demonstrate the ways in which 'geography matters' to the contemporary economic transition.

Rather narrower in focus than the regulationist school is the argument that capitalism is passing through a 'second industrial divide' *en route* to a future of flexible specialisation (Hirst and Zeitlin, 1989). The argument opposes mass production, using semi-skilled workers to produce standardised goods, to flexible specialisation, based on skilled workers producing a variety of customised goods. It is suggested that there are historical turning points, or 'divides', involving conscious choices between mass production and standardisation. The first such divide, in which mass production limited the growth of craft industries, occurred at the beginning of this century. We are now, it is claimed, witnessing a crisis of mass production constituting a second industrial divide, in which flexible specialisation will prevail; this reflects greater diversity in consumer demand, and the growth of flexible manufacturing technologies. The result is that small-batch production is feasible without loss of scale economies; the implication is that a reversion towards craft production might become possible. There are numerous criticisms of this model, perhaps the most important of which relates to the rose-tinted view of the future presented by its more evangelical advocates (Amin, 1994, 16). It is unrealistic, critics claim, to imagine that it will be at all straightforward to return to such a model, since the existing structures of Fordism will most likely adapt to new circumstances. In any case the grip possessed by multinationals on finance, marketing and distribution will enable them to continue to exert their dominance.

There are, then, a range of approaches to understanding contemporary transitions in the organisation of production. These share some common ground in interpreting the era that is passing but there is far less agreement about a putative successor. There is probably general consensus that technological change will have profound impacts on patterns of production organisation. There is much less agreement about how such changes will be managed and integrated with other elements of production reorganisation. This is why one sees a degree of hesitancy in identifying 'Fordism's unknown successor', evident in the terms 'post-Fordism', 'neo-Fordism', and 'after-Fordism'. The three prefixes indicate very different views of what is happening, as well as scepticism about the extent of empirical support for claims about new forms of production organisation and about whether a new mode of regulation has come into existence.

Thus, several important general tendencies in the mode of accumulation can be observed. These include: restructuring of employment based on new technologies and leading to greater labour force segmentation; an industrialisation (or intensification) of the service sector; greater mobility of labour provoked by new geographies of employment; heightened social polarisation between high-productivity/high consumption strata, and low-wage or no-wage strata; and an individualisation and pluralisation of lifestyles as a consequence. Similar emblematic developments may be identified in the mode of regulation: a triple displacement of powers of governance away from the national state downwards, upwards and horizontally (Jessop, 1993); a reorientation of welfare services away from universalism towards selectivity (both socially and spatially); and an abandonment of corporatist arrangements for governance (Burrows and Loader, 1994; Jessop, 1995). However, while such tendencies certainly can be identified, it is questionable whether a stable solution to the crises of Fordism can be said to have arrived. Jessop argues that this is so because of the absence of a strong ideology which secures a general acceptance of the 'rules of the game'. For example, the Blair government's rhetoric about the 'third way' is vacuous, and it is not yet clear how stable the support for this government will prove; on the other hand, however, some commentators discern sufficient common ground (around an ideology of 'Blaijorism') to justify the suggestion that a consensus around a 'one-vision polity' is emerging (Hay, 1996a; see chapter 3).

Theorists of flexible specialisation have also engaged in futurology and have suggested that this model offers considerable progressive potential, mainly through enhancing worker autonomy and participation and increasing the skill content of jobs. Examples are drawn from several paradigmatic locations (Silicon Valley; the 'third Italy') and firms (especially certain innovative large firms in Germany, Italy and Scandinavia) but it is questionable how generalisable these are. It is also debatable, suggest critics, whether solutions to the crises of Fordism could not be found through other means. Such solutions might well take very different forms depending on particular socio-cultural and political circumstances. Thus distinctions are made between 'high' and 'low' roads to flexible accumulation, with the high road being typified by progressive initiatives such as those found in some exemplar 'new industrial spaces' (see chapter 5) and the low road exemplified by the British approach to labour-market deregulation, viewed by some as a 'palaeoliberal hire-and-fire conception' of a labour market (chapter 6). The geography of flexible specialisation is also crucial since the central message of 'flex-spec' theorists (that a geographical clustering of productive activity will occur) has been adopted by influential commentators (e.g. Krugman, 1996), and turned into a soothing mantra for policy-makers. The important debate here concerns tensions between tendencies towards localisation and those pointing towards globalisation (chapter 2).

Regardless of their theoretical point of departure, theorists generally agree that what follows Fordism will not be stable in the absence of appropriate regulatory institutions. However, while there is a widely shared awareness of the limitations of both state planning and free-market liberalism, the appropriate regulatory structures for a new era are the subject of much debate. Some propose the creation of institutional networks facilitating relationships of 'trust' and stimulating the formation of 'social capital' which are regarded as essential prerequisites

for regional competitive success (Putnam, 1993). There is also debate about the kind of decentralised, pluralistic structures of governance which might be most suitable for emergent flexible forms of production organisation (Hirst, 1994; chapter 11). These questions implicitly pose challenges to spatial policy as conventionally practised (chapter 12), and more generally to the character and institutional form of state intervention (chapter 3). However, for some commentators (Peck and Tickell, 1994), no stable regulatory order has yet emerged and we are currently witnessing a desperate search for a 'new institutional fix'.

Such a search will have to confront and seek to resolve two further issues: social polarisation and exclusion, and environmental sustainability. There is much evidence that contemporary economic changes have had markedly uneven impacts on different social strata, producing a divergence in socio-economic prospects (chapter 8). This has implications for social stability through the costs of crime and of providing services to dependent populations; it also tends, potentially, to undermine support for welfare expenditures from those whose taxes will pay for them. Given evidence as to the reluctance of the better-off to pay higher taxes and/or their willingness to exit to the private sector, reconciling these demands will continue to be problematic. Second, sustainability is an issue rarely dealt with in the literature on emergent forms of socio-economic organisation, yet (as the experience of Britain's sunbelts illustrates), this will be a major constraint on growth, and its management a vital challenge (chapters 12 and 13). This in one sense brings the book full circle, back to the environmentalist legacies of British human geography.

1.4 OUTLINE OF THE BOOK

I begin with an attempt to assess Britain's 'place' in the world (chapter 2). This involves a consideration of Britain's relative economic and political decline, the impacts of globalisation, and the nature of 'British' national identity. It is argued that these connect logically with the discussion of the changing character of state intervention (chapter 3), in which the strategies pursued by successive governments are very much informed by – and attempt to respond to – perceptions of decline. This chapter reviews the so-called 'post-war consensus' on economic and social policy, the nature of 'Thatcherism', and the evidence for and against arguments that the election of the Blair government symbolises a new era of consensus politics in which the parameters of debate are very limited.

Four chapters follow which may be classed as 'economic geographies'. The first two relate to production: chapter 4 deals with deindustrialisation and new geographies of manufacturing, and chapter 5 with geographies of economic recovery. Key issues in the former chapter relate to the impacts of globalisation on state policies, and the impacts of foreign direct investment (particularly as an exemplar of production reorganisation). If chapter 4 is concerned with the decline of industrial employment, the focus in chapter 5 is on what replaces it. Issues covered include high technology industry, new enterprise formation, and geographies of the service sector. To contextualise this there is a discussion of the literature on new industrial spaces and the evidence for such spaces in the UK. This is followed (chapter 6) by an examination of emergent geographies of labour, paying particular attention to the extent of labour-market flexibility (in various senses) and polarisation, at individual, household and regional levels. Chapter 7 then examines geographies of money, bringing together research on several disparate topics: household incomes; the impacts of taxation policies; the availability of finance for enterprise; and the spatial impacts of ostensibly non-spatial expenditure programmes.

There follows a discussion of contemporary research on social polarisation and on the geography of the welfare state. The central point made here is that growing social polarisation has made possible divisive welfare policies which, in turn, have contributed to a broadening of socio-economic divides. In chapter 8 I first review the adequacy of conventional concepts of social class before examining the evidence for social polarisation and the argument that the key contemporary divide is between a discrete 'underclass' and the remainder of society. Discussion of geographies of the welfare state focuses on processes of privatisation (in a variety of forms) which mean that, at both a macro and micro-level, divisions are emerging between those reliant on public services and those able to avail themselves of private alternatives. The marketisation of public services appears likely to accelerate this process by producing divides within the public services.

What are the political consequences of these developments? Without necessarily reading off patterns of political mobilisation (such as voting behaviour) from patterns of uneven development, a case can certainly be made that (in the 1980s and early 1990s) there is strong evidence of association between these variables (chapter 10). However, if a wider range of 'political' behaviour is considered, the kind of quantitative explanations advanced by conventional electoral geography become less plausible. Much mobilisation in recent decades has cut across traditional alignments and draws motivation from a range of sources; environmental and other protest movements are a case in point, while nationalist and regionalist movements also seem to be on the rise. In turn, such movements have prompted calls for greater decentralisation and participatory democracy, and the changing character of subnational governance is consequently taken up (chapter 11).

The final two chapters consider, first, how best to manage the challenges of emerging forms of uneven development and, second, whether *sustainable* development is feasible. In chapter 12, the limitations and criticisms of previous forms of spatial policy are considered, and recent developments are assessed. Current initiatives seek to create the kind of 'institutional tissue' allegedly associated with successful new industrial spaces, and the prospects for such developments are considered. The question of the sustainability (or otherwise) of emergent forms of uneven development is the subject of chapter 13 which considers the management of a range of pressures (from agriculture, transport and urbanisation). A concluding chapter looks forward, considering forces which will shape the UK's human geography in the early years of the twenty-first century.

Throughout, the approach taken seeks to integrate important debates in contemporary human geography and the social sciences with major developments impinging on the UK's economic, social and political landscape. Inevitably this has, for reasons of length, involved selection and exclusion and readers will have to judge the merits or otherwise of the topics considered. In addition, because many readily available sources are updated annually, I have not relied on extracting the latest available statistics from Regional Trends or other such sources, which are easily accessible in any case. For more detailed evidence the numerous atlases available (Champion *et al.*, 1996; Dorling, 1995) will repay examination.

A PLACE IN THE WORLD 1

2.1 INTRODUCTION

This chapter is about *placing* Britain: understanding its location in relation to circuits of power in the global economy and polity; examining its position historically, where the key issue is the celebrated feature of British decline; considering what is particularly distinctive about British capitalism; and identifying whether there is a distinctive national identity which enables individuals and governments to make sense of their place in the world.

There are several reasons why these issues are important. Economically the pressures of globalisation impart renewed urgency to debates about national competitiveness and pose the question of whether distinct national economies can be said to exist. Explanations of economic decline are central to the formulation of strategies for its reversal and particularly for the construction of hegemonic projects aimed at mobilising consent. In addition, the rise of supranational institutions, such as the EU, raises questions about the capacities of the nation-state, and in turn about sovereignty. All these combine to raise profound questions about national identity.

Several authors have demonstrated the relevance of these issues to geography. Massey (1986) argues that we are used to explaining how the internal spatial structures of former colonial states reflect their interna-

tional role and position – for example, an export orientation superimposed on an indigenous economy. This kind of argument is also relevant to the UK. Heavy industrial regions were exposed in the nineteenth century to the vagaries of an apparently prosperous industrial system based on exports rather than indigenous demand (thereby being potentially unstable), while attempts by the British state in the inter-war years to maintain its international position (by propping up the value of sterling) had disastrous effects (Massey, 1986, 34–5), by making exports seriously uncompetitive and undermining the markets for basic commodities. Previously, of course, much of the prosperity of the same regions had followed from exporting to captive colonial markets. Taylor (1989) makes a similar argument with respect to Britain's excessive commitments to defence and to maintaining the Empire, which have 'severely distorted adjustment to the country's changing material base' (p. 19). The implications for the regions of the UK are considered by Lovering's (for example, 1990a and 1995a) analysis of defence-related industries. However, the fortunes of important cities such as Bristol and Liverpool, and particularly their prosperity in the eighteenth and nineteenth centuries, owe much to their key role in the Empire. Bristol, for instance, 'grew fat on the proceeds of slaves and sugar' (Fraser, 1979, quoted in Boddy *et al.*, 1986, 6). The relative success with which these two cities have adjusted to decline speaks volumes for the profound influence of such international orientations on the UK's human geography; Liverpool appears increasingly peripheral in an economy oriented principally towards Europe rather than the Atlantic (Lane, 1987, quoted in Meegan, 1989).

Perspectives on British decline and the nature of British capitalism also help to illuminate enduring features of uneven development, notably the disjuncture between financial and industrial capital and of the contrasts between the metropolitan hub of finance and commerce, and the industrial provinces (for example, Nairn, 1977, 23, Fn. 16; Marshall, 1987, 228). Hallsworth (1996) has elaborated on the implications of the apparent 'short-termism' of the UK economy for patterns of urban and regional development. The overcentralised British state, and the associated institutional geographies, are criticised by many writers, and clearly have implications for the concentration of particular forms of development in the South East of the UK (see chapters 11 and 12).

Finally the question of national identity – if there is one – must be discussed because much contemporary writing questions whether such identities can be ascribed a fixed, invariant character. It is argued here that British national identity is rooted in a white, southern England and is largely backward-looking – a kind of 'village-green nostalgia' for which former Prime Minister John Major was taken to task. The question of whether a more progressive sense of national identity can be constructed is also discussed, as this will facilitate or constrain political strategies, especially with respect to Europe.

These, then, are the arguments for this chapter. There are three main sections. The first discusses explanations of Britain's economic decline and diagnoses of the nature of British capitalism. Then there is a discussion of evidence for and against globalisation, with a view to determining whether politicians are justified in their fixation with this issue. The third section explores concepts of national identity and their geographical implications, with particular reference to Englishness, race, and rurality.

2.2 MAPPING THE BRITISH DISEASE: ACCOUNTS OF ECONOMIC DECLINE

Perhaps a central problem for historians of Britain is the explanation of economic decline; that for politicians is reversing it. The accounts and diagnoses of decline help 'place' Britain; perceptions of decline shape political strategies. The two together may help explain paths of uneven development in the UK and particularly the persistence of certain forms of development. The empirical evidence for decline is perhaps less significant than the *idea* of decline, and its apparent stranglehold on the psyche of politicians (Supple, 1994). Cannadine (1997), for example, comments on Thatcher's messianic sense that a reversal of decline was essential. Yet the period of loudest clamour about decline has witnessed growth rates greater than at any time since the 1870s. British GDP averaged growth of some 3 per cent between 1951–73 but, even as affluence became more widespread, other states pulled ahead. Thus while there might be a 'British disease, there is also a British rate of growth'. It is the comparison with other nations that produces 'fear of failing' (Supple, 1994, 442). This relative decline has both qualitative and quantitative elements.

Quantitatively, Coates (1994) argues that all major comparative indicators demonstrate a weakening of the UK economy's international standing and performance. Some of this was an inevitable effect of Britain's early industrialisation; other countries were bound to begin to catch up and the UK was overtaken by the USA and Germany in the late nineteenth century. The relative decline of manufacturing has attracted most attention: Britain's share of world manufacturing output has halved since 1960 while its share in world trade in manufactures has fallen from one-third in 1899 to under 10 per cent today. The argument that these figures are only of arithmetical significance is undermined because Germany and Japan have succeeded in maintaining their shares of trade and production at a higher level.

Perhaps of greater concern are the qualitative elements of decline. Newton and Porter (1988) chart the UK's declining share of what they term 'leading edge' (rather than declining) sectors of the economy. They discern an emerging structural weakness as the UK's economy failed to develop new products (particularly chemicals and electrical goods) in the 'second industrial revolution' of the late nineteenth century. By the 1980s, the competitive strength of the UK economy was concentrated in a limited set of low technology 'clusters' of related industries. Notable features were, first, a strength in the areas of military-related production (aircraft, defence goods, motors/engines, radios) which in turn had spatial consequences because of the dependency of certain regions on defence expenditures. Second, even in these areas of comparative strength, the competitive position of various 'clusters' of activity showed little depth. Finally there was a general weakening in terms of the UK's share of world exports. Taken together this indicated a persistent gap between UK-based manufacturing and its overseas competitors. The *de facto* devaluation of sterling in 1992 eased the position somewhat but the pound's subsequent strength has been a problem for exporters.

Gamble (1994, 30–8) summarises four theses on decline (table 2.1). The *imperial* thesis blames Britain's attempt to maintain a world role incommensurate with its economic resources. The *supply-side* thesis emphasises the relative backwardness of British manufacturing, explained in terms of the inadequacies of forms of state intervention or a stalemate between capital and labour. The *state* thesis focuses on the institutional characteristics of the British state which render it neither effective nor representative, while the *cultural* thesis prioritises a deep-rooted anti-industrial political culture which resists modernisation and innovation. These accounts vary in the emphasis placed on labour, the state and capital, as the following discussion shows. Recent political debate has been dominated by Gamble's 'market' perspective coupled with a neo-liberal critique of state intervention, but numerous counter-arguments have been put forward.

Labour

Blaming decline on organised labour usually entails the claim that trade unions have become a defensive obstacle to the free play of market forces, due to their restrictive practices and propensity to strike. The actions of unions in major industrial disputes in the 1970s are said to have had a crucial influence on the Conservatives' post-1979 economic policies. Connections between the unions and the Labour Party are said to have sustained Labour's commitment to full employment and welfare provision thereby contributing to the burden of unproductive public expenditure. The unions were also blamed for the economic difficulties experienced by particular regions, with Conservatives drawing attention to spatial coincidences between high unemployment and unionisation (see also Minford and Storey, 1991). The prescriptions which follow emphasise the need to increase flexibility in labour markets and restrain union power.

A counter-argument to accounts which blame unions is provided by Szreter (1997) for whom one cause of British decline has been a failure to invest in human capital to the same degree as competitor nations. Similarly, Reich (1992) and others argue that human capital is an essential precondition for competitive success in service-dominated, information-processing economies. There have been signs that this is recognised by New Labour, in contrast to the market-led approach to labour supply adopted by the Conservatives (chapter 6).

The state

The British state is criticised from both Left and Right. For the latter, an overactive state produces an excessively large burden of public expenditure which the private sector cannot support. Voters lack a budget constraint and the state is excessively open to special interest groups; both of these create a 'political marketplace' in which public expenditure is not easily checked. The end result is institutional sclerosis as the state simply cannot cope with the demands placed upon it. The policy prescription which follows is to roll

TABLE 2.1 Theories of decline

	Market	State	Class
World system the imperial thesis: Britain's world role undermined the domestic economy	legacies of empire i lack of exposure to international competition ii protectionism, subsidy iii abandonment of gold standard	overextension of British state i military ii financial iii foreign investment iv misguided foreign economic policy	character of British capitalism i divided capitalism city/industry split ii global capitalism
Economy the supply-side thesis: state/economy relations weakened the manufacturing sector	economic management i wrong macro-economic policy – Keynesianism ii public spending public sector too large; taxes too high iii too interventionist industrial strategy iv trade union power	developmental state i inadequacy of Keynesianism ii lack of coordination between government, industry and finance iii poor industrial relations iv no consistent industrial strategy v bias and size of spending on infrastructure	state/economy relations i incomplete Fordism; unsuccessful corporatism; inadequate investment coordination; poor industrial relations ii class stalemate
Polity the state thesis: Britain's state was too weak to promote modernisation	weak state i deficiencies of political market ii excessive expectations iii distributional coalitions; institutional sclerosis iv political business cycle	weak state i overload ii adversary politics – lack of effectiveness – inadequate representation	weak state i ancien régime ii dual crisis of the state iii exhaustion of social democracy
Culture the cultural thesis: Britain's political culture held back modernisation	anti-enterprise, anti-capitalist culture i welfare dependency ii elite culture and attitudes iii egalitarianism	anti-industrial culture i institutions of British establishment ii liberal ethos	anti-modernisation culture i aristocratic ethos ii Labourism

Source: Gamble, 1994, 31.

back the state, opening the economy as far as possible to market forces; this is usually linked to a monetarist critique, which suggests that the growth of public expenditure 'crowds out' productive private investment (Bacon and Eltis, 1976).

The international orientation of the British state also attracts criticism. Britain has attempted to play a variety of roles on the world stage in order to preserve its image as a world power, but has lacked the economic power to sustain extensive overseas commitments. There has consequently been a tension between these commitments and the need for economic modernisation. This was most evident in the economic 'stop–go' cycles of the 1950s and 1960s. There have been reductions in overseas commitments consequent on the withdrawal from Empire and Britain's reduced world-power status, but these have not released substantial resources for redeployment elsewhere. In addition, the priority given to maintaining sterling's value, as part of this international orientation, has not been beneficial to British industry.

Against such views there are also arguments that the state has been insufficiently proactive. Britain allegedly lacks a 'developmental state' (Gamble, 1994, 35); this would be active, interventionist, and geared to promoting continual modernisation. There is poor coordination between government, industry and finance, and there has been no consistent or effective industrial strategy. Spending on infrastructure, R and D, and training have all been neglected and adversarial relations between state and capital, on the one hand, and trade unions, on the other, have meant a failure to realise the full productive capacity of the economy. Where Britain arguably has had an industrial policy it has involved substantial defence expenditures (Lovering 1990a and 1995a) but these crowded out skilled personnel from civilian R and D in favour of military production. Moreover, 'soft' government orders for military equipment protected industry from competition and inhibited innovation. Other 'industrial' programmes involving the state (rescuing 'lame ducks', support for nationalised industries, or regional assistance) do not add up to a coherent policy.

These arguments flow through into policy prescriptions: for the Right, monetarism and rolling back the state; for the Left, a more active, interventionist state, taking a long-term view of investment and attempting more corporatist forms of economic management. In an attempt by the Blair government to avoid being tainted by association with old paradigms, much debate currently focuses on the rather inchoate concept of the 'third way' between socialism and *laissez-faire* capitalism (see chapter 3).

Capital

By far the most vigorous debates on economic decline have focused on the nature of British capitalism and on the part played (or *not* played) by capital therein. One prominent argument is that there is something peculiar and specific about English culture, which meant that industrialisation and modernisation were pursued with less vigour and less far than elsewhere. A second relates to the split between industrial and financial capital. Both these arguments have been reworked in Hutton's (1995) recent analysis of short-termism in the British economy.

Wiener (1981), Cain and Hopkins (1993) and Barnett (1986 and 1995) have all suggested that English culture and values produced, through the education system, individuals less capable than their international competitors of coping with technology, science and economics. This reflected a general unease with industrialism and its values which inhibited the quest for expansion, productivity and profit (Wiener, 1981). Drawing on Cain and Hopkins (1993), Hutton (1995) argues that there was an almost innate desire to be seen to be living a 'gentlemanly' lifestyle, producing income without effort. Property and finance, the obvious sources of such income, therefore soon appeared more attractive than industry as a focus for economic endeavour.

However, it is questionable whether concepts like 'culture' or 'gentlemanly capitalism' can shoulder the explanatory burden placed on them. 'Culture' surely cannot, on its own, explain a pattern of economic performance which varies geographically and temporally. It *might* explain the limited impact on the UK of social democratic traditions, so that British capitalism has never been accommodated to the idea of the state as an ally in programmes of economic modernisation (Coates, 1994, 148), but other influences are at work there as well. Nor does Cain and Hopkins' concept of 'gentlemanly capitalism' help; apart from problems of definition it is difficult to see what difference the 'gentlemanly' status of British capitalists made to the character of the UK's development (Ingham, 1995, 342).

What may be much more significant is the separation between industrial and finance capital which dates back at least to the nineteenth century. On this argument, British industry has never had the support it needs from financial institutions to enable viable programmes of modernisation to take place. Moreover political strategies have been quite deliberately formulated so as to prioritise finance capital over industrial capital. From this perspective, what needs explaining are the institutional structures which created and then sustained the advantageous position of financial over industrial capital. For while, according to the New Right, firms are simply Darwinian creatures responding to pre-given competitive conditions, 'there is nothing "natural" about the institutional, social, legal and cultural context' in which firms operate (Hutton, 1995, 111). Crucially it is the nature of the financial system in which ownership is located that is central to the character of any given economy's firms. This directs attention not to generic concepts – such as capitalism and markets – but to the specificities of *particular* capitalisms (Crouch and Streeck, 1996).

Hutton's analysis shares important features with the work of Perry Anderson and Tom Nairn in that it prioritises 'the destructive relationship between British finance and industry' (p. 111). The Industrial Revolution is often presented as a more-or-less spontaneous development, in which Britain's raw materials and labour reserves were combined in a deregulated economy which gave maximum scope for innovation and creativity. However, Hutton argues that the Industrial Revolution was almost an accidental by-product of an economic system geared, from well before the advent of industrialisation, to the development of trade, commerce and finance rather than industry. Britain's commercial and financial interests were cosmopolitan, searching to maximise rates of return; Imperial expansion served to open up channels of exchange. The City had minimal involvement in supporting industry, which relied on local sources (often, investors known to individuals or their families) for funds, though the profits from industrial growth were so high that companies could finance themselves without recourse to external finances.

What was distinctive about British industrial development, then, was that 'the state was not involved and ... the financial system remained distant from large-scale industrial lending'. Attempts were made to launch industrial banks but these never became established. Consequently the financial system was responsible for Britain's failure to exploit the wave of new inventions of the late nineteenth century. A pattern was thus firmly established:

> a national banking system disengaged from production; a risk-averse London stock market focused on international investment; equity finance made available only on the most onerous terms, heaping large dividend demands on British producers; a Bank of England concerned to preserve price stability and the international value of sterling; and an industrial base losing ground to foreign manufacturers with high productivity – and having to respond by bidding down wages to maximise retained profits, the only reliable and cheap form of finance.
>
> (Hutton, 1995, 123)

The City, the Bank of England and the Treasury promoted a rigid financial orthodoxy which survives to the present day. Only in exceptional circumstances have there been efforts to rectify this and such efforts have been short-lived. Arguably the current preoccupation with European Monetary Union, with its emphasis firmly on *financial* criteria, will do nothing to change this.[1]

For Hutton and those of similar views

(notably Ingham, 1984) manufacturing is in a very subordinate position and is kept there by an endemic short-termism in which maximising shareholder value takes precedence over everything else. The global preoccupations of finance capital block industrial modernisation because, first, banks were never very heavily involved in the prospects of their immediate localities. Lacking an interest in local prosperity, they never developed strategic leadership as was the case in Germany and the USA. Second, UK manufacturing firms had to rely on the stock market rather than banks as a source of capital. This left them facing a financial system which had a clear preference for short-term, rather than long-term lending, was reluctant to lend venture capital, and was insensitive to small business needs. More recently the short-termism of the British economy has been linked with the dominance of pension funds as sources of investment. Demands for high rates of return – linked in part to the privatisation of pensions, in part to the need for high rates of return to fund endowment policies purchased in the housing boom of the 1980s – exacerbate short-termism and impose impossible payback periods on investment for British firms (Minns, 1996; Hallsworth 1996). Thus the difficulties of the British economy are interlinked with – and exacerbated by – the pursuit of individualism and privatism in the British political culture, assiduously encouraged by the Conservatives (see chapter 9). Hutton traces various social and political consequences of short-termism. The demands placed on companies for a payback on investment give them no alternative to a vigorous attack on labour costs. 'Flexibility' and insecurity are becoming the norm in labour markets (chapters 6, 8), potentially undermining the foundations upon which assent for a comprehensive welfare system can be secured (chapter 9). Responding to these challenges presents formidable obstacles; Hutton talks of a 'stakeholding' society which would place greater obligations on firms and on the financial system, thereby working to rein in the power of the City, but there is little sign that such a programme is being pursued by Labour (see chapter 3; also Leys, 1995).

2.3 GLOBALISATION OF THE ECONOMY

For some commentators, globalisation has long historical roots. Beveridge, for example, is said to have discussed the implications of the possible relocation of routine manufacturing activities some 75 years ago (Harris, 1977, 2). However, the globalisation of economic activity has only become transparently obvious in recent decades. Most writers on globalisation emphasise the ways in which globalisation undermines the sense in which we can speak of 'national' economies, conceived as closed systems, and of the nation-state, conceived as a geographically bounded set of political, economic and social relations. Roughly speaking, the crude version of the argument is that there will be a global shift of routine assembly work into South East Asia as the 'tiger economies' sweep all before them (an argument rather weakened by recent events in those economies). The inevitable consequence will be the collapse of male, full-time employment in the Western 'rustbelts' with their 'smokestack' industries. People will have to adjust to a much more flexible labour market by pricing themselves into jobs. Such ideas, popularised by authors such as Ohmae (1990) or Kennedy (1993), seem to have become a widely held orthodoxy – and this despite their pessimistic vision of nation-states powerless to do anything, a vision in which domestic strategies of economic management are largely irrelevant. Where does the UK stand in relation to globalisation?

First, there are few examples of direct relocations of routine productive activities from the UK to, for instance, South East Asia. Individual firms may have done so (for examples, see Lloyd and Shutt (1985) or Peck and Dicken (1994)) but capacity reductions, closures and job losses have more often been attributed to technical change than to the New International Division of Labour (NIDL) (e.g. Fothergill and Guy, 1990). Where companies have closed capacity in the UK and expanded it abroad, this has often been within Europe where the states of the former Eastern Bloc offer potential for large labour cost savings.

Instead of seeing globalisation as a one-way shift it is more helpful to look at the net inflows and outflows of investment. Hirst and Thompson (1996) show that some 91 per cent of FDI is located in countries containing only 43 per cent of the world's population. The largest flows are between the core capitalist nations; 93 per cent of EU investment and trade takes place within EU states. FDI is significant to the British economy: foreign-owned firms account for over 10 per cent of GDP and inward direct investment for about 12 per cent of domestic fixed capital formation. However, *net* flows are typically in an outward direction and British investments overseas are much higher, proportionately, than their German or American equivalents. Though net flows of FDI are large, the statistics on flows of sales and asset volumes do not appear to show that the UK's economy is internationalised to a much greater extent than its competitors (with the exception of Japan). Assets and sales are also spatially concentrated: in a global marketplace one might expect an expansion of sales abroad but the manufacturing multinationals analysed by Hirst and Thompson do not conform to that expectation. Instead, sales in their 'home region/country' comprise at least two-thirds of their total sales (Hirst and Thompson, 1996, 80–3). A similar pattern is evident for sales and assets in both services and manufacturing (Allen and Thompson, 1997). Finally, considering the distribution of subsidiaries of UK firms, the UK appears to have a greater *spread* of subsidiaries than comparator states, but there is nothing particularly novel about this. There has also been an expansion in the number of international strategic alliances between firms since the early 1970s (Kozul-Wright, 1996, 155).

One inference might be that 'pure' globalisation has not occurred, but we cannot easily dismiss the idea of globalisation: there is considerable evidence for globalisation tendencies and processes. What may be more novel and influential than the globalisation of production is the *scale* of the globalisation of finance. Two separate developments are crucial: the rise in oil prices during the 1970s, which created a vast increase in the quantity of money in circulation; and technological change, which permitted a rapid acceleration in the circulation of money as exchange rates and commodity prices could be adjusted almost instantaneously. These changes permitted a rapid expansion in financial services as surpluses of cash had to be recycled; various new financial instruments were created to this end and trading in these (the Eurodollar market, for example) provided a further boost to financial centres. The scale of the financial system and the effects of technological change made it impossible to maintain fixed exchange rates or (as a result) capital controls. These were abolished (the latter by the Conservatives after the 1979 election) giving a major boost to London as an international financial centre. London was particularly well placed to benefit from the expansion of finance because of the multiplicity of markets in the City, which reduces the risk of launching new financial products (Hutton, 1995, 144). The subsequent deregulation of the City (the 1986 'Big Bang') gave a further boost to the City of London; this was designed to help secure London's position as a premier trading centre in financial services. The expansion of the City also gave more grist to Hutton's mill; he argued that the rates of return available in the financial sector made the position of industry even less tenable.

The Conservatives' reaction to the globalisation of finance was a straightforward attempt to work with the grain of the market, by deregulation. This raises a wider debate concerning what can or cannot be done about globalisation. Hirst and Thompson insist that there remains scope for action by the nation-state but others are more sceptical. In pessimistic accounts, nation-states are reduced to 'local authorities of the global system', whose job is simply to provide the 'infrastructure and public goods that business needs at the lowest possible cost' (Hirst and Thompson, 1996, 176). However, Hirst and Thompson argue that very few firms would welcome a genuinely globalised open economy: firms benefit from being located in a specific national base; embeddedness within national business cultures is important to firms; the stability provided by national legal and regulatory frameworks also offers benefits. Hence firms have a strong interest in the continued public governance of the economy (Hirst and Thompson, 1996, 188; Kozul-Wright, 1996, 160–1). Nation-states therefore continue to have a role, albeit as one political agency in a system of power from world to local levels. In this system some powers would be ceded to supranational bodies, to rein in the worst excesses of globalisation, while others are devolved to regions or localities, in what some term a hollowing-out of the state.

Such arguments have been challenged as politically reformist, since recent economic summits seem more focused on reducing to a minimum such obstacles as still remain to globalisation (Harman, 1996, 21–2). Even if this were not so, the veto power available to international capital markets is pivotal to the macroeconomic strategies of nation-states, forcing convergence between governments of quite different political colours. This, coupled with the impact of low-cost production in the East, is the real meaning and impact of globalisation: not a direct and visible global shift, but an emergent process with an insidious capacity to alter the behaviour of firms and states (Hutton, 1995, 312).

This restricts the state's scope for manoeuvre. One argument (the 'low road' to flexibility) is simply to cut labour costs in an attempt to ward off competition. There are plainly limits to such approaches unless a government is prepared to sacrifice its welfare state. The alternative is to accept, perhaps within a more integrated Europe, that the most feasible course of action is to invest in the skilled labour that will attract mobile capital. The Blair government's approach appears to accept the terms of globalisation (and thus the need for wage restraint) while arguing for the latter, though it remains to be seen whether the result will be an advance on previous failed modernisation strategies.

2.4 GEOGRAPHIES OF IDENTITY AND CITIZENSHIP

A nation's place in the world is not determined solely by its economic and political clout; nations are social constructions and political strategies are built on discourses about the character of nations. In Anderson's (1983, 15) words, attempts are made to create 'imagined communities': even in the smallest nation citizens will only know a minute proportion of their fellows, and so attempts are made to create a sense of community and belonging, by appealing to defining features of nations. These issues are illustrated with a discussion of the construction of nation and national identity in Britain, placing particular stress on the way these processes interact with geography.

The implicit assumption, in political discourse, of the homogeneity, stability and 'naturalness' of nations has been contested by a corpus of work which has argued that 'nations' are social constructions rather than pre-given, immutable categories. Indeed some commentators argue that nations, and national traditions, are very recent inventions (Hobsbawm and Ranger, 1983). Constructing national identities entails emphasising attachment to place, conceived of as a 'repository of socially and politically relevant traditions and identity'; place is not just a context for action but a source of identity. But how does the process of constructing national identity work?

For Penrose (1993, 29) there are three components of the concept of 'nation'. Nations comprise a distinctive group of people, and these groups are natural and immutable. Second, they occupy a specific territory; we have come to accept as 'natural' the idea that specific social groups occupy distinctive regions. Third, there is a conception of a quasi-mythical 'bond' between people and place, which 'melds, "naturally", the other two components into an immutable whole'. Penrose goes on to argue that because the general category of 'nation' is the central organising unit of the world order, the convincing construction of a specific nation is an essential prerequisite to the assertion of political power. Hegemonic political strategies therefore must attempt to create a credible image of 'the nation' around which to mobilise loyalty. There are at least three problems associated with this in a British context, relating to: the relationship between Englishness and 'Britishness'; the problems posed for the concept of an essentially 'white' national identity by non-white British citizens and immigration; and the extent to which national identity can encompass both urban and rural elements.

'Englishness' and 'Britishness'

The dilemma here is well illustrated by Samuel's admission, in the major edited volume on *Patriotism: the making and unmaking of British national identity*, that 'British' was substituted for 'English' in the title at a late stage, indicating the relative ease with which contributors had tacitly accepted the parameters and associations of 'Englishness'. If nations are defined in terms of the relationship between a social group and a specific territory, then the absence of agreement on terminology like this indicates a problem. Seen from England, there might be an equivalence between 'Englishness' and 'Britishness' but from Scotland or Wales, national identities would be defined very differently, in dual terms (Scottish and British; Welsh and British) (Osmond, 1988). Partly as a consequence of this ambivalence there is a reluctance to admit that nationalism exists at all (Chambers, 1993, 146; Nairn, 1977, 293), which is ironic given the ways in which the Anglo-British are enthusiastic celebrators of national identity (evidenced in pageantry, obsession with royalty, etc.). Instead patriotism takes the place of nationalism so that Taylor (1991a, 148) interprets Anglo-British identity as a 'deformed nationalism hiding behind its self-ascribed patriotism'. There are a number of reasons for this (perhaps best explained by Nairn (1988)) mainly to do with the peculiarities of the British state and Britain's incomplete modernisation in the age of revolutions. The monarchy was preserved and used as a symbol in national mobilisation, becoming a substitute for a democratic nationalist identity (Osmond, 1988, 189).

This is, of course, a specifically *English* nationalism; 'Britishness' simply 'stands in for the quaint authority and organic continuity of an English conservatism' (Chambers, 1993, 146). In fact, this Englishness is rooted

in a comparatively small proportion of the national territory. In an amusing demonstration of this point, Taylor (1991a, 150–4) maps places which, over time, have acquired associations with royalty. The geography of royal places is dominated by the Home Counties, and this pattern also applies for the other key elements in the social hierarchy such as the major public schools and important cultural and social events and venues. It is no surprise that patterns of economic development are so centralised when these sources of social power are so spatially concentrated. He also maps the distribution of places pictured in a photographic collection of images of England (Figure 2.1). The principal images represented say much about the components of Englishness and they, like his royal places, display an overwhelming southern concentration, which is deliberately accentuated by the unusual orientation of the map.

The ambiguities of national identity became highly visible under the post-1979 Conservative governments. By the 1980s the symbols of Britishness – the Empire, the monarchy – had lost much of their integrative force (McCrone, 1992, 210). However, the Conservatives demonstrated that it was not necessary to appeal to a national interest, but only to the 'old metropolitan heart of the Empire' (Gamble, 1988, 214) to secure a comfortable majority. It became increasingly evident that their vision of national identity – an entrepreneurial, individualistic one – found few supporters in the celtic fringe. However, the emergence of a nationalist alternative was inhibited by the fragmented nature of Welsh and Scottish identities. Osmond (1985, xix–xviii) questions homogenising views of Wales, arguing that at a minimum there are three distinct political areas, which limit the prospects for the emergence of a coherent national identity (see also Giggs and Pattie, 1992). The same could be said for Scotland where there is continual internal rivalry – between Highlands and Lowlands; between older industrial areas in the west and the oil-based economy of the east; and even between Glasgow and Edinburgh (Harvie, 1994; McCrone, 1992). In Northern Ireland, too, attention has been drawn to the need for more diverse and pluralist conceptions of

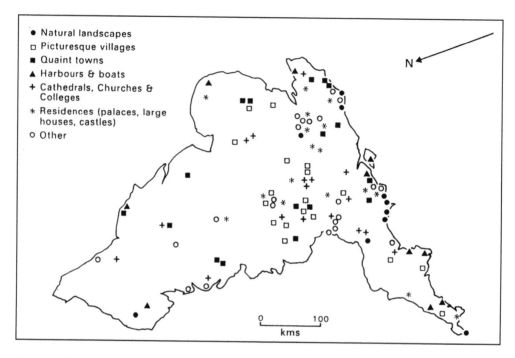

- • Natural landscapes
- □ Picturesque villages
- ■ Quaint towns
- ▲ Harbours & boats
- + Cathedrals, Churches & Colleges
- * Residences (palaces, large houses, castles)
- ○ Other

N

0   100
kms

FIGURE 2.1 Landscapes of Englishness (Taylor 1991a)

identity than those constituted purely in sec-tarian terms (Graham, 1997; Graham and Shirlow, 1998). Even within England there are growing signs of the assertion of regional distinctiveness, in the context of debates about regional government (chapter 11).

However, Miller (1998, 192–4) argues that identities are not exclusive categories; people have problems choosing between, say, a British and a Scottish identity. The 1994–95 Local Governance survey asked respondents to indicate how much they identified with their local government district, their 'region' (including Scotland, Wales and eight regions of England), Britain, Europe and various other sites. Interestingly, the balance between British and regional identification varied. In the English South and the Midlands, identification with Britain ran well ahead of identification with region; in Yorkshire and northern England, identification with region was about equal to identification with Britain; while in Scotland and Wales identifi-cation with the region ran well ahead of iden-tification with Britain. Nevertheless, there was strong evidence of *multiple* identifica-

tion: of those identifying strongly with place (scoring at least 6 on the scale used in Miller's study), at least 60 per cent in every part of Britain identified strongly with *both* Britain and their region. There are clear dif-ferences as between Scotland and Wales (with strong regional identification) and the English South (with a strong identification with Britain alone) but the most consistent feature is the scale of multiple identification indicating that identities are typically both plural and contingent upon particular circum-stances (Miller, 1998, 193).

These ambiguities are problematic from the perspective of political strategies. 'Britishness' is, on the one hand, a more mod-ern and inclusive concept than 'Englishness', since it does not presume a common culture; it thus allows for a more pluralistic under-standing of the nation, which sees it as a citizenry. However, it is tainted by its associ-ations, particularly with military conquest, Empire and the nation-state. Englishness, on the other hand, has several positive associa-tions evoking a people rather than a state, a literary tradition, and an image of rural tranquillity (see p. 32). But it is equally to be regarded as regressive, xenophobic and exclusionary. Reconciling the tension between 'Englishness' and 'Britishness' as competing visions of identity remains a key problem.

'Englishness' and 'whiteness'

Similar difficulties attend the concept of Britishness in respect of multiculturalism. Conventional visions of British identity are overwhelmingly white, with little space for immigrants or non-white British citizens, a view vigorously challenged by Gilroy (1987). He shows convincingly how symbols of 'race' are mobilised in Conservative rhetoric about national decline which is 'presented as coinciding with the dilution of once homogeneous and continuous national stock by alien strains' (p. 46). Discourses about 'race', furthermore, are often conducted through metaphors which draw analogies between war and conquest, on the one hand, and immigration, on the other. Terms such as invasion, occupation and alien territory figured strongly and frequently in the rhetoric of Enoch Powell and his followers, for example. Speaking geographically, such language directs attention to policing the boundaries of citizenship and determining who is entitled to entry and residence. Smith (1993) argues that Britain has always seen itself as a country of emigrants, has never positively welcomed immigrants, and rarely accommodates them in significant numbers. Immigration policy therefore seeks to protect the integrity of the national character. In the 1960s legislation sought to restrict the entry of people without a demonstrable claim to an 'Anglo-Saxon heritage'. By the mid-1970s immigration had been reduced to a trickle via a strict exclusionism, in which the economic benefits which might have accrued from a more relaxed policy could not offset a concern for the nation's cultural heritage. During the 1980s the Thatcher governments pursued largely symbolic policies which emphasised the exclusion of non-whites, regardless of the economic costs of doing so (Smith, 1993, 61–2). Policies towards refugees exemplify this; immigration control is one

area where Britain seems to wish to maintain independent policies from the rest of the EU, effectively double-locking the door on immigrants. Whereas economic migrants from the EU can enter the UK freely, this is in stark opposition to asylum seekers, who must prove that their motivation is *not* economic (p. 67). Smith also shows how policies operate so as to exclude those whose family background is not spatially rooted in the UK. She therefore argues that nations are only partially described by lines on a map; the policies governing access to territory are equally significant, and territorial boundaries may be less significant for what they physically contain than for what they symbolically exclude. Smith (1993, 71) concludes that immigration policy places the preservation of a particular type of society before the well-being of the economy, and prioritises parochial aims of nation-building over a more universal code of human rights.

Claims to citizenship and discourses about identity are not restricted to arguments about who shall overcome immigration hurdles. Whatever the promise held out of citizenship rights, in practice there has been consistent evidence of racial segregation in Britain, and this in turn structures the opportunities open to immigrant groups, in terms of housing and labour markets, and access to welfare services. When New Commonwealth immigration began after the Second World War, it was initially anticipated that the residential clustering of immigrant groups would be a short-lived phenomenon, and that dispersal and integration would eventually solve the highly visible 'problem' of spatial concentration. The Notting Hill riots of 1958 effected a transformation in attitudes here, drawing attention to the numbers and spatial concentration of non-white immigrants. This was interpreted as a cause of environmental decay and a 'harbinger of competition for scarce resources between "black migrants" and "white indigenes"' (Smith, 1989a, 158). Crucially this was not seen as an effect of racism but as a cause of prejudice. This cleared the ground for Enoch Powell's speeches which represented racial segregation as a threat to the national territory; for instance, he referred to the 'transformation of whole areas ... into alien territory' (quoted in Smith, 1989a, 159). One response was a tightening of immigration controls. Another possible response would have been the pursuit of a more pro-active housing policy aimed at achieving greater integration. However, responsibility for dispersal was decentralised to local authorities and no clear policy guidance was given. Despite some pro-active policies (Atkinson and Moon, 1994), processes of individual and institutional discrimination still circumscribe the housing opportunities open to black people, who do not enjoy access to the basic services of the welfare state on the same terms as the white majority. There is considerable evidence of concentration and residualisation of the non-white population into areas of poorer-quality housing and schools (Phillips, 1998). Thus, the boundaries of the nation are restricted in at least two senses: first, through restrictions on membership of the national community expressed in immigration controls; and, second, through processes of discrimination in the housing and education system which reproduce inequality for non-white citizens. Smith (1989b, 170) therefore concludes that racial segregation indicates black people's limited access to citizenship rights and symbolises the force of white supremacy in Britain.

'Englishness' and rurality

The approved and dominant images of England are pastoral and green (P. Wright, 1985, 87); on being asked what is 'really' England, 'many urban children reply with images of the monarchy, the flag and ... most interestingly, of the countryside, the "green and pleasant land"' (Williams, 1983, 181). Just what the characteristics of Englishness are, and where they are located, defy easy definition, so that Wright speaks of 'the vagueness of deep England' (1985, 81–7). There is an awareness that the virtues of Englishness comprise 'tradition, order, religion, idiosyncrasy', which are 'located in the South, and especially the rural South, and embodied in a matching symbolic scenery' (Matless, 1994, 78–9). However, pinning these down has reduced writers to producing

eclectic examples of definitive images, ranging from Stanley Baldwin's 'sight of a plough team coming over the hill' to John Major's 'warm-beer-and-village-green nostalgia', in attempts to communicate the incommunicable and define the indefinable. Once again, though, as with Taylor's geography of key places in the British establishment, the Home Counties are hegemonic.

This identification with rurality and the virtues of rural life raises three issues. First, and connecting back to the literature on decline, it underpins a nostalgia for a lost world and works against modernising impulses (Wiener, 1981). Attempts to recreate an idealised past therefore account for a disproportionate share of the nation's resources (Hewitson, 1987). Second, it is associated with an anti-urbanism, which has impinged on attempts to regulate and plan urban development for many years. The influence of anti-urbanism is evident in the extent of conflicts over development, particularly in (but not confined to) the South East, reflecting the difficulties of accommodating those fleeing the cities in search of a rural idyll. Allen *et al.* (1998) regard such desires not solely as individual preferences but as social constructions: aspiring to and achieving a particular rural lifestyle is seen as a key indicator of social status. This construction of Anglocentric rurality is an exclusive one: it draws upon the 'networks of privilege which make up the established southern middle class' (Allen *et al.*, 1998, 29). Finally, it might be tempting to dismiss all these rural metaphors as 'harmless romanticism' but, as Osmond (1988, 157–9) observes, when they are combined with 'a focusing of imagery almost exclusively on southern England there is a distinct sense of a split opening up' both within England (see Steed, 1986) and between England and the rest of the UK.

2.5 CONCLUDING COMMENTS

Identifying Britain's place in the world is difficult because, as the foregoing discussion has shown, cartographic exactitude is impossible: some enduring features of the character of the British economy might be identified but economic decline is always relative (the events of 1998 may force a reappraisal of the comparative position of South East Asia, for instance) while identities are always unstable and in flux.

What we can say is that political strategies are usually formulated as attempts to deal with relative economic or political decline. They therefore start from a diagnosis of the causes of such decline, of which there are many available possibilities; moreover (though not developed in this chapter) accounts of economic decline are clearly applicable to places within the UK as well as to the national economy. Discussions of decline also point to alternatives; the most relevant recent illustration is the analyses of the institutional diversity of capitalism (e.g. Crouch and Streeck, 1996).

In turn, analyses of decline cannot be separated from contemporary concerns about globalisation and, in particular, whether there remains scope for relatively autonomous economic management on the part of nation-states. Even if the reality and extent of globalisation are disputed, it is a crucial parameter determining state policies (chapter 3): the major political parties both appear to accept that there is little states can do but acquiesce to the pressures of globalisation. Related to this, though not considered here, is Britain's position *vis-à-vis* the European Community, since there is debate about whether such supranational institutions, perhaps with a subsidiary federal structure, offer more effective protection against globalisation.

This in turn raises questions of identity, because under the joint pressures of globalisation and of greater political integration, the nation-state is plainly under threat. In these circumstances, mobilising consent for a political project is becoming harder because there is no available alternative sense of British identity to the 'Anglo-British' one that has dominated British politics for many years. According to Samuel (1998, 43), 'no less striking than the collapse of British power ... is the unravelling of any unitary idea of national character'. The Conservatives have swum against this tide more than once: military adventurism in the Falklands provided one (successful) opportunity, but it is less clear that current attempts to unite around an anti-European identity will be successful. Labour seems still to be groping towards a unifying identity – witness Blair's attempts to sell his vision of a 'young country' and proclaim the virtues of 'Cool Britannia'. There have been calls for alternatives. Osmond (1988, 266; see also Dodd, 1995; Haseler, 1996) asked us to imagine 'a different kind of Britain, one containing a patchwork of self-governing communities relating not only to each other but to a wider European context as well'. Whether this is feasible raises important questions about the character of state intervention and political strategy, to which we now turn.

NOTE

1 For assessments of Hutton's work see Barratt Brown (1995), Leys (1995), Hay (1996a) and Mohan (1996).

3

THE RECONFIGURATION OF
STATE INTERVENTION

3.1 INTRODUCTION

Explaining Britain's human geographies without reference to its place in the world economy is clearly no longer defensible. The neglect of state policies has also been a significant omission from many texts. Yet it was really only with the Thatcher years that serious attention was paid within British human geography to the impacts of state policies.

These impacts are highly visible but theorising the state has been contentious: precisely how, for what purpose, and in whose interests the state carries out its tasks has been extensively debated. Few would now adhere to the view from traditional pluralist political science that the state is merely a neutral arbiter between competing interests. Nor can the state be viewed simply either as the agent of the capitalist class or as an 'ideal collective capitalist' which somehow takes responsibility for those issues (for example, law and order, welfare, or infrastructure) which offer insufficient potential for private profit. One response to such deterministic approaches has been to 'bring the state back in', emphasising its autonomous powers (Driver, 1991); certainly during the Thatcher/Reagan decade it appeared that the state had the capacity decisively to shape the political agenda (Krieger, 1986).

Critics of such views have argued from a Marxist perspective that one must not lose sight of the fundamental constraints on state activities imposed by the process of capital accumulation (Jessop, 1990). A key contribution of regulationist accounts of contemporary capitalist societies has been to emphasise the ways in which this process can be stabilised. Theorists emphasise the importance of a relatively stable aggregate relationship between production and consumption, and this necessitates a correspondence between regimes of accumulation and modes of regulation (Amin, 1994; see chapter 1). The concept of a *mode of regulation* refers to regulatory mechanisms, institutional ensembles, and a body of beliefs, habits and norms that are consistent with and supportive of a regime of accumulation (Jessop, 1990, 310). In this context, the role of the state is, broadly speaking, to ensure that changes in the conditions of production and consumption are consistent and compatible (for example, via taxation and expenditure policies). The state also seeks to ensure that the mode of regulation is successfully reproduced (for example, via education and health policies). State intervention is predicated on three elements: an accumulation model, a hegemonic project which has the support of an appropriate social coalition, and particular forms of statecraft (Martin, 1992, 126–8).

An *accumulation model* implies a politically and ideologically grounded vision of: the pattern of economic growth; how that pattern can be sustained; how the benefits from accumulation should be distributed across the social base; the position, or insertion, of the nation in the international economic system; and the role and scope of state intervention in promoting or regulating the accumulation process. Here, commentators typically identify a decisive break between Keynesianism and Thatcherism, contrasting the respective priorities given to the various sectors of the economy, the extent of state intervention in the economy, and the extent to which the benefits of growth were equally distributed. Successful accumulation strategies also rely on *hegemonic projects* of political, intellectual and ideological leadership, which in turn rest on an appropriate social coalition. Thus, Thatcherism crucially sought to entice the electorate with the prospect of reversing national decline, ending the dependency culture, and revitalising the spirit of self-help and enterprise. Attracting the support of key social strata (the skilled manual working class, historically strong supporters of Labour) was central to this. Distinctive styles of state intervention, or *statecraft*, appear to be associated with particular historical periods; a central claim of the Conservatives, again, was that the post-war social democratic state apparatus had had its day. In the process both business and labour interests were marginalised as state power came to be implemented in an authoritarian, centralised way (Gamble, 1988; Jessop, 1990, chapter 7; Hudson and Williams, 1995; Martin, 1992, 128).

Of course, state intervention is not unproblematic and various crisis tendencies may test the limits to forms of intervention and, *in extremis*, lead to a search for a new mode of regulation. The state may experience a legitimation crisis, as it is unable to deliver on its agenda, a fiscal crisis as expenditures exceed revenues, or a rationality crisis as the competing imperatives to which the state is subject make the pursuit of policy impossible without negative, unintended consequences (Painter, 1995, 85–8). These are tendencies but occasionally, when such tendencies are actualised, they can at the very least bring down governments and even pose questions about the character of state intervention itself. Consequently turning-points can be identified in the character and extent of state intervention. Thus, in the UK the era of 'con-

sensus politics' has been contrasted with the subsequent two decades of Thatcherism. The post-war consensus reached a number of limits, including fiscal crises as the Fordist regime was exhausted and legitimation problems as the state was unable to deliver on its commitments. It is questionable whether subsequent events have seen the emergence of a stable successor to the era of consensus politics. Thatcherism reached limits of its own, though some now claim to discern a one-vision polity because of the extent to which the major parties now share the same assumptions.

These arguments are illustrated here by examining the replacement of consensus politics by Thatcherism, the reasons for the downfall of the Conservatives, and the rise of Blair's New Labour. The aim is to draw broad-brush contrasts and attention should be drawn to inevitable problems of determining when particular epochs began and ended (exactly when did 'consensus' break down?), to differences within periods of ostensible consensus (e.g. Harold Wilson's move towards economic planning in the 1960s, or the shift towards a more radical Thatcherism after the 1987 election) and to tensions within parties (e.g. the tension between neoliberal and neoconservative philosophies: Hay, 1996b, 131–5).

3.2 THATCHERISM AND MAJORISM: FROM ONE NATION TO TWO, 1979–1997

The three decades of consensus politics after 1945 may now seem like ancient history, but some discussion of that era is necessary for two reasons: first, and most importantly, to illuminate what was novel about Thatcherism; second, because the economic policies of the 1945–75 era consciously sought, through expenditure and taxation programmes, to reduce interregional disparities.

One nation politics, 1945–1975

Most commentators would date the end of the era of consensus, 'one-nation' politics to

1975, the year when massive public expenditure cuts were implemented by Labour in response to demands from the IMF. Until then, in terms of an accumulation strategy, both main parties accepted the mixed economy, a substantial degree of public ownership, and the use of Keynesian techniques of economic management to stabilise the economy and promote balanced growth. The major political parties drew support from all regions of the country by pursuing 'One Nation' politics. This involved an inclusive commitment to full employment and a comprehensive welfare state with the intention of convincing the electorate that there would be no return to the inequalities of the pre-war years. In terms of statecraft this era was notable for corporatist structures for policy-making in which attempts were made to involve labour and business interests in decision-making.

It is easy to exaggerate the degree of consensus and the integrity of the post-war settlement. For Pimlott (1989) the notion is a product of retrospective assessment; the main parties were some distance apart in policy terms in the early post-war years, and the Conservatives only accepted the new agenda following their massive electoral defeat in 1945. Others point out that the Conservatives consistently explored the possibility of reversing post-war initiatives (e.g. in health care: Webster, 1990; see also Timmins, 1995). More fundamentally, Jessop (1992, 16–17) suggests that the era of consensus contained the seeds of its own destruction because of the unresolved tension between economic and social priorities within it. Put more crudely by Barnett (1995), Labour prioritised building the 'new Jerusalem' over reconstructing the industrial base.

A distinctive feature of the consensus era was the attempt systematically to reduce spatial inequalities. Martin and Sunley (1997) argue that, under the Keynesian welfare state (KWS), the national economic space was the essential geographical unit of economic organisation, accumulation and regulation. Keynesian politics required a high degree of closure of the national economy in order that domestic policy measures could have the

desired effect. This is in contrast to the circumstances which obtained during the 1980s and 1990s as exchange controls were abolished and the economy became increasingly open. Second, Keynesianism required a degree of spatial centralisation of the political regulation of the national economy. This was because demand management (Keynesianism's key policy innovation) required centralised powers of coordination and manipulation of fiscal and monetary policy. Third, there was a high degree of spatial *integration* of society, for example, via welfare policies designed to foster consistent national standards across all regions of the UK. This contrasts sharply with more competitive modes of regulation such as that prevailing in the USA. Fourth, the KWS's model of economic intervention was essentially redistributive and stabilising, with the net effect of *reducing* inter-regional income differentials (MacKay, 1994 and 1995), through public expenditure programmes and public employment (though see chapter 7). It is debatable whether the pursuit of *geographical* equity via Keynesian policies was a deliberate choice rather than a by-product of other policies, but by comparison with the 1980s and 1990s the Keynesian era was one of socio-spatial convergence and the Thatcher years were ones in which it seemed that spatial inequalities were quite deliberately used as an element in a political strategy. Privatisation programmes, for example, were motivated by strategic calculations as to the constituencies and territories that stood to benefit.

However, the basis of Keynesian policies was undermined, in two ways. First, Keynesianism proved unable to resolve economic rigidities, as evidenced by inflation, unemployment and the performance of 'lame duck' industries; the rising costs of welfare provision provoked a fiscal crisis; and the government's inability to control the demands of organised labour ultimately undermined its legitimacy. Second, the possibilities for national economic intervention were undermined by globalisation, because of the porosity of national boundaries (chapter 2). These conditions arguably give states no option but to withdraw from extensive regulation and intervention and to minimise the tax burden, thereby ceding economic power to global markets and corporations. Such arguments raise the historical question of the differences between the Conservative governments and their predecessors, and also the pressing contemporary question of the differences, if any, between the Blair government and the Conservatives.

Two nation politics, 1979–1997

While the Callaghan government were eventually forced to abandon some key commitments and principles of the consensus era, the Thatcher governments ushered in what came to be known as 'two-nation' politics. The Thatcher governments diagnosed the causes of relative decline in terms of an overactive state; the overwhelming priority was therefore to roll back the state and thus, in true neoliberal fashion, re-establish the primacy of markets as a steering mechanism. While Thatcherism was often (initially) equated with monetarism, there were several elements to Thatcherite economic strategies: liberalisation, deregulation, privatisation, recommodification of the residual public sector, internationalisation, and tax cuts. These were implemented against the background of a tight fiscal policy aimed at controlling inflation. These diverse policies had a number of aims.

First, manufacturing productivity and competitiveness were to be boosted; a tight fiscal policy, and constraints on borrowing, necessitated restructuring of production. This would produce a 'leaner and fitter' industrial base, by promoting a shake out of excess labour and unproductive capacity. Second, in order to renew corporate profitability, the public sector was cut back; it was claimed that public expenditure had 'crowded out' productive investment, and that profitability had also been eroded by excessive inflation and wage settlements. Third, productive investment had to be reconstituted in order to shift resources from declining and traditional industrial sectors into new, high-technology sectors, processes and products. There were also attempts to stimulate enterprise through market deregu-

lation and the encouragement of 'popular capitalism'. Fourth, the boundary between public and private sectors was (selectively) redrawn via numerous privatisation initiatives (Feigenbaum et al., 1998). Within what remained of the public sector, policies of selectivity and targeting became the norm as the fiscal squeeze intensified, while market mechanisms were introduced as and where politically feasible. Finally, there were attempts to reduce the power of organised labour. Unions were regarded as monopolistic organisations which distorted the operation of markets and held down productivity growth (Graham, 1996; Jessop, 1992, 30–1; Martin, 1992; Hudson and Williams, 1995).

This is a broad summary of Conservative strategy, but there are important qualifications to be made. First, this was not a blueprint, implemented systematically once in government. Commentators draw attention to the varying emphases on different policies over time, most notably the identification of a (brief) heyday of 'radical Thatcherism' after the 1987 election (Gamble, 1988). Second, political considerations often overrode economic rationality – for example the continued fiscal privileges of owner-occupation, the search for tax cuts, and the simultaneous reluctance to reduce various tax breaks which benefited the middle classes. Thus 'the government's concern to reward its supporters [interfered] with the pursuit of a rational economic strategy' (Jessop, 1992, 36–7; see also Jessop et al. 1990). Third, Thatcherism was always far more than a narrow economic project: instead, Hall argued that Thatcherism was a hegemonic project, seeking to mobilise popular consent through a range of rhetorical devices (S. Hall, 1985; see also Jessop et al., 1984 and 1988). For Hall, the post–1979 Conservative governments offered a distinctive combination of populism and authoritarianism (Hay, 1996b, 136–51).

It was populist in its appeals to self-interest and independence, in opposition to the dependency culture of a decaying welfare state. Thus, there was a deliberate and systematic attempt to shift the pattern of social interests and values away from the inherited culture of state support, subsidy and dependency, towards a new ethos of self-help, enterprise and individual initiative. This was articulated rhetorically in various ways: slogans such as the enterprise culture, popular capitalism, and rolling back the state gained widespread currency (Hay, 1996b, 149–50). It was supported, in material terms, by several strategies of which privatisation, and the sale of council houses, accompanied by tax cuts, were arguably the most significant. The first two of these gave the government more scope for the latter, while one effect of privatisation policies was a weakening of trade union power. The effects were ultimately that, through share ownership or owner-occupation, many more individuals now had a material stake in the success of the Thatcherite project (Feigenbaum *et al.*, 1998).

However, Thatcherism did not stop there, eventually extending market principles into the core services of the welfare state. The post-1979 governments had in effect reintroduced a division between the 'deserving' and the 'undeserving' in welfare, so that services used mainly by the poor became residualised; this was especially true of housing. Education and health care had been largely immune from this, but after the 1987 election the Conservatives felt confident enough to embark on market-led reforms in education and health care. The rhetoric of freedom and choice appealed strongly to voters – particularly in respect of schools, where local housing markets came to be driven by the quality and perceived success of schools, leading to greater social segregation (chapters 8 and 9).

Thatcherism could consequently be described, in contrast to the era of 'One Nation' policies, as a 'two nation' politics of inequality. Rhetorically, this worked through a number of oppositions: between dependence and independence, public and private, state and market, 'North' and 'South', manufacturing and services, collectivist versus individualist, and so on. In Gamble's terms the aim was an essentially geopolitical one: the 'South' would ultimately swallow the 'North', conquering the remaining bastions of outmoded economic, social and political structures (Gamble, 1989).

Expanding the scope for the operation of market forces required the powers of a strong state. The British state is uniquely centralised, which grants exceptional leverage to a government with a large majority. The Thatcher years were characterised by systematic and determined use of the powers available to the central state; this was perhaps most visible in the coal dispute of 1984–5, where the government were plainly determined to defeat the miners by any means necessary, even to the point of creating a *de facto* national police force (Beynon, 1985). It was also evident in a succession of centralising measures which Jenkins (1995) termed the 'Tory nationalisation of Britain'. The disciplinary powers of the central state were used to sweep away obstacles to reform, as well as to coerce and contain those marginal to the Conservative project.

To what extent did the Thatcher era represent and produce a decisive transformation? There will always be a gap between what a government would like to see done and the effects of specific legislation designed to bring about desired outcomes. Consequently some have pointed to continuities as well as discontinuities, and areas where radical change has been hindered, to argue that the Thatcherite revolution was more rhetorical than real (Marsh and Rhodes, 1992). Against this, a distinctly Thatcherite strategic agenda *can* be identified; with respect to privatisation, union reform, local government reform and the welfare state, Hay (1996b, 152) contends that the Thatcher governments 'presided over a period of structural transformation perhaps unprecedented in the post-war period'. He argues this on the basis of the limited likelihood of any initiatives in these areas being reversed.

Leys (1990) presciently suggested that the Thatcherite settlement could prove to be as long-lasting as that of the KWS, while Kavanagh (1997) argued that Major's 1992 election victory indicated the extent to which Thatcherism had transformed the political landscape and ushered in a new consensus. For while Major may have attempted to present Thatcherism with a human face, and thus soften its radical edge, in practice the 1992–97 Major government continued to pursue Thatcherite policies (Hay, 1996b, 166–79) most notably in a gradual shift towards a 'workfare' rather than a 'welfare' state, and the extension of privatisation via the Private Finance Initiative (PFI), a *de facto* commercial approach to infrastructural provision (e.g. roads, but also health care). Major also sought to create a niche for Britain on the edge of Europe by opting out of the Social Protocol of the Maastricht Treaty, but this served to construct the UK as an offshore assembly line for Far Eastern and American multinationals.

Nevertheless, in certain senses the Conservative governments did not deliver the desired results. Productivity and competitiveness improved mainly because the 'tail-end batsmen (the least efficient plant) were shot' (Coates, 1994), while public expenditure remained high due to the social costs of mass unemployment. Moreover, the Conservatives' accumulation strategy and hegemonic project ultimately proved self-defeating and ran up against several limits.

First, the *laissez-faire* approach to economic management eventually produced diseconomies and externalities because of the economic boom in the South East, the costs of which became evident in rampant house price inflation, skill shortages, congestion and pressures on the environment and public services. The consequence was a resurgence in inflation from the early 1990s but the corrective action taken by the government (a substantial hike in interest rates) tipped the whole economy (not just the South East) into recession, choking off an emergent recovery in the peripheral regions. In this sense regional imbalance was both an effect of, and a constraint on, national economic renewal (Martin, 1989a). This imbalance was not unconnected with government policies – principally, tax cuts – which had disproportionately benefited the South East, fuelling a consumption- and import-led boom.

Second, Thatcherism's labour-market policies produced greater segmentation and failed to deliver a more highly skilled workforce. Labour markets were to be regulated

purely by price competition in a 'low road' to flexibility, but there are definite limits to how far wage levels can be cut in this way, while the jobs so generated are of poor quality and lead to an erosion of the skill base. At the same time, training policies were left to the market, producing poaching of skilled personnel because there was little incentive for individual firms to invest in training. Skill shortages continue to act as a constraint on growth, and the pursuit of this 'low road' has become discredited as a labour-market strategy (chapter 6).

Third, Thatcherism was ultimately self-undermining as a political project. By unleashing free markets and diminishing the power of organised labour, Thatcherism undermined economic security among those social groups who were its initial target audience and beneficiaries. Those sections of the working class aspiring to upward social mobility:

> emerged not in the sunlit uplands of bourgeois security, but onto a desolate plateau of middle class pauperdom. The poignant irony of Essex man and woman struggling up the economic escalator only to meet the bedraggled figures of the professional middle classes staggering down is a narrative of our times.
>
> (Gray, 1996, 11)

The increase in insecurity – expressed most vividly in rising levels of indebtedness and house repossession – came to be blamed on a government whose neoliberal programme offered little by way of relief (Gamble, 1996, 19).

Finally, Thatcherism arguably overreached itself in its attacks on local democracy and intermediate institutions. The Poll Tax fiasco is the most obvious example: in pursuit of a way of financing local government drawing on public choice theory (Hepple, 1989), the government generated a uniquely unpopular solution of flat-rate charging, which bore no relation to people's ability to pay. There was also a widespread sense that local democracy had been undermined by a battery of centralising measures, and that a quango state had emerged in which appointments depended on political patronage. Such perceptions of sleaze, corruption and centralisation were particularly important in the final collapse of the Major government, but they were undermining the legitimacy of the Conservatives well before that. Consequently, by 1997, Conservatism was exhausted and discredited. The government's reputation for competence in economic management had been undermined by the events of the early 1990s and although the economy recovered as a result of corrective action taken after the fiasco of 'Black Monday' in 1992, this was not enough to restore credibility. Economic insecurity and the difficulties experienced in the housing market alienated many of the Conservatives' core supporters. Finally, divisions over European integration threatened to split the party.

3.3 BLAIJORISM, THE BLAIR GOVERNMENT AND THE 'THIRD WAY'

Just as the Thatcher governments drew inspiration from intellectual attempts to provide a critique of the failures of 'planning', the Blair project can (broadly) be characterised as a critique of the failures of neoliberalism and an attempt to find a 'third way' between *laissez-faire* and state planning. This starts from the premise that as relatively closed national economies no longer exist, there is little scope for national-level economic management (see chapter 2). New Labour finds much common ground with the Conservatives; in particular, there is agreement that competitive advantage in the global market place lies in 'low non-wage labour costs, low levels of social protection, and deregulated labour markets' (Marquand, 1997, 336). There has been an inexorable shift of power towards the financial markets, which has resulted from an almost global economy, the deregulation of finance, and advances in technology which facilitate rapid switching of funds, so that actors in the financial markets are becoming a 'new global dominant class' (Crouch, 1997, 358). New Labour therefore finds itself in the familiar position of emphasising the role of

social policy in adjusting to market forces, while simultaneously proclaiming that nothing can be done about the dominance of finance capital, or about globalisation. For evidence of this one need look no further than Blair's response to the decision of Fujitsu to close a substantial plant in his constituency.

Labour may share common ground with the Conservatives on globalisation but there are differences with respect to the role of the state. There is a broad acceptance of the primacy of markets as steering mechanisms for the economy and (most clearly evident in debates about nationalisation) a concern to distance the government from 'old Labour' public ownership. Such forms of intervention have become impossible in an era of globalisation, flexible specialisation, social reflexivity and risk (Beck, 1992; Giddens, 1994a, 24–6).

Acknowledgement of past failings does not mean there is no role for the state: intervention may have been discredited but it is still necessary. There has been widespread recognition that if decisions are left purely to market forces, failures will result. Individuals or corporate entities operating in a market may take decisions which are rational from their point of view but which produce suboptimal results from a society-wide perspective. The unleashing of market forces during the 1980s demonstrated the paradoxical character of the strategies of the Conservatives: an attachment to traditional bulwarks of society (the family, the 'community') clashed with a neoliberal desire to roll forward market forces. The writings of John Gray exemplify this: originally a (qualified) supporter of the Conservative project, he then identified some of these internal tensions before ultimately concluding that Thatcherism was unable 'to anticipate that freeing markets would fracture communities, deplete ethos and trust within institutions, and finally mute or thwart the economic renewal which free markets were supposed to generate' (1996, 41; see also Gray, 1993).

The implication here is that successful capitalist economies require more than price signals and contractual relationships to make them function effectively. Free markets do not exist in a pure form; instead, markets are social constructions, shaped by particular institutional arrangements (Hindess, 1987). This directs attention to the institutional diversity of the world's capitalisms. There is evidence (e.g. Crouch and Streeck, 1996) that the more successful capitalist economies differ from that neoclassical vision in several important respects. The 'Rhenish model' of capitalism, for example, attempts to incorporate 'some sense of collective interest, long-term commitment, trust and cooperation, and recognition of the role of interests other than those of shareowners'. A generalisation of such a model could have included: institutional reforms, encouraging forms of cooperative, high-infrastructure, collectively oriented enterprise; measures at the international level to replace national attempts at regulating capitalism; and strategies to improve skill levels (Crouch, 1997, 356–7). Indeed some writers insist that in a globalising economy state intervention is even more necessary. Boyer and Drache (1996, 4–5) argue that the economic success stories of the Fordist era show that 'markets work best where the state is a strong regulator', while Castells (1996, 86–8) states that the 'informational global economy is . . . highly politicised': success depends on the active involvement of national states.

It is debatable whether it would be desirable (or feasible) to replicate other variants of capitalism in the UK. Nevertheless some elements of the institutionalist critique seem to have had an impact, notably Hutton's (1995) damning criticisms of short-termism and his proposals for a 'stakeholder' society. His argument is that Anglo-American capitalism is dominated by an emphasis on contractual property rights, deregulation and flexibility, with transactions mediated solely through price. This is a low-trust society in which the maximisation of returns to shareholders takes precedence over the interests of all others with a 'stake' in a firm. Hutton argues that a broader conception of governance is necessary in which employees, suppliers and consumers of firms would have an influence on decision-making. He cites German and Japanese evidence to show how long-term relations of trust, coupled with a demand for

reasonable financial returns over the medium term, can secure the future of companies which, under a more short-term and demanding regime, would have ceased to exist. Such relationships can also help protect the interests of other stakeholders.

Thus the advocates of stakeholding suggest that it confers two types of economic benefits (Gamble and Kelly, 1996, 25–6). First, instead of treating as many costs as possible as externalities, and disclaiming responsibility for them, stakeholding firms would seek to handle those costs through internal negotiation. Second, stakeholding is said to encourage better flows of information which will lead to the generation of mutual commitment. In turn, this is said to facilitate awareness of and response to risk, and in particular determining who is to bear what elements of risk. Set against these arguments, if each and every group affected by the activities of a company were said to have a stake in it, the idea of stakeholding would be unworkable. The firm could become deadlocked if too many stakeholders had decision-making powers; consumer or environmental groups might easily be in conflict with shareholders. However, these arguments do not invalidate the idea of stakeholding; they simply raise the question of whether or not it is better than its alternatives.

Clearly, the extent to which the state can achieve this kind of innovation is debatable. Here, instead of calling for socialisation, Hutton argues that the challenge is to devise ways of steering markets in ways which create 'a better economic and social balance, and ... a culture in which common humanity and the instinct to collaborate are allowed to flower' (Hutton, 1997, 64–5). This task will not be achieved by establishing a 'web of administrative bureaucracies', nor by top-down planning. Instead the state has to find points of 'maximum leverage on the private sector which will set in train the dynamics [the state] wants to see, while protecting and advancing ... the public interest' (Hutton, 1997, 66). To date, as far as corporate governance is concerned, there is little sign of the pursuit of stakeholder politics and Crouch (1997, 358) argues that, far from a coherent public philosophy, stakeholding is 'little more than commending to employers the value of consulting their workforces'.

In the absence of a more positive commitment to stakeholding, Labour's policies are focusing particularly on supply-side reforms: incentives for longer-term investment and a 'University for Industry' to promote modernisation, are two obvious examples. Greater investment in education and training, including the welfare-to-work programme, are also relevant. There are also initiatives designed to promote more co-ordinated action by the state on policy areas which cut across departmental boundaries ('joined-up thinking'). The establishment of the Social Exclusion Unit is the clearest illustration here. The novelty of New Labour here is thus in its claim to be developing a more intelligent state. There is also the prospect of change in the area of devolution and constitutional reform. Here, almost anything would represent an improvement on the degree of centralisation of the Conservative era. Proposals for regional development agencies, elected assemblies in Scotland and Wales, and elected mayors could all begin to recreate important intermediate institutions between centre and locality (chapter 11). Electoral reform of some sort is also on the agenda.

Where the Blair government do claim distinctiveness is in the proposition that they have discovered a 'third way' between capitalism and socialism. The terminology is arguably vacuous, because all social systems are positioned somewhere on this spectrum. Furthermore, the Blair government is not one heavily laden with ideological baggage and so the concept signifies a pragmatic brand of trial-and-error politics. Nevertheless some distinctive principles have emerged, which are well summarised by White (1998). These are the concepts of opportunity, responsibility and community.

The concept of opportunity must mean more, White argues, than just formal equality of opportunity. Rather it entails a commitment to substantive or real opportunity for basic goods such as education, jobs, income and wealth. Conversely, social exclusion (chapter 8) can be understood in terms of patterns of power that deny access to

minimally acceptable shares of these goods, sufficient to allow individuals to participate fully as citizens. Now while Labour has accepted certain elements of this view (for example, in a commitment to a national minimum wage (albeit one set at a fairly low level)) as yet there is no prospect of a substantial increase in benefit levels. This has been the subject of acrimonious debate between 'old' and New Labour with the former arguing for more redistribution and the latter insisting that cash alone will make little difference to the socially excluded.

The concept of responsibility features strongly in New Labour thinking, being coupled with rights or entitlements. Those in receipt of benefits from the state need to recognise their reciprocal obligations. The most obvious example here is the development of proposals which tie eligibility for state benefits not just to active job search, but even (in the case of the 'New Deal' for the young unemployed) to accepting work or training on almost any terms.

This marks a significant difference of principle between New and 'old' Labour. 'Old' Labour arguably saw the state as having enforceable responsibilities towards citizens, but this downplayed the idea that individuals have reciprocal and enforceable responsibilities towards the state. However, emphasising such obligations sails close to the wind of American-style 'workfare', in which individuals must undertake menial tasks to be eligible for any assistance. There must be a parallel obligation on the state to ensure that responsibilities are reciprocal – for example, guaranteeing that 'New Deal' training opportunities are of a high quality.

Blair is also arguing that citizens have responsibilities to a community of which they are part. Here New Labour is drawing (loosely) on communitarian values imported from America (e.g. Etzioni, 1993). The creation of stable communities appears an increasingly urgent task given growing evidence of social polarisation and segregation (chapter 8). This discussion has been much influenced by Putnam's (1993) concept of social capital. Putnam argued that in contrast to much social science which accounts for variations in welfare provision in rather economistic terms, the causal connections operate in the opposite direction. Prosperity derives from strong traditions of civic participation and community association. This theme featured strongly in the Commission on Social Justice report (CSJ, 1994, 307–9). The problem is that it is difficult to legislate to ensure that people behave in a communitarian fashion. Rich networks of social capital do not come into existence overnight. Critics of communitarianism rightly argue that it is a punitive philosophy which enforces certain codes of behaviour under pain of negative sanctions (e.g. withdrawal of benefits). Apart from changes to benefit regulations, a good example would be measures to evict anti-social tenants from social housing. At whom are such strategies aimed?

One plausible answer would be that they are aimed primarily at middle-class taxpayers, whose support for the Blair project was crucial, but whose continued bearing of the tax burden cannot be presumed. Britain has a 'culture of contentment' like that of the USA (Galbraith, 1992); it is principally suburban and rural taxpayers whose taxes support benefit recipients in inner cities and deindustrialised areas. Workfarist policies offer some reassurance that their taxes are being well spent. However, whereas the Blair government demands responsibilities of those in receipt of state benefits, there is no parallel demand for responsibility from the rich, for whom existing levels of taxation seem sacrosanct. Still less is there a demand for civic responsibility from corporations. The point of such communitarian strategies, then, seems to be to reassure the middle classes and multinational capital that they can have their cake and eat it. Such initiatives are indicative both of the way Labour has, in seeking to broaden its electoral appeal, to buy off the middle classes *and* impose its own self-denying ordinance on taxation, which will limit the scope for government-sponsored modernisation. For by accepting various parameters laid down by the Conservatives, Labour has sought and obtained electability by default rather than devising a genuine alternative to neoliberalism – despite the availability of alternatives (Hay, 1997, 372–4) which have

largely been marginalised by Blair's disciplined new model party. The reasons are not hard to find: New Labour's social coalition, which propelled it into office, will not take kindly to measures which will result in increased taxation.

3.4 THE CHANGING NATURE OF THE STATE

As well as the foregoing shifts in political strategy, the character of state intervention has changed substantially over the past quarter century. Key developments have been the gradual shift to a regulatory role for the state, a 'hollowing out' of states as their powers are displaced in various ways, and a shift (resulting from both of the foregoing tendencies) from *government* to *governance*.

A corollary of vigorous programmes of privatisation and market reform of public services has been the emergence of a qualitatively different role for the state. In the welfare state local authorities and health authorities now hold the purse strings for purchasing services from a disparate body of statutory, voluntary and commercial providers. This results from the separation of their purchasing and providing roles, which resulted from critiques of bureaucracy and of the tendency to prioritise producer rather than consumer interests. In the principal utilities the process has gone much further: electricity, gas, telecommunications, water and transport are all substantially in private hands. The consequence has been the rise of the regulatory state: the state has been rolled back in one sense and rolled forward in another. The utilities provide a good example. As public services, extension throughout the land and the operation of egalitarian charging mechanisms were in part financed through an implicit cross-subsidisation from wealthy communities to unprofitable (rural and disadvantaged) locations. Since privatisation, utility companies have sought profits by a combination of cherry-picking and social dumping. Profitable markets (the City of London) see the benefits of competition as companies seek a share of the action. Poor communities, where consumers use services sparingly, are less favoured (Drakeford, 1997). The response of utility companies has been disconnections of defaulters or (which is more insidious) demanding high deposits or forcing people into prepayment systems, concealing the politically sensitive issue of disconnection. Graham and Marvin (1994) argue that this is actually creating ghettos in which the most rudimentary existence (cooking meals, phoning relatives, flushing a toilet) becomes less and less universal.

Yet this is possible because of a comparatively lax regulatory regime. It has been argued that privatisation has benefited the wider economy by facilitating the international expansion of British companies. However, the corollary has been that the socio-spatial consequences of utility company decisions have been neglected. Clark (1992) therefore argues for a focus, in understanding the state, on its administrative frameworks and practices; the establishment and enforcement of regulatory frameworks are an important element of this. The British case has seen relatively weak regulation which has been unable to counter the effects of competition on marginal consumers in marginal locations. The 'windfall tax' levied by Labour on the utilities was a response to perceptions of excessive profiteering but this has not changed the regulatory regime at all. The extent to which universal access to utilities can be achieved through the regulatory state will continue to pose considerable challenges to regulators and governments.

The 'hollowing out' of the state has attracted much attention. The state has, of course, lost powers and functions and autonomy through privatisation, but 'hollowing out' is particularly associated with changes in the spatial constitution of the state. Jessop (1993) discerns a triple displacement of power from the nation-state. In an upward direction elements of power and sovereignty are ceded to supranational bodies: the EC's influence, for example on environmental policy, is especially important here. In a downward direction, the thrust of central government policy is to devolve responsibility to local organisations or to local tiers of national organisations: such

decentralisation and localism are implicit in numerous welfare programmes such as health care or labour market policy (Mohan, 1995a, Chapter 9; Peck, 1996). However, decentralisation is often more apparent than real, being accompanied by rigorous central control and monitoring. Third, there is evidence in some locations of greater cross-national cooperation between subnational organisations, which Jessop regards as a horizontal displacement of the powers of the nation-state. Examples might include networks of local authorities facing similar challenges in the sphere of economic development. Jessop's claims are abstract but there is considerable evidence for the tendencies he describes. A consequence of this hollowing out has been the de facto transfer of power to unelected quangos and public–private partnerships (chapter 11); another has been a 'regulatory deficit' in the management of uneven development (Peck and Tickell, 1995a). We should also remember that hollowing out is not an autonomous process but results from conscious and deliberate political decisions.

Finally, some writers claim to discuss a shift in state practice from *government* to *governance*. The implication is that we must consider state power much less as a hierarchical, top-down exercise than hitherto; images of a sovereign state are misleading. In a narrow sense governance relates to the transformation of the state from a direct producer of goods and services to an overseer of their production: this involves the state in 'steering, not rowing' (Osborne and Gaebler, 1992). More broadly, governance signals an acknowledgement that the political system is increasingly differentiated. Consequently the task of government is to 'enable socio-political interactions [and] to encourage many and varied arrangements for coping with problems, and to distribute services' (Rhodes, 1997, 51). Such interactions involve a range of public and private agencies, which overlap in complex networks: commercial and voluntary agencies delivering services on a contractual basis on behalf of (and in partnership with) government agencies; public–private partnerships in regeneration; agencies and quangos.

Some authors argue that changes in the character of state intervention have been associated with new *governmentalities*, or new ways of thinking about governing. Specifically, more responsibilities are devolved to individuals and communities, and as a result new relationships emerge between government and forms of self-government. What is distinctive about this is the new freedoms allocated to individuals and communities, who become autonomous actors who are deemed to be responsible, through their individual choices, for their own welfare. Government seeks to act upon individuals through their supposed allegiance to particular communities (Murdoch, 1997, 112). Central to this process are discursive strategies which seek to construct particular representations on which communities and individuals must act. Discourses of localism and self-help are familiar in the welfare state (chapter 9) but Murdoch shows how these were deployed in respect of rural policy in a 1996 White Paper. The narrative of this document emphasises the diversity and local character of the countryside, which provides a justification for the appropriateness of local decision-making. However, such decisions are not to be made by the state (whether central or local) but by the (small, tightly-knit and self-reliant) communities that make up rural society, who will devise flexible responses to local needs. The corollary is that central government plays a much reduced role: the White Paper is littered with references to partnerships, an enabling role for government, and creating the frameworks in which businesses can prosper. Murdoch (1997, 116–17) argues that the state is moving towards more selective and indirect modes of intervention, and using 'discourses of community, locality and diversity in order to promote a profound shift in state/society relations'. Much the same can be said of many pronouncements by New Labour; the discursive strategies of the state will continue to attract attention.

3.5 SUMMARY

The past two decades have witnessed the hegemony, and then conclusive failure, of neoliberal economic policies, accompanied by an anti-democratic centralism in politics. The limitations of this are clear but there remains profound distrust of state intervention, while economic globalisation and technological innovation will prevent a return to 'old Labour' politics. This is in essence the position hypothesised in Jessop's (1993) prescient arguments about the kind of state that would be required to cope with the putative post-Fordist economic future. This would have three principal characteristics. First, its principal tasks will be promoting innovation and enhancing competitiveness, largely through a range of supply-side initiatives. Second, social policy will become subordinate to the success of states in a global economy. Third, the nation-state will be 'hollowed out' as its powers are displaced upwards, downwards or horizontally (Jessop, 1993, 9–10). Put together these constitute an emerging Schumpeterian workfare state (SWS): Schumpeterian after the economist, Joseph Schumpeter, whose work emphasised the relationships between innovation and competitiveness; and workfare, because of the concern to promote flexibility and competitiveness in social policy. There are extensive debates about these propositions (Jessop, 1992, 1995; Burrows and Loader, 1994) but the general point of Jessop's argument is that the conditions under which the KWS operated no longer obtain.

Consequently, according to Hay (1996b), we now have a 'one-vision polity', in which there is substantial agreement on certain principles, such as an open, globalising economy, and a minimal (and different) role for the state. This might be regarded as Jessop's 'neo-statist' variant of an SWS, inasmuch as it seeks a more active role in managing technical change and upgrading the skill base. However, the indications are clearly that this remains a centralising government and there remain deep tensions over the extent to which redistributive social policies will be pursued. For these reasons interpretations of the nature of 'Blairism' will remain contested and elusive. The 'third way' at present remains just that: a route, not a final destination.

4

GEOGRAPHIES OF PRODUCTION I

DEINDUSTRIALISATION AND REINDUSTRIALISATION

4.1 INTRODUCTION

As recently as the early 1970s, one of the major texts on the UK's human geography accorded about 350 pages to a sector-by-sector description of the key heavy industries and the factors which underpinned their geographical distribution – a discussion couched firmly in the terms of classical industrial location theory (Stamp and Beaver, 1971, 277–624). Changes to the geography of production, driven by forces not even considered by conventional texts, have prompted far-reaching debates. This chapter is intended as an overview of the main elements of the changing geography of industrial activity, and of the competing interpretations of these phenomena. The key feature is of course the geography of deindustrialisation, which has decimated the productive base of certain places; within this, the major single element of deindustrialisation – certainly in employment terms – has resulted from changes in nationalised industry policies from 1979. This is followed by a discussion of the continued salience of the urban–rural shift in manufacturing, and finally there is a discussion of attempts to regenerate the industrial base through foreign direct investment.

The processes of deindustrialisation, reindustrialisation and tertiarisation (chapter 5) need to be examined not in isolation but as part of a wider set of economic changes oper-

ating on a long timescale. Thus, various theories of 'long waves' in economic development suggest that major technological innovations supersede old products and processes and/or revolutionise production in existing industries; these accounts are often associated with the idea of 'Kondratieff waves' (cycles of innovation over a 30–50 year period) (Hall, 1985; Martin, 1988b; Bull, 1991). Likewise some accounts of deindustrialisation see it as an inevitable outcome of long-run sectoral shifts in economic activity (section 4.2) in which deindustrialisation and tertiarisation are (potentially) complementary processes. However, such exogenous influences are mediated by social and political arrangements and structures. Thus an understanding of deindustrialisation, or of the growth of foreign direct investment, would be seriously hampered by a neglect of state policies; and arguably no understanding of Britain's economic landscapes is satisfactory without acknowledgement of the key role played by the City and the practices of financial institutions. It is erroneous simply to understand geographies of production in terms of the operation of market forces alone, without drawing attention to the institutional context within which markets operate.

The focus of the chapter is on the absolute and relative decline of manufacturing, but this decline has of course prompted efforts to modernise manufacturing. It is worth considering whether the measures taken to achieve this were the only ones available, whether they were inevitable or not, and whether they have succeeded. This raises numerous questions about the social and economic costs and benefits of economic change (Turner, 1995a). The chapter begins by briefly putting deindustrialisation, and the explanations for it, in context. Key dimensions of deindustrialisation are then discussed: the geography of recession, in the early 1980s and 1990s; the impacts of nationalised industry policies and of privatisation; and the urban–rural shift. There follows a discussion of new forms of production organisation within manufacturing, concentrating on the impacts of foreign direct investment and assessing whether contemporary changes indeed signify a novel departure or whether they merely represent efforts to preserve old modes of accumulation. Throughout, attention is paid to the contrasting methodological standpoints of the authors whose work is reviewed.

4.2 DEINDUSTRIALISATION IN CONTEXT

Since at least the 1960s there has been a steady decline in the relative and absolute contribution to the British economy of manufacturing, regardless of whether this is measured in terms of output, investment, exports or employment. However, the weak performance of the industrial economy has been a source of concern for over 100 years. In this sense deindustrialisation is simply a new name for an old problem (Martin and Rowthorn, 1986).

The post-war 'long boom' to some extent shielded the UK from the regional industrial problems of the inter-war years, and helped disguise the poor competitive performance of manufacturing. However, even before that boom ended in the 1970s, deindustrialisation was already under way. Key dimensions of this include: a cessation of productivity growth; a decline in output; a trade deficit in manufactured goods from 1983; declining profitability; and reductions in employment. Manufacturing employment peaked at 8.7 million in 1966; it is now around 4 million. In terms of relative shares, industry (manufac-

turing; energy and water supply; construction) accounted for 48 per cent of British civilian employment in 1955, which made the UK one of the most highly industrialised states in history (Hall, 1991); by contrast, by 1996 that share was approximately 23 per cent and today there are now more than two service sector employees for every worker in industry. Only in the West and East Midlands does industry's share of employment exceed 30 per cent, but the figure for London is 11.6 per cent, while manufacturing alone now accounts for under 10 per cent of employment in London and Surrey. Only in Derbyshire, Leicestershire and Mid-Glamorgan are over 30 per cent of the workforce now in manufacturing (Figure 4.1). By contrast as recently as 1981 the only regions recording less than 33 per cent engaged in production industries were London (25.3 per cent), the South East as a whole (29.7 per cent, though the figure for the rest of the South East, excluding London (RoSE) was 33.9 per cent), and the South West (32.7 per cent). The West and East Midlands had 46.6 per cent and 47 per cent of employment in industry.

There are several contrasting explanations for deindustrialisation (see Rowthorn, 1986). One account emphasises that deindustrialisation results from economic *maturity*. Over time there is a secular shift from agriculture, through manufacturing, to services; the share of services in total employment expands inexorably, mainly because of the changing structure of demand as incomes increase. Given Britain's early industrialisation it is an obvious candidate for this explanation (see also Kitson and Michie, 1996, 35). This argument is clearly relevant to the UK from the 1950s since not only was industrial employment relatively large, agricultural employment was limited at some 5 per cent of the workforce. Any expansion in services could therefore only be at the expense of manufacturing, in contrast to other states which could absorb labour from their larger agricultural sectors. The maturity thesis therefore explains both the early fall of industrial employment in Britain, and the intensity of decline; Britain's economy was already close to maturity by the mid-1950s. Other states followed Britain in the 1960s and 1970s as

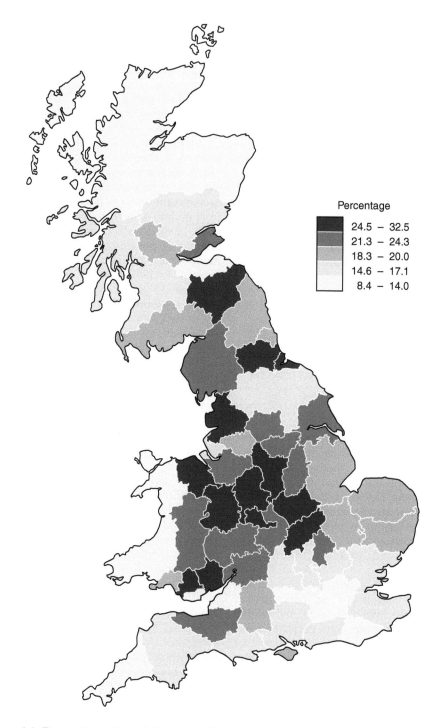

FIGURE 4.1 Proportion of workforce employed in manufacturing, 1996 (NOMIS)

their agricultural labour reserves were depleted. Despite the superficial descriptive accuracy of this account, it does rest on the proposition that employment in services rises inexorably.

A second set of arguments coalesces around the notion of *trade specialisation*, concentrating on foreign trade and the UK's role in the international division of labour. In the 1950s the UK was a workshop economy, importing food, raw materials, and oil, and consequently running a substantial trade surplus on manufactures (to compensate for the deficit in non-manufactured goods). There has, however, been a massive shift in the structure of trade (for reasons explained in Rowthorn (1986, 15–19)): the deficit on non-manufacturing trade has been eliminated and with it has gone the need for a substantial surplus on manufacturing. No other country had quite the necessity to devote such attention to manufacturing, which explains the intensity of the decline of manufacturing in the UK. Related to this explanation are arguments which stress the influence of shifts in consumption patterns, which cause labour to be diverted to services (Kitson and Michie, 1996, 35).

A third set of explanations focuses on the *failure* thesis – the failure of British manufacturing to compete successfully or to produce the output required for a prosperous and fully employed economy. There are two elements of change here. There was a relative decline up to 1973: productivity and output were rising but not as quickly as in other states; and an absolute decline after 1979, with output falling dramatically (and only in the early 1990s recovering to pre-1979 levels). There are numerous accounts of this failure (see Coates, 1994). However, despite the attractions of these arguments, Rowthorn (1986) concludes that the maturity and trade specialisation theses account for most of the decline of British industry: even if productivity had been sustained, employment would still have fallen. Thus, deindustrialisation was inevitable largely because of the stage of development which the UK had reached, but it is arguable that the responses (and deliberate policy choices) of governments meant that the impacts were more severe in Britain than else-

where, not least because no attempt was made to steer the process of deindustrialisation. Instead, market forces were given free rein.

4.3 KEY DIMENSIONS OF INDUSTRIAL CHANGE: GEOGRAPHICAL PERSPECTIVES

Geographies of industrial decline have attracted the attention of authors working within quite different research traditions (Massey and Meegan, 1985). For convenience one might identify divisions between those who emphasise 'location factors' and those drawing on a 'political economy' tradition. This distinction signals differences in empirical focus and methodology. There is a contrast between those who emphasise extensive approaches to the study of location factors, seeking empirical generalisation, and those who stress intensive investigations of a limited number of case studies. This spills over into geographical contrasts, as Townsend (1993) points out; much scholarship has focused on major employers in a limited set of locations (coal mines, steelworks) to the neglect of small enterprises in whole swathes of small-town Britain. Because of the quite different explanatory perspectives adopted, both schools can be accused of one-sidedness. Thus political economists stress (and sometimes condemn) external circumstances (globalisation, state policies) failing to acknowledge that the particular circumstances of individual plants make them differentially vulnerable to changing circumstances (Sayer, 1995). Likewise a one-sided emphasis on location factors fails to ask important questions about the structural contexts in which firms operate. Despite apparent differences in assumptions and methodology, these contrasting approaches may have more to offer one another than is often acknowledged (Massey and Meegan, 1985).

The emergence of the North–South divide

The period of accelerated deindustrialisation from 1979–81 had severe long-term consequences. Manufacturing output did not recover to 1974 levels until 1988; levels of investment in 1978 were not matched until ten years later, and even then were about half the level of investment in distribution and in financial and business services. The post-1979 period was the most intensive phase of industrial retrenchment and rationalisation in British history. Even without the Thatcher government there would have been a serious recession, but almost half the losses in employment in the 1980s are attributable to the effects of monetary and fiscal policy against a background of underlying recession (Martin, 1986). These policies were highly deflationary; they heavily squeezed industrial output and profit margins; and high interest rates (and the short-term demands of lenders) constrained firms' responses, leading to drastic shedding of labour. The Conservatives welcomed the scale of the slump. A short-term recession would not force reorganisation of production; instead the forces of creative destruction had to be unleashed, producing a 'leaner and fitter' industrial base and one which would be capable of sustained recovery. Less stringent policies would no doubt have been possible – the degree of deflation imposed on the economy was heavily criticised – and might have given firms scope to adjust in ways other than wholesale capacity reductions.

The effects of the recession of the early 1980s were severe: the slimdown and shake-out of manufacturing were achieved almost entirely by cutting jobs, productive capacity, and investment. Investment in manufacturing fell by a third between 1978 and 1983; firms sought drastic reductions in employment; when these were insufficient, company liquidations rose to an all-time high of 13 000, mainly reflected in the disappearance of small, single-plant firms. Within large corporations, the most obvious response was rapid labour shedding – exceeding 50 per cent of the workforce in many cases – with adverse consequences for the affected localities.

Champion and Townsend (1990) observe various patterns within these broad changes. One was the decline of the large firm: the number of firms employing over 1000 people dropped from 1132 in 1971 to 452 in 1988 (see also Fothergill and Guy, 1993). That decline in large plants has continued; in 1995, there were 409 manufacturing establishments with over 1000 employees (table 4.1) while the number of plants of all sizes has shrunk rapidly, the only exceptions being growth in firms employing under 20 people. The effects of such closures on particular localities were notable, particularly given the government's declared intention of promoting economic revival through small business formation: the numbers of new businesses needed to replace employers of more than 1000 individuals was an obvious challenge to the burgeoning economic development industry.

Second, the plants most vulnerable to closure were branch plants of multiplant (if not

TABLE 4.1 Size distribution of manufacturing establishments, United Kingdom, 1971–95

Employment size	Number of plants		Change, 1971–95	
	1971	1995	No.	%
10–19	16 617	19 927	3310	+19.90
20–99	27 408	22 083	−5325	−19.40
100–199	6140	3684	−2456	−40.00
200–499	4850	2376	−2474	−51.00
500–999	1650	748	−902	−54.66
1000+	1132	409	−723	−63.90
Total 10+	**57 797**	**49 227**	**−8570**	**−14.80**

Sources: 1971: Champion and Townsend, 1990, table 5.2; 1995: Office for National Statistics (1998) table 5.

multinational) concerns. In Fothergill and Guy's (1993) study, of some 515 000 jobs shed by named corporations, 320 000 (over 60 per cent) were lost in the Assisted Areas; on a population basis the (then) Special Development Areas lost twice their share. Interviews with managers of externally controlled manufacturing plants suggested that reductions in capacity were due not to characteristics of plants *per se* (i.e. productivity) but to the peripherality of plants in corporate hierarchies at a time when corporations were losing market share. Many such closures were regional policy factories attracted in previous rounds of investment, emphasising the weaknesses of such an approach to policy (Stone, 1995, 182). A full understanding of precisely which plants were closed, and why, can arguably be provided only through studies of the position of plants within corporate hierarchies (e.g. Townsend and Peck, 1984 and 1985).

Third, given earlier comments on globalisation (chapter 2), to what extent were reductions in capacity in the UK accompanied by overseas expansion? Several studies suggest that domestic contraction was accompanied by overseas expansion. Lloyd and Shutt (1985) found that, between 1979 and 1982, the top 50 manufacturing companies in the UK increased overseas production from 36 per cent to 44 per cent of their output; simultaneously they cut their UK-based workforce by 400 000 while expanding it abroad by 40 000. This might appear to indicate support for the new international division of labour (NIDL) thesis (chapter 2), in which relocation of production is driven by a search for cheap labour, but Peck and Dicken's (1994) study of Tootal, the textile firm, emphasises changing market relations rather than the company's labour relations *per se*. Such examples call into question the concept of a *national* economy and of *British* deindustrialisation. Rationalisation of capacity was not simply a process of – or driven by – efficiency-raising slimming; instead, British capital was bound up with the international restructuring process. The Thatcher governments regarded this internationalisation as evidence of economic recovery and strength, but its negative impact, on employment and visible exports, was given less prominence.

No regions escaped deindustrialisation but the impact was very uneven. Absolute job losses were substantial: Dunford (1997, table 20.3) shows that in the 10 travel-to-work areas (TTWAs) experiencing the greatest decline in employment in 1981–91, a total of 710 000 jobs were lost in energy/water supply and manufacturing combined. Major cities featured prominently: London lost 271 000 manufacturing jobs, and Birmingham (72 000), Manchester (68 000), Glasgow (51 000) and Liverpool (47 000) were also prominent. The percentage decline in manufacturing output between 1979 and 1983 in the South East, East Anglia and the South West of England was less than half that which occurred nationally and one-third of that which occurred in Wales, the West Midlands and the North West. Consequently the gap between the more buoyant South and East, and the former industrial heartlands, widened greatly. This was only partly an effect of variations in industrial specialisation, though clearly such factors had an influence: the North West and the West Midlands suffered particularly from international competition in textiles and metal manufacturing, and from the decline of various forms of engineering (Flynn and Taylor, 1986; Lloyd and Shutt, 1985); other regions (South Wales, NE England) experienced major contraction in basic industries such as coal or steel (Beynon *et al.*, 1991). Regardless of variations in industrial composition, the peripheral regions suffered a higher rate of job loss across most industries; the rate at which manufacturing capacity was reduced differed, and so did the manner in which it occurred. The greater the rate of industrial decline, the greater the proportion of compulsory redundancies in the discharge of labour. Townsend (1983) showed that Scotland, Wales and the North East of England experienced substantially more redundancies than would have been expected if each region had experienced redundancies in line with its share of national employment in each industry. Champion and Townsend (1990, 76) show, using shift-share analysis, the extent to which higher-than-expected rates of employment contraction were a particular feature of the 1978–84 period in the peripheral regions. The net effect was a widening of inter-regional unemployment disparities and the emergence of concern about the North–South divide. This period saw the expulsion from the labour force of substantial numbers of men on a more-or-less permanent basis (chapter 6). Townsend (1993) demonstrated the regional impact of recession in the early 1980s by maps indicating the time at which there was a marked upturn in unemployment: on this indicator the North slid into decline much earlier than the South of England. It is understandable that this relative decline is presented simply as the working-out of market forces, but this view would be an over-simplification, since a key influence on the decline of manufacturing employment in the northern regions has been the impact of state policies, notably those relating to nationalised industries and privatisation.

Redrawing the boundary between public and private sectors: denationalisation and privatisation

Regardless of which government had been in power in the 1980s and 1990s, the nationalised industries (principally steel and coal, but also shipbuilding) would have faced severe challenges due to global overcapacity and the ready availability of cheap imports (Sadler, 1992). However, it is at least arguable that there were alternatives to the drastic programmes of capacity closure and labour-shedding that were set in train after 1979. Hudson's (1989a and b) analysis of nationalised industry policies is drawn on here. It exemplifies a political-economy approach which relates the decline of key industries to the impact of the international political and economic environment mediated by national-level political decisions.

The history of nationalised industries raises important questions about the limits to public ownership as a strategy for regional development. The original purposes of nationalisation were twofold. From the point of view of capital, nationalisation offered the

prospect of securing the production of key inputs from sectors which were technically backward, with a history of inadequate investment; these were hardly the 'commanding heights' of the economy. This helps explain the absence of substantial opposition to nationalisation. From the point of view of labour, nationalisation appeared to offer better wages and working conditions, as well as helping to control trajectories of local, regional and national development; crucially, nationalisation could help ensure there was no return to the mass unemployment of the inter-war years in localities dependent on the production of basic commodities.

However, the goals set for nationalised industries were arguably at variance with securing high levels of employment. In order to underpin the competitiveness of UK manufacturing, nationalised industries were generally operated according to strict efficiency criteria. Hudson (1989a) argues that this was not compatible with the political need to maintain employment. Although successive governments at least attempted to reconcile these competing imperatives, by 1979 the consensus about the operation of nationalised industries no longer held. The Conservatives viewed these industries as bastions of outmoded working practices and industrial relations, and had never forgiven the unions – particularly the National Union of Mineworkers – for their role in ousting the Heath government in 1974. As a result they sought to operate the nationalised industries on *de facto* commercial lines, aiming ultimately to privatise them. The pursuit of profitability led to savage cuts in employment and to a redefinition of terms and conditions of employment for the remaining workers (see Beynon *et al.*, 1991). The government were also well aware that the costs of such a strategy would fall on those areas with a strong Labourist tradition as part of their wider 'two nations' political project (chapter 3). This meant that they were generally indifferent to the social impacts of plant closures, with rare exceptions such as the (temporary) reprieve for the Ravenscraig steel plant in central Scotland in 1982.

Several changes in the international political and economic environment facilitated this switch in policy. First, in the 1980s global overcapacity and overproduction of coal and steel meant that nationalisation was no longer essential to guarantee production of such commodities within the UK. Second, the growing internationalisation of the economy made a switch to overseas sources feasible and provided competition, notably in steel and shipbuilding. Third, the recession of the early 1980s was an ideal backcloth against which to introduce anti-union legislation – particularly legislation designed to limit the extent to which unions in different sectors of the economy could act in concert. Finally, privatisation policies in related areas, such as energy production, both exposed the coal industry to additional competition and broke the links between, for example, coal production and electricity generation, thus threatening the viability of individual pits. Preparations for privatisation of coal also involved closing any remaining 'unprofitable' pits. Hence, as a prelude to privatisation, the Conservatives announced the closure of 31 pits in late 1992. The ensuing political outcry initially forced the Government to reprieve 12 pits, but British Coal subsequently reverted to a policy of incremental closures, so that within a couple of years more pits shut than the 31 originally earmarked for closure (Turner, 1995b, 27).

The net effects have been substantial. Between 1978 and 1987 alone, 431 000 jobs were lost in coal mining, iron and steel, and shipbuilding combined. Of these job losses, the Northern region, Scotland and Wales lost 89 000, 60 000 and 59 000, respectively. Coal mining had employed 107 000 people in these regions in 1978; in 1996 less than 10 000 remained. British Steel shed 129 000 jobs between 1979 and 1993 (Townsend, 1997, table 4.3). Because of the spatially concentrated nature of this decline, communities suffered substantial downward multiplier effects. The collapse of these industries affected both firms supplying services and components to them, and businesses depending on consumption by their employees (Stone, 1995, 180–1; Turner, 1995a). Moreover, the nature of many settlements affected – one-industry towns – was such that the localised impacts were severe; unemploy-

ment and poverty had an adverse effect upon health and social cohesion, with some communities recording widening inequalities in health during the 1980s and 1990s (Phillimore et al., 1994). What was also being jettisoned were human resources (skilled labour) which would be difficult to replace; for example, efforts to sustain ship repair on the Tyne have been bedevilled by difficulties in obtaining the requisite numbers of skilled workers.

These developments reveal much about the limits to nationalisation as a political and economic strategy, for while the nationalised industries helped preserve employment in the generally favourable economic conditions of the 1950s and early 1960s, no real innovations in management or productivity were achieved. In the hands of a government determined to expose the economy to the full force of international competition, the nationalised industries, and those places which depended on them for employment, were particularly exposed, yet because of the locationally concentrated nature of those industries, the social costs of restructuring fell largely in locations not disposed to support the incumbent government. Not surprisingly some saw this as a central part of the overall 'two nations' strategy. Crucially, one cannot understand the origins and impacts of nationalised industry policies other than in terms of political decisions rather than the technical requirements of production. Opponents of these policies generally emphasised the broader costs of increased transfer payments to unemployed workers and their families, costs which generally exceeded the subsidies being paid to the industries concerned (Beynon 1985). Such objections cut little ice, as did the argument (made with particular force in respect of coal) that the UK could not afford to sterilise millions of tons of fuel. The Government's response, characteristically, was to assert that these sclerotic relics of Britain's industrial past needed – and benefited from – the shock of privatisation. On some counts British Steel is now one of the world's most efficient steel producers (Hudson 1994; Morris 1995), but the social costs of its transformation have been considerable.

Similar general points could be made about privatisation, a key policy of the Conservatives which has produced substantial downsizing and job losses in diverse areas, notably utilities, transport, local government and defence. Between 1985 and 1997, some 121 000 jobs were lost in defence-related manufacturing as a result of expenditure cutbacks, increased competition and privatisation (table 4.2). Because of multiplier effects, these can be trebled, as estimates suggest two additional jobs for each manufacturing job dependent on defence spending. In absolute terms South East England was hit hardest but Lovering's (1995a, figure 5.1) statistics demonstrate that the local impacts of some 70 000 job losses were greatest in communities in the North West (e.g. Barrow) which depended heavily on such employment. Despite optimistic proposals that this could

lead to a 'peace dividend' and to conversion of the defence industry to civilian uses, in practice savings have been used to cut taxes, almost no conversion has occurred, and a market-led restructuring has had severe impacts on affected localities. Nor has there been a demilitarisation of the economy: if anything, major corporations have retrenched *into* defence production, acquiring smaller defence-related companies and increasing their concentration on defence work (Lovering, 1998, 11).

Tickell (1998) shows that utility privatisation has also been associated with large-scale job losses. Since privatisation, there have been 203 000 job losses in the various companies subject to Labour's windfall tax, a figure which comfortably exceeds the numbers of opportunities to be created by Labour's welfare-to-work initiative, which is funded

TABLE 4.2 Employment change in defence-related manufacturing and utilities

	Defence-related employment				Utilities employment			
	1985/6	1996/7	Absolute change	% Change	1981	1996	Absolute change	% Change
North	15 000	5000	−10 000	−66.6	26 736	14 258	−12 478	−46.6
Yorks/Humbs	4000	5000	1000	+25.0	45 060	23 200	−21 860	−48.5
E Midlands	8000	3000	−5000	−62.5	32 520	19 016	−13 504	−41.5
East Anglia	4000	1000	−3000	−75.0	18 619	15 032	−3587	−19.2
South East	84 000	27 000	−57 000	−67.8	221 777	112 702	−109 075	−49.2
South West	22 000	18 000	−4000	−18.2	45 869	30 633	−15 236	−33.2
W Midlands	7000	5000	−2000	−28.6	50 572	25 974	−24 598	−48.6
North West	23 000	8000	−15 000	−65.2	55 091	32 730	−22 361	−40.5
England	167 000	71 000	−96 000	−57.4	496 244	283 545	−212 699	−42.9
Scotland	18 000	7000	−11 000	−61.1	47 886	29 657	−18 229	−38.0
Wales	3000	0	−3000	−100	29 578	13 128	−16 450	−55.6
N. Ireland	8000	0	−8000	−100	ND	ND	ND	ND
UK	200 000	79 000	−121 000	−60.5	573 708	326 330	−247 378	−43.1

Sources: cols 1–4 *Statement on the defence estimates, 1987–88,* table 6.9, and *Defence statistics,* 1998, table 1.12; cols 5–8 NOMIS: 1981 data are activity headings 1610 and 1620 (production/distribution of electricity and gas), 1700 (water supply) and 7902 (telecommunications). 1996 data are classes 4010 (electricity), 4020 (gas), 4100 (water supply) and 6420 (telecommunications).
Notes: There may be minor definitional differences (due to changes between the 1980 and 1992 Standard Industrial Classifications) but broad trends should be clear.
ND No data from this source.
Defence statistics are estimates rounded to the nearest 000.

by the tax. Regional estimates of the impact can be obtained for the electricity, gas, water supply sectors and telecommunications, which are broadly compatible with the totals for the companies discussed by Tickell. Some 250 000 jobs were lost in these between 1981 and 1996 (although some losses may be due to reclassifications in the census of employment). Much of this is due to pre-privatisation restructuring; 100 000 job losses took place between 1981 and 1991, for example. However, the pace of job loss has been steady in these sectors throughout the period in question, with most regions losing 40 per cent of jobs in utilities, though some of these jobs will have been transferred due to subcontracting. In mitigation, moreover, several privatised companies have relocated jobs to such regions (call centres and billing functions), which in qualitative terms contributes to the changing balance between industry and services and between male and female employment.

The urban-rural shift?

A significant recent trend in the location of manufacturing industry in the developed economies has been the relative growth of rural as opposed to urban areas. By 1991 the rural areas of Britain had 250 000 more manufacturing jobs than thirty years previously, representing a 45 per cent increase. This contrasted sharply with the collapse of manufacturing in urban areas. Fothergill and Gudgin (1982) first demonstrated an inverse relationship between settlement size and changes in manufacturing employment for the 1952–81 period. Much of this growth was concentrated in southern England, notably East Anglia (North, 1998). It is important to note that the shift reflects not so much direct relocation of production as differential growth (i.e. expansion/contraction of existing firms).

We might expect the shift to be weaker in the 1980s, when deindustrialisation accelerated. In fact manufacturing employment in remoter rural areas increased by 4.6 per cent from 1981–89, compared to a decline in all other types of location. The relative shift towards rural areas is even greater when allowance is made for industrial composition. Using shift-share analysis employment

change in a region can be disaggregated into three components: a *national* component (what would have happened if manufacturing employment had changed at the national rate); a *structural* component (the expected change given each area's sectoral mix) and a *differential* component (a residual, interpreted as reflecting local advantages and disadvantages). The column showing differential shifts in table 4.3 is of most interest. Remoter areas in southern Britain gained nearly 40 000 more manufacturing jobs in 1981–89 than would have been expected on the basis of national trends and their industrial mix. The same is true for rural areas in the 'north' which experienced an even larger differential shift.

By contrast, Townsend notes that London lost 170 000 more jobs than would have been predicted from its industrial structure. This negative differential was also experienced by other large cities, in both the North and the South. Note that differential shifts were generally better in northern areas than in the South. As Stone (1995, 178–9) observes, however, for most places of any size this simply meant that employment was not declining as rapidly as the national average. Other commentators put a more optimistic gloss on this, arguing that despite the substantial decline of manufacturing in London, productivity has increased (Graham and Spence, 1995, 897–900). Finally, the contrasts in fortunes of urban and rural areas were even greater for high technology industry although growth was often spatially concentrated (e.g. around Cambridge: Keeble, 1992).

Consequently, the urban–rural shift could not be regarded as a passing phase of British economic geography. This applied especially to Greater London and the principal cities, where the strong growth of financial and business services failed to arrest a decline in total employment. Some relative shifts of manufacturing activity to industrial areas took place, but the absolute numbers of jobs in factories did not rise. In non-metropolitan cities the growth of services and tourism was enough to offset a greater proportionate loss of factory jobs than in industrial areas, but the overall expansion of employment fell below the national level. The massive

TABLE 4.3 The urban–rural shift in manufacturing employment, Great Britain, 1981–89, in 000s

	Total		Total change 1981–89		National shift	Structural shift		Differential shift	
	1981	1989	No.	%	No.	No.	%	No.	%
South									
Inner London	282.6	180.4	−102.2	−36.2	−42.6	+20.1	+7.1	−79.7	−28.2
Outer London	401.3	263.3	−138.0	−34.4	−60.5	+13.3	+3.3	−90.9	+22.7
Non-metropolitan cities	416.0	323.6	−92.4	−22.2	−62.7	−13.7	−3.3	−16.0	−3.8
Industrial areas	358.0	322.5	−35.8	−10.0	−54.0	−2.2	−0.6	−20.3	+5.7
Districts with New Towns	200.4	178.1	−22.3	−11.1	−30.2	+3.5	+1.8	+4.3	+2.2
Resort, port and retirement	169.5	159.9	−9.6	−5.7	−25.5	+11.4	+6.7	+4.5	+2.7
Urban and mixed urban–rural	695.4	652.4	−43.0	−6.2	−104.8	+36.6	+5.3	+25.2	+3.6
Remoter, mainly rural	274.0	283.9	+9.8	+3.6	−41.3	+11.4	+4.2	+39.7	+14.5
Total	2797.2	2364.1	−433.5	−15.5	−421.6	+80.7	+2.9	−93.0	−3.3
North									
Principal cities	604.9	413.5	−191.4	−31.6	−91.1	−21.6	−3.6	−78.6	−13.0
Other metropolitan districts	1133.6	938.7	−194.9	−17.2	−170.8	−35.6	−3.1	+11.5	+1.0
Non-metropolitan cities	293.3	251.2	−42.1	−14.4	−44.2	−0.4	−0.1	+2.5	+0.8
Industrial areas	610.0	568.3	−41.8	−6.8	−91.9	−26.9	−4.4	+77.0	+12.6
Districts with New Towns	195.5	193.6	−1.9	−1.0	−29.5	−5.0	−2.6	+32.6	+16.7
Resort, port and retirement	48.9	45.7	−3.2	−6.5	−7.4	+2.9	+6.0	+1.2	+2.5
Urban and mixed urban–rural	207.6	195.3	−12.3	−5.9	−31.3	+4.9	+2.4	+14.1	+6.8
Remoter, mainly rural	159.7	169.6	+9.9	+6.2	−24.1	+0.7	+0.5	+33.2	+20.8
Total	3253.5	2775.9	−477.7	−14.7	−490.3	−80.7	−2.5	+93.0	+2.9

Source: Townsend, 1993, table III.

deindustrialisation of the industrial North was answered by a modest rural and semi-rural recovery, but with relative gains accruing to medium and small-sized labour market areas. There was a clear inverse relationship between deindustrialisation and tertiarisation, with those regions suffering most from the decline of manufacturing gaining least from the expansion of jobs within the service sector.

While not conducting a detailed shift-share analysis Turok (1999, table 2) shows that manufacturing decline between 1981–96 exhibited the familiar urban-rural gradient. London experienced the greatest decline (402 000 job losses; −59 per cent) followed by other conurbations (615 000; −41 per cent) and free-standing cities (185 000; −35 per cent). Towns and rural areas (748 000; −23 per cent) were the only area type to experience a decline in manufacturing less than the national average (−32 per cent).

Explaining the urban-rural shift has, over the years, generated much debate between those emphasising traditional 'location factors' (congestion, property prices) and those emphasising the role played by rural areas in particular spatial divisions of labour (see chapter 1). Townsend argues that the patterns evident in the 1980s reflected historically specific and short-lived circumstances. The propulsion of investment to greenfield sites resulted from the combined consequences of pressures on both land and labour in the 'core' regions of the UK. Townsend (1993) argued that such pressures offered more plausible explanations than putative moves towards flexible accumulation. Turok (1999), likewise, emphasises physical constraints which may retain their salience given pressures on available land in the South East (chapter 13).

There are conflicting assessments of whether this shift continues to reproduce stereotypical spatial divisions of labour. Conventionally much economic geography has been critical of the impact of external control on regional economies (particularly where multinationals are involved), seeing it as generating branch-plant economies in which peripheral regions are the location for routine, assembly-line jobs paying relatively low wages. Such spatial structures were criticised: local linkages and spinoffs were weakly developed; value added was retained within corporations; plants were highly vulnerable to capital flight (Pike, 1998, 882–3). Against this, some recent commentaries claim to identify positive local economic development effects from some plants, possibly because of the high standards they expect of suppliers and the demonstration effect of new working practices (Cooke, 1995). Potter (1995) argues that recent debates about flexible specialisation open up the possibility that the character of branch plants might change for the better. Flexible production may lead to increased product and process complexity, higher skill levels and greater local integration in branch and subsidiary plants. This might imply 'a better quality of recent branch plant investment' than in previous rounds of investment. Potter (1995, 174) claims that in his study of 140 externally controlled firms in Devon and Cornwall, there was 'widespread dynamism in products and markets, in production technologies and process organisation, and improvements in skill levels and R and D activity'. This is an optimistic view which contrasts with Pike's (1998) assessment of restructuring in the car industry. Pike saw little sign of the development of the ideal typical 'performance' branch plant; instead innovation was heavily constrained. It is likely, however, that the rise of flexible production systems will see more firms externalising labour-intensive activities, a trend likely to be facilitated by improved telecommunications (North, 1998, 184–5).

New forms of production organisation? The impacts of foreign direct investment (FDI)

The organisation of this book posits an opposition between 'rustbelt' and 'sunbelt' sectors which is arguably too neat, because one response to industrial decline has been efforts to reorganise production processes within established branches of production. These efforts have been most closely associated with FDI although they are not confined to foreign-owned production units. Large inward investments were held up by the

Conservatives as exemplars of their success in positioning the UK as the 'enterprise centre of Europe'. Major investments are also important as flagships for the localities concerned, and as demonstration projects: just as the nationalised industries exemplified outmoded working practices which had to be swept away, foreign investments are said to symbolise the nature of the industrial future. But these developments may not be an unalloyed blessing: what may be rational from the point of view of an individual firm may not be so for the national economy (due to technological dependence and the costs of attracting investments) or for the localities in question (due to the costs of engaging in a territorial competition for investment). Three key issues are considered here: the scale of new investments and the net inflows and outflows; the implications for places benefiting from (or competing for) investment; and the effects on the national economy.

The scale of FDI in the UK is substantial but must be put in context by a consideration of *net* inflows/outflows. Table 4.4 shows that, for the G10 countries, the overall level of new investments has risen sharply since the mid-1980s and the UK has been a major destination, receiving more FDI than Germany.

None the less France has recently received more inward investment than the UK. Conventionally, Conservatives argued that the UK's success in attracting FDI reflected its flexible labour markets and deregulated economy, but France's recent success suggests that flexible labour markets may not be the only factor in investment decisions. Instead, the level of FDI in the EU has been rising across the board, in anticipation of gains from the completion of the internal market, and this is probably what has made the UK an attractive destination. The key objective of Japanese investors, for instance, has been to locate within the EU to overcome trade barriers; particular locations *within* the EU were then chosen on the basis of local circumstances (Barrell and Pain, 1997, 52).

Turning to net inflows/outflows, UK firms have more assets overseas than foreign firms have in the UK, for both manufacturing and non-manufacturing. The UK is also a net investor in the EU; over one-third of UK FDI is located within EU states compared to one-fifth a decade ago. In 1994 the value of UK companies' assets in the EU exceeded that of assets held in the US. If the UK's flexible, deregulated business environment offered such advantages, why has this drift to

TABLE 4.4 FDI inflows and outflows, 1976–95

		1976–80	1981–85	1986–90	1991–95
		($ billion, annual averages)			
G10	outflows	39.3	41.3	154.0	193.5
	inflows	19.9	31.3	111.1	116.1
	net flows	−19.4	−10.0	−37.9	−77.4
UK	outflows	7.8	9.2	28.1	25.4
	inflows	5.6	4.3	21.7	17.2
	net flows	−2.2	−4.9	−6.3	−8.2
France	outflows	1.9	2.7	16.7	23.2
	inflows	2.2	2.2	8.1	19.0
	net flows	−0.3	−0.5	−8.6	−4.2
Germany	outflows	3.7	4.2	14.4	27.6
	inflows	1.2	0.8	2.7	3.7
	net flows	−2.5	−3.4	−11.3	−23.9

Source: Barrell and Pain, 1997, 50.

Europe happened? The implication is that labour regulations are not seen as insuperable obstacles. Barrell and Pain (1997, 54) suggest, however, that a sceptical attitude to European Monetary Union might be more damaging in that foreign investors whose primary markets are in continental Europe would have greater incentives to locate there.

There are other implications for the national economy of inward investment. One relates to the balance of payments. Depending on the balance between the distribution of sources and sales, the result of internationalisation could be that the British car industry produces record output while actually being a net drain on the balance of payments (Rhys, 1995), if the majority of sales are within the UK and the bulk of expensive components come from outside it. A second implication relates to technological dependence: is inward investment the best way to enhance a nation's industrial base? Major investments may merely be importing processes and jobs developed elsewhere, producing a 'screwdriver economy' in which the UK's role is reduced to that of an assembly line for companies orientated towards a European market. This might be part of a 'Hong Kong' model of economic and political development, as an offshore free-trade, low-wage zone, but this hardly seems sustainable.

Moving from the national to the regional scale, FDI may have various implications for patterns of uneven development. First, in terms of regional shares, Hill and Munday (1992) show that Wales, Scotland and the Northern region have consistently attracted the largest shares of inward FDI, which they explain in terms of wage levels, grants and infrastructural provision. Though there have been losses as well as gains, the job creation effects should not be dismissed lightly. In the northern region 22 per cent of manufacturing employment is in foreign-owned plants while on Wearside the figure is over one-third (Stone, 1995); similar figures were reported for Scotland and Wales. While foreign-owned manufacturing units in Scotland represented under 5 per cent of the total of manufacturing units in 1989, they accounted for 24 per cent of wages and

salaries, 30 per cent of gross value added and 39 per cent of net capital expenditure (Collis and Noon, 1996, 5).

Stone and Peck (1996) analysed stocks and flows of firms and employment in the foreign-owned manufacturing sector (FOMS) for four regions (Northern Ireland, Northern England, Wales and Scotland) for 1978–93. Although there were substantial new investments in all these regions, and the total number of plants increased everywhere, total jobs in the FOMS declined in Scotland and Northern Ireland although they increased in Wales and the North (table 4.5). In the latter two areas the numbers of new creations exceeded the number of closures. The reverse was true for Scotland and Northern Ireland, where closures removed jobs equivalent to 39 per cent and 49 per cent of the original stock of employment in FOMS. Additionally much of the apparent gain in FOMS employment is an artefact, resulting from foreign acquisitions of UK-owned companies. The different performance of the four regions has in part resulted from different patterns of ownership. Scotland and Northern Ireland were rather more dependent on US-owned plants and the period was marked by a steady retrenchment in the US-owned sector (over 66 000 job losses, 1978–93; Stone and Peck, 1996, table 5). By contrast Wales and the North were much more successful in attracting Far Eastern investments.

Given the importance of FDI, regions engage in vigorous competition to attract investments, through promotional activities and financial incentives. Nissan, for example, were apparently offered grants totalling some £100 million; for Lucky Goldstar the package amounted to some £30 000 per job created, implying a total commitment of around £200 million. This compares with previous criticisms of capital-intensive regional policy grants (see chapter 12). However, to what extent can this be regarded as an industrial strategy? There are benefits in terms of productivity: Keeble (1991) contends that FDI was an important contributor to the Welsh economy's comparatively good economic performance. Other commentators speak approvingly of the demonstration effects of inward investments. However, the touchstone

TABLE 4.5 Components of foreign-owned manufacturing change by region, 1978–93: number of plants (and associated employment)

	Initial stock	New creations (+)	Closures (−)	Acquisitions (+)	Divestments (−)	In situ (+)	In situ (−)	Net change	Stock at end of period
Northern Ireland	68 (30 100)	27 (4162)	28 (14 653)	24 (12 610)	8 (1660)	18 (1992)	13 (4648)	+15 (−2197)	83 (27 903)
Scotland	352 (108 065)	144 (19 219)	158 (41 760)	84 (19 272)	46 (9261)	55 (8407)	79 (18 148)	+24 (−21 976)	376 (86 089)
North	154 (49 201)	89 (14 618)	50 (10 366)	79 (15 295)	16 (3294)	36 (3307)	53 (13 105)	+102 (+6455)	256 (55 656)
Wales	217 (58 900)	129 (17 000)	59 (6700)	118 (22 600)	53 (12 400)	52 (5500)	64 (16 800)	+135 (+9100)	348 (68 000)

Source: Stone and Peck, 1996, table 4.
Notes: Scotland 1979–92; Wales 1979–93.

for academic commentary has been the extent to which FDI generates spinoffs and linkages in the local economy. Conditions are usually attached to grants, such that firms have to agree that their output will incorporate a certain proportion of components derived locally. This is dictated to some degree by new systems of production, such as 'just-in-time' systems, a key feature of which is the delivery of components as and when they are needed (thereby avoiding the costs of maintaining stocks). Thus it was suggested that Nissan had agreed that between 60 to 80 per cent of components were to be sourced locally; but for these purposes it was agreed that 'local' meant within the EC. Sadler (1997, 314–19) describes the build-up of Nissan's linkages. Initially links to local firms were small-scale (Amin and Tomaney (1991) reported that only 18 of Nissan's 177 suppliers were based in the North East) but as Nissan developed relationships with suppliers they increased local purchasing. Purchases on components averaged *c.* £800 m. in the early 1990s; of this some 77 per cent was spent within the UK and 43 per cent in North East England (Sadler, 1997, 315). Munday *et al.* (1995) discern evidence of the development, over a period of time, of subcontracting complexes in Wales: they argue that the long-term nature and stability of supply linkages have enabled firms to engage in longer-term resource and production planning. Thus, they suggest, plants are not just 'warehouses' with final assembly lines piecing together goods imported from the Far East. On the other hand Stone (1995) suggests that newly-established Japanese plants have largely sourced from pre-existing component suppliers; those suppliers located near the new car plants have tended to be those making bulky and/or relatively low-value items. Other work on electronics plants in South Wales suggests that most key inputs are drawn from England or beyond, with Welsh inputs being limited or minimal (Phelps *et al.*, 1998, 131–2).

Commentators disagree on the extent to which these developments signify innovation in production organisation, or simply intensification of previous practices. The blurred divide between the old and the new is emphasised by Hudson (1997a). Changed external circumstances have prompted demands for a more flexible reorganisation and relocation of production. This typically involves enhancing labour productivity via: new fixed capital investment in automated production technologies; more intensive ways of organising work and the labour process; and more efficient ways of processing material inputs. These have been key preconditions for experimenting with new ways of producing in large quantities, such as just-in-time (JIT) production systems. These seek to combine the benefits of economies of scope, small-batch craft production, and a greater flexibility in responding to demand, with the benefits conferred by scale economies. These are often production methods which combine elements of old (mass production) and new approaches. There are sharply differing accounts of the power relations between firms resulting from these initiatives. The persistence of vertical integration and hierarchical relations between companies in many sectors (food, footwear, electronics, cars) suggests that in many respects characteristically Fordist relations between companies are by no means a thing of the past. Hudson's (1994) work on the automobile industry emphasises that to date there is little evidence of territorial production complexes founded on new transplants and foreign investments. At best, he argues, new geographies of production in traditional sectors offer the possibilities of branch plant investment with only limited linkages into the regions where investment takes place. This is essentially an attempt by capital to preserve old modes of accumulation (high-volume production). The more optimistic accounts of the impacts of FDI on regional economies acknowledge this but argue that local linkages take time to build up as suppliers adapt to higher standards. As the closure of the Siemens factory on Tyneside demonstrates, of course, even major investments are vulnerable to changing circumstances; they can be mothballed even before supply chains have become established.

4.4 CONCLUDING COMMENTS

The past two decades have witnessed a transformation of the UK's industrial landscape, which has largely taken the form of the removal of many of its most striking physical symbols from locations once celebrated as workshops of the world. To what extent has this dramatic process of restructuring produced an economy equipped to compete internationally, and to what extent, theoretically, does this conform with the views of those analysts who argue that the economy is tending towards a 'post-Fordist' future?

There is scepticism regarding the extent to which an economic miracle has occurred and even optimistic assessments are divided on how much credit can be given to the actions of governments (since they see the post-1992 period of growth as resulting largely from the effects of the 1992 devaluation of sterling, a devaluation resisted to the last ditch by the then government). There is evidence that economic growth rates in Britain in recent decades have converged on those of West European states, but this is partly due to a slow-down elsewhere rather than accelerated growth in the UK. Moreover, even though productivity has increased in British industry, it still lags behind that of Germany and the relativities between the two have been maintained for over 25 years (Currie, 1996, 44–5). This leaves an absolute gap in levels of economic performance. The source of the recent increase in productivity in the UK is, of course, the more stringent industrial relations climate during the Thatcher years, but this is generally recognised to have had only a one-off effect on productivity. Much of the growth in productivity evident during the 1980s and early 1990s resulted from the elimination of the least-efficient productive capacity and through a harsher and more exploitative regime in the workplace; over 50 per cent of the growth in manufacturing productivity between 1979 and 1989 was attributable to reduced employment while over the same period output grew by only 12 per cent (Glyn, 1992, 79). And even if productivity had increased, industrial production in

1994 was hardly higher than in 1979 (Bonefeld et al., 1996, 193), causing considerable concern to those who regard trade deficits on manufactures as an indication of economic failure because of the difficulties of trading in services. The wholesale destruction of industrial capacity has left the UK ill-fitted to expand production in manufacturing.

Increased profits could have been channelled into higher levels of investment, but several commentators show that this has not happened. Glyn (1992) suggests that in the 1979–92 period, the asset base of the economy continued to deteriorate, with investment largely going into property (specifically housing, retailing and office development) rather than into modernising the UK's productive capacity. As a consequence there were only two years during the 1980s in which manufacturing investment climbed above its 1979 level and the UK's capital stock was shown to be older and smaller than that of its major competitors, with German industry having about 30 per cent more machinery per worker-hour than British manufacturing (Hutton, 1995, 98). In sharp contrast, the share of profits paid to dividends rose dramatically. It followed, finally, that as a consequence of capacity-shedding and low investment, capacity constraints in the economy were reached comparatively easily, posing the ever-present threat of resurgent inflation. Imports were sucked in as part of the 1980s' consumer boom, contributing to a weaker balance of payments position. As was acknowledged by a House of Lords Committee (1985), this *does* matter: deindustrialisation *can* damage your wealth. It does so in three ways. First, the service sector ultimately depends on the size of, and rate of growth in, the manufacturing sector. Second, cumulative industrial decline restricts firms' spending on training, thereby constraining one of the processes necessary to shift resources within the national economy into new, more productive sectors. Third, a deteriorating trade position in manufacturing creates dangers such as deflationary macroeconomic policies which follow from any resulting balance-of-payments deficit (Kitson and Michie, 1996, 46). Acknowledging these points implies an investment strategy that will enable the UK to close the capital stock gap with other major industrial countries; the alternative is an acceptance that the UK's comparable competitors are the NICs.

Given these remarks it is clear that in quantitative terms some scepticism must be expressed as to whether the British economy has been transformed. This also applies in *qualitative* terms, with reference to the claim that a distinctively post-Fordist economy has been produced. There is of course evidence of a degree of 'Japanisation', but the extent to which modernisation within particular sectors has occurred is highly variable (see for example the various sectoral studies in Turner, 1995a). But despite widespread acknowledgement that Fordism has had its day, it remains unclear precisely what has supplanted it. For example, efficiency- raising strategies in the British economy (labour- and capacity-shedding) appear as classically Fordist (or capitalist) efforts, and it is not clear what is new about that. The economy is leaner, and meaner, but not necessarily post-Fordist. There is evidence of the adoption of some innovative practices, but it is debatable how widespread these are. Consequently there is uncertainty about how far a *stable* post-Fordist regime of accumulation has emerged. Perhaps the most plausible interpretation is that, at least within manufacturing, what we are witnessing is *neo*-Fordism: efforts to preserve, often in not particularly novel ways, the Fordist regime. Such efforts, however, seem predicated on continued degradation of standards of labour regulation (chapter 6); as such, they will reach definite limits, sooner or later.

5

GEOGRAPHIES OF PRODUCTION II
SUNBELTS AND NEW INDUSTRIAL SPACES?

5.1 INTRODUCTION

The focus here is on the extent to which new geographies of economic activity are emerging in the UK and on the extent to which they entail a reworking of spatial divisions of labour. Contrasts are drawn with the geographies of industrial decline considered in the previous chapter. The prominent geographer Peter Hall remarked that 'tomorrow's industries are not going to be born in yesterday's regions' (P. Hall, 1985): to what extent is that assertion still valid? Of course, the implicit opposition between 'rustbelt' and 'sunrise' sectors is slipshod. It's not possible simply to sail off into the sunrise, leaving rustbelts to rust, and thinking in these terms encourages a narrow view of developmental trajectories and possibilities (Massey *et al.*, 1992, 247–8). In particular it encourages the view that the only feasible way forward is slavishly to copy existing regional success stories.

There is much rather imprecise discussion of emerging geographies of production in the UK as the various references to silicon landforms (glens, fens) or high-tech geometric shapes (arcs, crescents) makes clear. This imprecision seems to indicate uncertainty about precisely why developments take the form they do. The emphasis on intangible features of places and chance circumstances by some authors is symptomatic, perhaps, of this uncertainty. Explanations range from some fairly conventional multivariate statistical exercises (e.g. Keeble's work on new enterprises) to approaches which emphasise non-market forces and institutional features of the British economy and those which seek to explain the emergence of 'new industrial spaces' through an understanding of tacit knowledge and transaction costs.

Three key elements of recent economic change are considered: high technology industry, new business formation, and the service sector. Some important trends cut across these, such as the growth of computer services. These are sectors seen as crucial to a successful transition away from the mass-production Fordist era; they have also been seen as emblematic of the economic transformations seen as desirable by the Conservatives. All have been the subject of state policies designed to promote a selective restructuring of the economic base, whether in the form of direct subsidies (e.g. tax breaks for investment in particular sectors), deregulatory measures to promote competition (e.g. the 'Big Bang' of 1986 in financial services), or a generalised politics of enterprise, designed to encourage entrepreneurship. The extent to which these various schemes have been successful raises important questions about the attractions of, and limits to, the various neoliberal policies pursued since 1979 and about the extent to which they can be generalised across the UK. Finally, the chapter considers current debates in geography about the contribution to regional development of 'new industrial spaces', and assesses whether there is evidence for such spaces in Britain.

5.2 HIGH TECHNOLOGY INDUSTRY

'High technology' has many resonances and tremendous symbolic importance. It has, of course, been used as a symbol of modernisation by national governments, as in the Wilson government's talk of the 'white heat' of technological change in the mid-1960s. Equally it is deployed by representatives of particular places keen to promote their attractions to such developments. High technology offers a new image of economic success and – it is to be hoped – a new recipe for regeneration, being portrayed as the answer to the decline of 'smokestack' industries (which themselves were once at the leading edge of technology). However, the regional success stories of a few favoured locations cannot easily be emulated. If we accept the arguments that there are long waves of economic development and that technological change through creative destruction helps economies out of recession, then ensuring the conditions in which innovation and new technology can thrive is evidently crucial. However, these long waves themselves are not immutable forces of nature to which places must bow down, and it is evident that some regions have been much better placed to respond than others.

It is helpful to begin by looking at some definitional questions, including the important distinction between high technology processes and products. Almost all offices, for instance, will now use high-technology products (computers) but these products may have relatively low-tech production processes. This is a point commonly made about the computer industry, since silicon chip production and the assembly of computers are relatively low-tech activities and thus the value added by such activities within a particular national economy may not be great. Similarly the growth of telephone call centres in peripheral regions is made possible by technological advances but the character of work in such centres is very routine and standardised.

Discussions of high technology activities routinely refer to the difficulty of arriving at an agreed definition of what counts as high technology. Definitions include variables such as the proportions of highly-skilled employees, high ratios of R and D expenditure to sales, worldwide markets for products, and a fast growth rate, but clearly these do not exhaust the possible definitions nor are they watertight. Hall *et al.*'s (1987) work therefore identified the proportion of a sector's labour force engaged in technical occupations. On this definition 36 industries had higher than the national average (for manufacturing) of engineers, computer scientists or researchers. Even this does not solve the problem since, for example, not all parts of the production process are necessarily high technology while R and D activities, and corporate HQs, may not be located within the UK. Hall *et al.* therefore emphasised the capacity of an industrial sector to employ scientific and technical practices, and the generation of technologically advanced products. This generally led to a focus, in their analyses, on the electronics, aerospace, computing, pharmaceutical and biotechnology industries.

Analysis of the geography of high-technology industry using location quotients (LQ) (for the same group of industries considered by Hall *et al.*) reveals that by 1996 little had changed since their (1987) work (figure 5.1). The national share of employment in hi-tech was 4.8 per cent but three counties had more than double this (Berkshire (12.7 per cent, LQ = 2.65), Isle of Wight (11.8 per cent, 2.45), and Hertfordshire (9.7 per cent, 2.02). Cambridgeshire, which had not featured strongly in the early 1980s, was fourth (8.75 per cent; LQ = 1.82). Other counties with a share of hi-tech employment at least 50 per cent greater than the national average were Mid-Glamorgan, Surrey, Buckinghamshire and Hampshire. Conversely, those counties with LQs of under 0.5, indicating a share of high-tech employment less than half the national average, were typically in the rural periphery or deindustrialised areas such as South Yorkshire, West

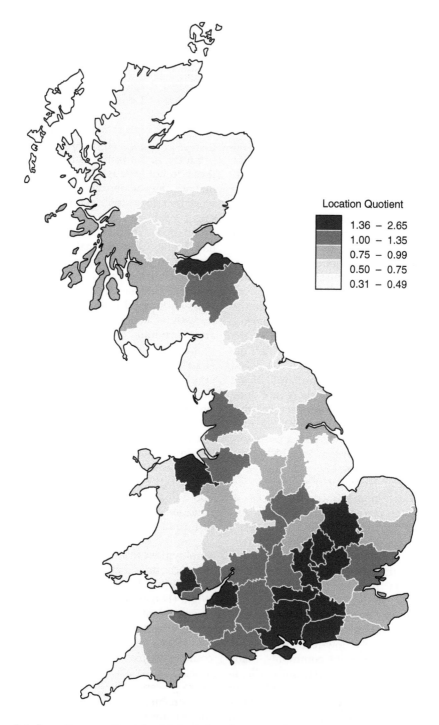

FIGURE 5.1 Location quotient for high-technology employment, 1996 (NOMIS)

Glamorgan and Staffordshire. Recent patterns of growth between 1993–96 do not indicate that this situation will change much. Berkshire, Cambridgeshire, Surrey, Oxfordshire and Buckinghamshire all experienced growth of around 40 per cent in their total number of hi-tech jobs; nearly one-third (45 000) of the 141 000 additional jobs in hi-tech were in these five counties. There were some indications of growth spreading to the East Midlands. Elsewhere particularly rapid percentage increases in Mid Glamorgan (70 per cent), Lothian (52 per cent) and the Borders (61 per cent) must presumably be attributed to large inward investments.

Hall's account of the origins of this belt is interesting for several reasons. He argues that the process was kick-started by the post-war decentralisation of manufacturing industry from London, in response to the need for space for physical expansion. This produced individual entrepreneurs moving out of existing plants and forming new enterprises. Crucially, these entrepreneurs responded to the substantial presence of the public sector in the area west of London, notably the defence research establishments (DREs) which massively expanded during the 1950s, largely because of the Cold War. The DREs placed contracts with key suppliers who developed prototypes and then obtained subsequent production contracts, leading to the emergence of related clusters of subcontractors and the generation of agglomeration economies. The role of Ministry of Defence procurement expenditures is crucial: the MoD is the largest customer of British industry; more than 50 per cent of national R and D expenditure is channelled through the MoD and there is a significant spatial bias to such expenditures with roughly half going to the South East (see table 7.2 below). Subregionally procurement expenditures tend to be distributed to contractors closest to the DREs and Ministry of Defence contact points.

Infrastructural investment decisions are also identified by Hall et al. as contingent factors in the growth of the M4 corridor: the choice of Heathrow as the major London airport, the development of the M4 and the priority given to the high-speed railway to

Bristol – all these contributed greatly to the improved accessibility of the area. In this context Hall et al. note that the route chosen for the M4 – to the south of Reading – passed much nearer to the DREs than would have been the case had it gone north of Reading.

Considerations of labour supply assume greater significance, particularly those deemed necessary to the attraction of highly-educated professional staff. Hall et al. contend that the environmental and lifestyle attractions of locations such as the M4 corridor are central to the development of that area. However, Massey et al. (1992) argue that such considerations are not decontextualised, innate desires; they are socially constructed images which appeal to particular social strata, and such lifestyles are only achievable by those on high incomes. Agglomeration economies are also thought to be very significant in terms of explaining the concentrations of high-tech industry, which is held to be innovative and – at least in its early stages – dominated by small firms which depend on obtaining external economies, especially in the recruitment of skilled labour. From this perspective firms beget firms – there is a cross-fertilisation of ideas and movement of personnel between companies. This may be useful in explaining the localised existence of a few concentrations of high-tech firms, but it is less valid in relation to interpreting the expansion of high-tech into previously non-industrial areas.

It is difficult to see how this favourable combination of circumstances could easily be replicated. One candidate might be a putative M11 corridor, with Cambridge at one end (boosted by the reported involvement of the computer entrepreneur, Bill Gates) and the London Docklands at the other: just as government investment helped sustain the M4 corridor, it could be argued that infrastructural developments – Stansted Airport, the M25, the M11 – are having a similar effect (Keeble, 1989). Cambridge is cited by Castells and Hall (1994) as one of their 'technopoles of the world' though this seems exaggerated given the comparatively small numbers employed there. But it is fallacious to suppose that there is much potential for the wider replication of the 'Cambridge phenomenon'.

Thus, while growth of high-technology industry in locations such as South Wales and Scotland is welcome, the regional development implications may be somewhat limited. They are not all stereotypical branch plants – for example, those in Scotland are not all engaged in assembly-line occupations, and there is some development work (e.g. adapting products to the nuances of the European market). In Scotland, early developments followed a familiar pattern, with the industry being established following wartime dispersals, notably of Ferranti, through which much MoD work was channelled. Initially there were comparatively limited spinoffs directly from Ferranti, but the company did recruit key engineering skills and this created a pool of labour which was subsequently to prove attractive to foreign investors. The main impetus behind the development of Scotland's 'silicon glen' has been inward investment by large UK and overseas companies. Although employment in hi-tech industries in Silicon Glen now exceeds that in traditional sectors (such as coal or steel) in Scotland, it still accounts for only 6–10 per cent of the workforce. Turok (1993) argues, furthermore, that Silicon Glen is not deeply 'embedded', in the sense of developing local linkages and spinoffs. He can identify only 12 per cent of purchasing taking place locally in the electronics industry; his study found few high value-added, technology-based linkages between firms. Instead, subcontracting linkages were of a fairly conventional kind, largely on the basis of price. Moreover, he found few signs of indigenous development, although the operations of some firms did include some development work. In general, however, as Henderson (1987) makes clear, the rationale underpinning firms' location decisions is a familiar one: wage levels are relatively low; reproduction costs are low because of the tax environment and because of the incentives offered to firms; the concentration of new entrants in Scotland's new towns is related to an apparent desire to draw on new reserves of 'green' labour and avoid the highly unionised urban centres.

Similar arguments have been advanced by Morgan and Sayer (1988) in their study of South Wales. In their view, high-technology industry has produced development *in* rather than development *of* regions: they point to the limited spinoff effects and the ways high-technology investments are used by firms simply to rework existing divisions of labour – to deploy new working practices or introduce non-union environments for example. Thus, the electronic engineering industry in South Wales was characterised, first, by a very high degree of external control, and, second, by very low-status activities within corporate divisions of labour, weighted towards older, mature and less science-intensive subsectors. However, their plants did not conform totally to the branch-plant stereotype; many plants played a comparatively unique role within their parent firm, and were therefore less dispensable, and by no means devoid of R and D functions (1988, 163).

On the other hand most jobs in the plants they studied were semi- or unskilled, and the plants often had difficulties recruiting professional and managerial staff. A key difference between their two study areas was evident in the respective proportions of managerial and unskilled manual staff. In Berkshire these proportions were 13 per cent and 12 per cent, respectively, whereas in South Wales the figures were 3 per cent and 66 per cent (Morgan and Sayer, 1988, 221). They also found little evidence of multiplier effects (sourcing of components usually involved overseas links because of an absence of indigenous suppliers) and most of their plants occupied a precarious position within corporate hierarchies.

There are, as a consequence, reasons for scepticism about the form and sustainability of Britain's high-technology industries. The M4 corridor, for example, has arguably developed because of a uniquely favourable conjunction of circumstances, and similar developments on anything like the same scale elsewhere are highly unlikely. Current employment trends seem set to reinforce the pre-eminence of South East England. The form of high-tech development elsewhere is more likely to be that described by

Henderson (1987) in Scotland. Finally, a general point is that the international performance of British high-technology industry may be relatively weak, lagging behind other states where there is more of a dirigiste attitude to promoting leading-edge sectors. Commentators also argue that the excessive dependence of hi-tech industry in Britain on military (as opposed to civilian) R and D, may restrict its development (Lovering, 1995a).

5.3 DIFFUSING AN ENTERPRISE CULTURE? GEOGRAPHIES OF NEW ENTERPRISE FORMATION

Reviving the enterprise culture was seen by the Conservatives as a crucial part of their political project, and small businesses and individual entrepreneurship played a crucial symbolic role in this. Contrasting the rhetoric of dependency associated with the Keynesian welfare state with that of independence and entrepreneurship, the ideological message was that the greater the extent of new business formation, the more likely the prospects of economic success and of a reduction in unemployment. At one level the former is entirely plausible – who can argue against 'enterprise' (however vaguely formulated it is as a concept)? The connection with the latter rests on a somewhat selective reading of the evidence (from the USA) as to the employment-creation effects of small enterprises – though it was this rather selective reading which clearly influenced the government in the early years of their administration. A range of policies were pursued over and above the consistent exhortations about the enterprise culture. Policies included a range of allowance and incentive schemes, available on a national basis, which were designed to promote the diffusion of an enterprise culture throughout the country. There was a reallocation of aid to businesses away from grants and capital subsidies, towards a more general enterprise politics (chapter 12). Note, however, that the availability of private capital for new enterprises exhibits a marked spatial bias in favour of the

South of the country, reflecting the role of venture capital firms in lending money in locations where they believe returns are most likely to be guaranteed (see chapter 7).

New enterprises have diverse origins, a fact sometimes forgotten by their advocates, for whom any new enterprise formation is inherently good and a sign of imminent economic recovery. Some are artefacts of accounting systems: for example, changes in VAT thresholds encourage firms to subdivide their activities into separate trading elements whose turnover individually is less than the threshold at which VAT is payable. Others are properly considered not as elements of individual entrepreneurship but as opportunistic responses to the process of privatisation, notably the highly favourable terms on which certain public assets have been disposed. Still others could be termed 'forced entrepreneurship' as companies respond to adversity by offering jobs to individuals only on condition that they accept the status of self-employment. This has financial advantages to companies (relieving them of the obligation to pay employers' National Insurance and pension contributions) but its benefits to workers are less clear. There is, as a result of these definitional issues, some overlap between potential indicators of the enterprise culture.

There is a further methodological point to consider. Small or new businesses are arguably a 'chaotic conception': enterprises are considered together simply on the basis of their size or turnover. This may not be helpful in discussing a very heterogeneous sector: there is a large difference between a self-employed hairdresser or shopkeeper, and a highly-capitalised biotechnology business which has spun-off from a university science park or a multinational corporation. Indeed, Taylor (1999) questions the whole concept, arguing that identifying discrete enterprises is almost impossible. Instead, we should think in terms of small firms as 'networked temporary coalitions' (p. 16) which come together for short periods of time and then disband. Within these coalitions individuals engage in a range of discrete activities rather than concentrating on one element of their business. These cautionary remarks should be borne in mind in discussing the patterns of new enterprise formation.

Keeble and his co-workers (e.g. Keeble and Bryson, 1996; Keeble *et al.*, 1991; Keeble and Walker, 1994) have provided the most comprehensive documentation of the geography of the enterprise culture and its spatial and temporal impacts. They rely on data relating to registrations for VAT purposes which, despite some limitations, provide a valuable index of the extent of activity and its variability between regions. New registrations rose from 158 000 in 1980 to 256 000 in 1989, though subsequently falling back to 239 000 and 206 000 in 1990 and 1991, respectively as recessionary conditions affected enterprise formation. The net result was a growth in the total number of businesses in the UK from 1.29 million in 1980 to 1.71 million in 1989. The fastest rates of growth were observed in some expected sectors: finance, property and professional services (the latter dominated by computer services and management consultancy). These accounted for 54 per cent of net growth in new business formation. In the production sector the numbers of businesses increased by 30 per cent while the number of retailing businesses declined, reflecting the growing dominance of large supermarkets and the decline of independent retailers.

There were substantial regional variations within these totals. The South East recorded 850 000 new businesses, a net increase of 200 000, or 46 per cent of the national total. Considering net increases (births minus deaths) the traditional manufacturing regions had net growth rates of less than 40 per cent of the South East. The high death rates were notable as well: in some regions net growth was only one-seventh of the new businesses being formed. But, as figure 5.2 makes clear, the enterprise culture was firmly rooted in the South East, where new firm registration rates were typically at least double those in peripheral regions. Regional disparities have remained substantial throughout the 1990s with the South East remaining well ahead of other regions (CSO, *Regional Trends*, various dates). There are also spatial variations between sectors. The growth of new firm formation in the production industries

is relatively dispersed, whereas new firm formation in the non-production sector, not surprisingly, is closely allied to the concentration of such activities in the South East.

In seeking explanations for the geography of new business formation, Keeble and his co-workers adopt a multivariate statistical methodology, seeking to explain new firm registration rates with reference to a range of independent variables. On the demand side they include indicators relating to previous growth in local income or GDP, and changes in population, which they plausibly argue were key indicators of potential stimuli to new enterprise. On the supply side – as determinants of the potential number of entrepreneurs – they include variables relating to local occupational, firm size and sectoral structures, capital availability, and changes in unemployment. There are some obvious problems of multicollinearity which they acknowledge: thus, on the supply side, occupational structure and capital availability (indexed through house prices as a surrogate for collateral) are not independent of each other since, *ceteris paribus*, higher house prices are associated with particular occupational structures.

Their general conclusions were that the rate of population change in the preceding five years was a key determinant of new business formation: this reflected the demand-side pull of growing market opportunities, and the supply-side push of an increasing pool of potential local entrepreneurs, especially in less-urbanised and rural counties. This is tied in with discussions of the importance of counter-urbanisation and selective migration – namely, that many individuals are departing from cities to set up new enterprises in rural environments for largely 'lifestyle' reasons (for further discussion see North, 1998). Population change as an independent variable is clearly being made to operate on both the demand and supply side here.

Other influences are not so direct in their effects. Despite their assertions about the significance of counter-urbanisation, concentrated urban demand remained important, they suggest, because a concentrated local

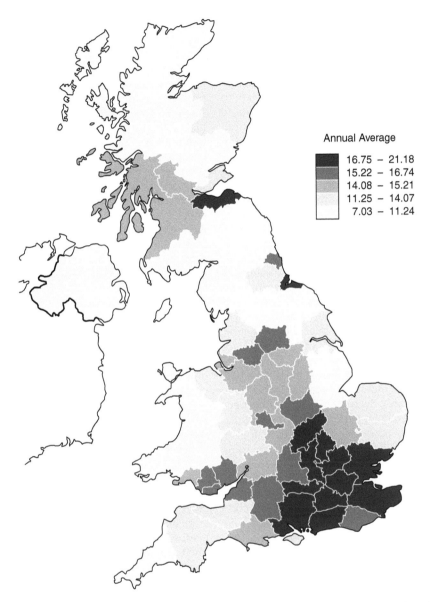

Annual Average

■	16.75 – 21.18
	15.22 – 16.74
	14.08 – 15.21
	11.25 – 14.07
	7.03 – 11.24

FIGURE 5.2 Spatial variation in new firm formation rates in the UK, 1980–90 (Keeble and Walker, 1994)

market and agglomeration economies are said to help stimulate new enterprise formation. This is a conventional 'incubator' hypothesis, in which cities are seen as crucibles for innovation and change. How far this is true is perhaps questionable. For example, given the scale of deindustrialisation, it is not easy to see how many cities can continue to perform this function, at least in the manufacturing sector, and it is possible that the growth of new businesses reflects, in part, the fragmentation of existing enterprises in response to changing external circumstances (e.g. privatisation or outsourcing policies).

Capital availability – indexed through

house prices – is also used as a measure, but again questions need to be asked about the processes through which this works. Surely the crucial indicator is not house prices but positive equity? Households with substantial outstanding mortgages are surely unlikely to be able to engage in new business formation, since financial institutions will be reluctant to lend to them. More generally, a discussion of this topic surely implies a consideration of the role of financial institutions in supporting new business formation. Numerous commentators refer to the difficulties faced by entrepreneurs in attempting to secure venture and other sorts of capital (see chapter 7), yet Keeble *et al.* did not consider this although they referred to 'policy effects', namely whether or not enterprise agencies (British Coal/British Steel Enterprises) were operating in particular places.

They also identify the pre-existing firm size structure as a potential causal factor, which operates in different ways in manufacturing and services respectively. For manufacturing they argued that a pool of small firms is very important because of the concept of linkages, the supply of large numbers of components, and the role of innovation. For services, however, their view is that a number of large firms are needed since individuals establishing themselves in new businesses will often have spent some time in larger firms, in which they acquire professional expertise, reputation, and client contacts. This appears plausible for some kinds of services – mainly high-level business and 'producer' services – but less so for others (e.g. tourist/leisure services), and again it seems to highlight some of the difficulties with aggregate, statistical explanations of this kind. For what Keeble *et al.* are seeking to do is explain an aggregate pattern which – as they themselves admit – is made up of many different and discrete subsets.

Finally they draw attention to the influence of local enterprise cultures as a further explanatory factor, but they index this by the political affiliation of local authorities. Presumably the intention is to analyse what it is about the character of 'place' and how this contributes to enterprise formation, but it is not obvious how political affiliation measures this, given that it changes from one election to another, except insofar as it is a measure of the propensity of the electorate to support a pro-enterprise policy. Understanding the character of local enterprise cultures probably requires studies based on more intensive methodologies which are less capable of being generalised (see for example, Pahl, 1984). This is an important issue, because studies of new business formation in peripheral regions, such as North East England and South Wales, indicate the difficulties of achieving the diffusion of the so-called enterprise culture. New business formation rates are generally lower than the national average, despite very considerable efforts to stimulate small firms and self-employment. In addition, many of those engaging in this activity have been aptly characterised as 'reluctant entrepreneurs', who see themselves as having little alternative but to attempt to set up on their own account, in the absence of other opportunities for paid employment (Rees and Thomas, 1991, quoted in Turner, 1995b). Thus there is comparatively little evidence of the establishment of an enterprise culture in the peripheral regions, and even where firms were established, Stone (1995) reported comparatively high failure rates.

This brings us to the overall impacts of attempts to stimulate an enterprise culture. First, given that the median size of new enterprises is five to six employees, clearly large numbers of new businesses are needed if substantial inroads are to be made into the unemployment figures. Second, one needs to assess the extent to which genuine *new* jobs are created. Much growth is almost certainly due to subcontracting of functions by large organisations – consultancy, marketing, data processing, IT services – and/or competitive tendering for 'ancillary' services (the latter process having in addition, transferred many jobs from the public to the private sector). This might increase the total number of enterprises without increasing the numbers of people employed – another form of jobless growth. Indeed, in some of these cases new firm formation will be associated with less stable and less well-rewarded forms of employment for those transferred to new employers. Third there is the question of sustainability – not in the environmental sense, but in the sense of the duration of the enterprises. Many enterprises established on the back of the consumption-led, credit-financed boom in the South East in the 1980s were of a short-term nature. As soon as interest rates began to rise, choking off consumption, many businesses in the property, retailing and leisure sectors failed. Plainly, the task of any strategy for stimulating enterprise must be to create more sustainable businesses which are less subject to cyclical fluctuations in the economy.

5.4 SERVICES, SPACE AND REGIONAL DEVELOPMENT

The service sector is crucial to understanding the changing economic geography of the UK. For some decades commentators have heralded the post-industrial economy (Bell, 1973), and, with the collapse of manufacturing in the 1980s, services had become the major – in some cases the sole – source of employment growth. But the service sector is very heterogeneous and defies one-dimensional analyses.

Debates about services involve both definitional confusion and arguments about the contribution of services to economic growth. On the former point, we can identify four possible uses of the term services: service industries, occupations, products, and functions. Service *industries* are defined as enterprises where the final output is in some sense 'non-material' regardless of the occupations that make up the labour force. Service *occupations* are present in all sectors of the economy: they include clerical and managerial personnel, sales representatives, and those providing financial and legal services. Service *products* may be provided by both manufacturing and the service sector; for example, manufacturers may provide service contracts or post-sales advice services. Finally, service *functions* are performed by all of us – for example, using household goods to carry out domestic tasks ourselves (DIY, laundry) (Allen, 1988).

Regarding the role services play in economic growth, commentators have generally rejected distinctions of the 'basic/non-basic' kind, in which manufacturing is seen as the core of the economy, in favour of analyses which (to varying degrees) accord independence or autonomy to services. Walker (1992), for example, sees services as largely fulfilling the role of serving a '*more* important economic activity – direct production' (Allen, 1992, 294). Urry, by contrast, contends that certain services are independent sources of growth, in that the strength of the service sector in certain localities is not dependent on, nor reducible to, a strong manufacturing base (Urry, 1987). Marshall and Wood (1992) feel that it is too simplistic to regard *either* services *or* manufacturing as the core productive activity. Instead, the functional *interdependence* between sectors should be stressed. For example, key manufacturing industries are heavily tertiarised, and the empirical evidence of corporate restructuring contains a strong non-production component (Marshall and Wood, 1992, 1262). Thus, rather than seeing services as dependent, a service-oriented view of economic change would recognise the role of specialist services in the revitalisation of certain industries (Marshall and Wood, 1992, 1266). This analysis does not privilege services over manufacturing but it does accord services a more prominent place in economic restructuring.

A key distinction in recent developments in services has generally been between producer and consumer services. Producer services can be thought of as supplying expertise which enhances the value of the output of other sectors at various stages of the production process. Such services are 'traded within companies, on the open market, and through their contribution to the competitiveness of other sectors'. Crucially, 'demand and supply need not coincide geographically', which is why such services are considered to be significant in terms of uneven development (Marshall *et al.*, 1988, 6). By contrast, consumer services supply final demand and are therefore closely related to the distribution of the population, although the distinction is not watertight: tourism, for example, usually entails interregional or international flows of money and is therefore 'tradeable' (Begg, 1993, 818).

The contribution of services to regional development is not confined to those services which form part of a region's economic 'base', for two other reasons. First, high incomes in a locality will induce additional consumption expenditures (on retailing, leisure or entertainment) thus supporting jobs. Second, services impact on the appeal or competitiveness of a locality; inward investment and indigenous enterprise may be attracted or stimulated by a suitable range of locally supplied business services. Recent literature points to the variations in the availability of business support services and the consequent need to strengthen that infrastructure (e.g. Bennett *et al.*, 1994; Wood, 1996). Put another way, a strong service sector may permit increasing returns in all sectors of the economy, by enhancing the ability of economic agents to learn and promote innovation (see also chapter 12, on regional policy).

Accounts of the rise of services differ in emphasis. Some commentators discern an inevitable shift towards a 'post-industrial' society sustained by the growing importance of knowledge, skills and information to the production process. They adopt an optimistic perspective on the potential for technological change to facilitate progressive developments, by enhancing job tasks and encouraging homeworking. In contrast, more pessimistic accounts see service growth as a corollary of deindustrialisation; this expansion is largely a consequence of slow relative growth in productivity. These views both find expression in contemporary patterns of regional development. London and SE England appear to exemplify a post-industrial structure, particularly given the growth of international finance and business service activities. By contrast in the deindustrialised North and West of the UK, the decline in production employment has not been offset, in general, by expansion in producer services and higher-order corporate activities (Marshall *et al.*, 1988, 7). However, such a simplistic division between 'post-industrial' and 'deindustrialised' regions does not capture the range of service sector geographies. These are illustrated through consideration of producer services (particularly the City of London's financial services), consumer services, and public services.

Producer services

Financial services have arguably become the most important element of the UK's economy in the 1990s, although they have been prominent for most of the twentieth century and some commentators argue that the primacy of finance is a structural feature that has been apparent since before the Industrial Revolution. The financial services sector as a whole has been well placed to benefit from recent technological and regulatory change. The latter has operated both internationally (in terms of the UK financial system's relationships with the rest of the world) and domestically.

Internationally the emergent New International Financial System (NIFS) has resulted in greater openness of national economies and nation states have had little option but to acquiesce and to maximise the benefits to their own economies. The driving force behind the NIFS has been pension funds and insurance companies, which have enormous funds at their disposal and conduct what are essentially global searches for investment opportunities. The NIFS has also been facilitated, if not driven, by technological change and it is also associated with the emergence of new financial instruments and markets. Despite the rhetoric of globalisation the NIFS is characterised by a *recentralisation* of activity in the core capitalist nations; three centres – New York, Tokyo and London – are dominant. In these circumstances, banking and commercial capital are becoming ever more international, and this undermines the capacity of national governments to control capital flows. Deregulation has been a common response. In the UK, this meant, among other things, scrapping of regulatory controls (e.g. fixed commissions on transactions), the introduction of automated dealing, and opening up membership of the Stock Exchange to non-UK financial institutions.

Domestic financial services have also been subject to a wave of liberalisation, with a substantial 'bonfire of controls' in the 1980s.

Measures taken included the removal of foreign exchange controls, removal of restrictions on consumer credit, the termination of the building societies' interest rate cartel, and the abolition of controls on banks and building societies. These changes, added to the encouragement of private pension schemes and other investment vehicles (e.g. endowment mortgages), stimulated considerable expansion in financial services.

The consequences have been particularly visible in the City of London (Pryke, 1994). The physical landscape has been transformed, via a wave of office development designed to cope with the requirements of technological change (better ventilation and space for wiring new information systems). The Docklands developments, notably the Canary Wharf prestige project, also exemplify this. The internationalisation of banking has produced a very rapid growth of foreign banks in London. In 1968, 125 foreign banks were directly represented, employing 9 100 people. By 1986 – that is, _before_ 'Big Bang' – there were 400, with employment totalling 53 800 (Leyshon and Thrift, 1997, 137). Allen (1992, 298) argues that this process of internationalisation has 'effectively tied London and the South East into a network of international relations which are very different from growth patterns across other regions'. Associated with these developments, there have been skill shortages and (partly because of a process of catching up with other financial centres) salary levels have risen dramatically. This has in turn had effects on property markets and also (through increased consumer spending) multiplier impacts in terms of job creation in retailing, leisure and other services. Socially, too, the rise of the 'Yuppie' and the 'loadsamoney' culture gave a new twist to regional stereotypes (Leyshon and Thrift, 1997, 136–63).

London's dominance of financial services remains unchallenged. In 1996 there were 313 000 jobs in financial services in London, representing 32.5 per cent of the national total. This is broadly comparable with Marshall _et al._'s (1988) figure of 34.5 per cent (changes in SIC classifications make direct comparison impossible). Employment growth in financial services from 1981–91

was most rapid in the outer South East with increases of over 50 per cent being recorded in 18 counties, most of which were in a belt encircling London but extending to Dorset, Avon, Gloucestershire and the south Midlands (figure 5.3). Northern outliers included Lothian, Cheshire and North Yorkshire. However, small reductions were experienced in four counties (Tayside, Cleveland, Lancashire and Durham), the latter losing one-third of its (small) total of jobs in financial services. This suggests that Leyshon and Thrift's (1989) predictions of the rise of provincial financial centres may have been over-optimistic. Townsend (1997, 140–7) argued that decentralisation from London was occurring mainly to Home Counties locations because firms still needed proximity to London but were searching for cheaper labour. As McDowell and Court (1994) observe, this process replicates a scenario identified by feminist geographers, namely the social construction of jobs as being particularly suitable for reserves of non-militant female labour willing to work part-time while bringing up children. There is, then, some support for Marshall's (1994, 41) contentions that metropolitan areas are 'increasingly becoming the base for higher-order … services, with routine administration being relocated to their hinterland'. Statistics on employment change for the 1993–96 period provide support for this view. In the context of generally static employment, Cambridgeshire, Wiltshire, Avon and Surrey experienced growth of over 10 per cent; Grampian and Glamorgan did likewise, albeit from a low base. However, London's continued strength was evident, with the 38 000 additional jobs there exceeding total growth in the rest of the UK.

While technological change ought to facilitate decentralisation of office employment to peripheral regions, this does not appear to have been happening on a large scale. The exception to this trend has been the growth of telephone call centres, which has permitted employment growth in peripheral locations such as Tyne and Wear (Richardson and Marshall, 1996). Large companies (British Airways; the AA) have relocated information, sales and inquiry services in search of

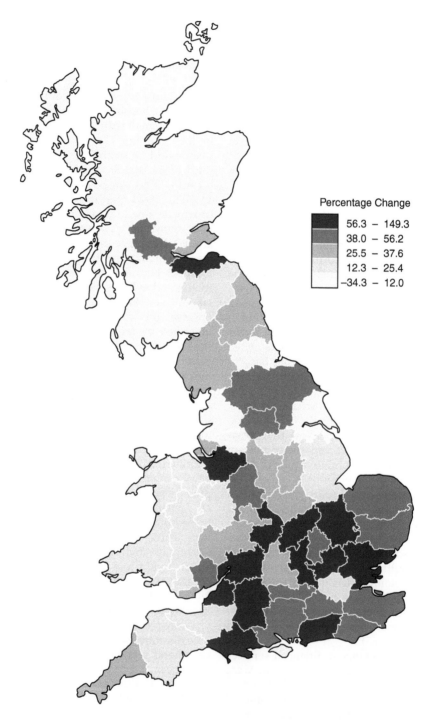

FIGURE 5.3 Percentage change in financial services employment, 1981–91 (NOMIS)

cheap labour and (arguably) better-qualified labour with low turnover. These centres now employ some 200 000 people and have been an important source of employment growth, though in many respects they reinforce existing spatial divisions of labour in services.

Broadly similar conclusions can be drawn from a study of the computer services sector, employment in which now totals some 279 000, up from 54 000 in 1981 (Coe, 1998). This expansion reflects the growing importance of IT and the tendency for many firms to outsource hardware and software provision and management to specialist firms. Computer services remain overwhelmingly concentrated in the South East: for 1996, LQs of at least 1.5 were found in Greater London, Cambridgeshire, Hampshire (2.09), Hertfordshire, Buckinghamshire (2.72), Surrey (2.99) and Berkshire (4.06). In total, 140 000 jobs, representing precisely 50 per cent of UK employment in computer services, were located in these areas. There are good reasons not to expect much dispersion of computer services. The main influences identified by Coe included the historical development of the computer industry, and the structure and hierarchy of the UK economy. This inevitably favoured the South East; by contrast, in peripheral regions growth was hampered by the small size of the indigenous IT industry and of the local market. Coe was sceptical of theories which stressed 'agglomeration economies created by close proximity to markets and suppliers' (p. 2065). He emphasised historical inertia and 'the overall attractiveness of the location to US software product companies' rather than inter-firm linkages and low transaction costs. In this sense his work reflects similar criticisms of the literature on new industrial spaces (section 5.5, below).

Consumer services: tourism, retailing and culture

Consumption dominates contemporary human geography, perhaps to excess. It also attracts an increasing share of attention from politicians concerned at the decline in traditional production industries. Few local authorities now lack a place promotion strategy, central to which is usually an emphasis on the desirable attributes of a place for various forms of active or passive consumption. But what are the regional development implications of all this activity?

A conventional view might be that consumer services are a dependent sector: they rely on surpluses being generated elsewhere in the economy. Increased attention to producer services has challenged the view of all services as dependent, and Williams (1997, 236) suggests that consumer services function as 'basic' activities, 'albeit by importing consumers rather than exporting products'. He writes approvingly of regional shopping centres (RSCs) importing shoppers to a region and thereby generating external income. His example - the Metro Centre, Gateshead - certainly attracts much of its trade from non-residents of Gateshead, and on this basis he argues that it brings substantial income to Gateshead. However, as this is a *regional* shopping centre what really needs to be assessed is the impact on other shopping centres in the region, rather than those within the immediate locality. RSCs are almost certain to draw trade from elsewhere even when there is steady growth in consumer spending, though as Williams points out they may have other spinoffs if visitors spend money on other leisure activities while visiting the RSC in question.

There are further reasons for questioning the regional impacts of retailing. It is a sector in which ownership and control are becoming increasingly concentrated and so profits flow from regional economies to metropolitan cores (Townsend, 1997, 157). Patterns of investment, favouring ex-urban development with all that that implies for urban sprawl and sustainability, reflect demands of institutional investors for short-term profits. Finally, although it is an area of employment growth, much of this is part-time and low-paid. On Williams' own figures geographical variations are minor and do not support the view that (at the regional level) consumer services are a major engine of growth. His table 4.2 presents location quotients for consumer services employment for regions, but all are very close to indicating minimal deviation from the national average share of employment.

Tourism has been criticised in similar terms, but this has not stopped national and local governments from making extensive efforts to develop new forms of tourism; at least half of all local authorities have pursued active policies (Townsend, 1997, 187). Coastal resorts have experienced a relative decline - many coastal towns, including some on the south coast, received Assisted Area designation in 1993 - and there has been a shift in promotional activities. There is, for example, an emphasis on historical and cultural attractions, on farm-based rural tourism, and on short-term tourism of a 'city-break' kind. Hewitson (1987) criticises the attempts to preserve almost any elements of a (sanitised) heritage as symbolic of a nation mourning its lost glories and unable to develop an alternative vision of its future (see also chapter 2, above), but this is the reality for many localities desperate for new sources of income. However, the multiplier effects have been exaggerated - there are benefits from tourist spending but once displacement effects are considered (due to diversion of tourists from alternative attractions) indirect multipliers are limited. Finally, tourist-related growth has added to the urban–rural shift, benefiting rural and semi-rural locations, small towns, and medium-sized cities due to the combination of environmental and cultural attractions possessed by such places (Townsend, 1997, 189).

The third set of activities considered here includes cultural industries, the media, entertainment and a range of consumption-based activities. Some authors argue that these can promote an urban revival, symbolising the vitality and creativity of cities and creating social cohesion by providing fora in which diverse groups interact. Glasgow and Birmingham have achieved some success in marketing their cultural attractions, and others are seeking to follow suit but it is questionable, again, how many and what sort of jobs these strategies create. Turok (1999) suggests that there is little sign (so far) of substantial growth in this sector.

Geographies of the public sector

Public sector services have received comparatively little attention in the regional devel-opment literature, but this situation is changing.

Jefferson and Trainor (1993, 1319) comment on a paradoxical situation: 'although many commentators abhor the national growth of the public sector, most regional agencies attempt to secure as large a share of the national public sector as possible'. Public sector employment has grown consistently throughout the OECD countries for most decades since the Second World War and, even when growth in *absolute* terms ceased, its *relative* share rose because of larger reductions elsewhere (e.g. as a result of deindustrialisation).

The definition of public sector employment used by Jefferson and Trainor includes the principal government services of health, welfare, public administration, education and defence, but not the public corporations (many of which, such as the nationalised industries, have by now been returned to the private sector). On this definition 5.3 million people were public sector employees in 1996, compared to 4.1 million in 1971. However, in proportionate terms, the public sector accounted for 24 per cent of employment in 1996 compared to 18.7 per cent in 1971, reflecting reductions elsewhere.[1]

Much of this employment is distributed broadly in line with population distribution; this is especially so with regard to education and health care. However, depending on the availability of other employment opportunities, there are substantial variations in the public sector's contribution. Thus, at the regional level, the Northern region, Scotland and (particularly) Wales have an over-representation of public sector employment, the respective location quotients being 1.13, 1.11 and 1.27. Eight counties have more than 30 per cent of their workforce in the public sector (Dyfed, South and West Glamorgan, Merseyside, Northumberland, Gwynedd, Devon, and the Fife region). At the scale of local authority districts, in Carmarthen (69 per cent) and the Rhondda (58 per cent) the public sector is the majority employer. In a further 15 districts the figure exceeds 40 per cent, including some southern university and administrative centres (Cambridge, Oxford,

Bath, Exeter); an additional 21 districts have at least 35 per cent of their employed workforce in the public sector, including Newcastle-upon-Tyne (36.4 per cent) and Swansea (35.8 per cent), where the figures reflect the presence of decentralised government offices.

These spatial variations raise the question of the role of the public sector in regional development. If the public sector draws into productive employment resources which would not otherwise have been used, then there should be increases in output, employment and income; if not, and factors of production are diverted from more productive sectors, there will be an overall fall in output. This would suggest relocation of public services to areas where opportunity costs are lowest (Jefferson and Trainor, 1993; Marshall, 1996). There are limits to such a strategy: many services are population-related; there may be operational reasons why some services (defence, for instance) must be in certain places (for ready access to communications and to government); and there are social and economic arguments for centralisation of services even within regions.

In this context we should consider the impacts of policies designed to defend, or to disperse, public sector services. First, several commentators have stressed the role of the public sector in providing access to a basic minimum of employment opportunities, which is why the privatisation of public services has been vigorously resisted, at least in some urban and northern local authorities, on the grounds of their local multiplier effects (Geddes, 1994).

The 'basic/non-basic' distinction is often invoked in discussion of the public sector, with services oriented towards the local population being regarded as dependent, or 'non-basic'. This may be an oversimplification, since it can be argued that certain minimum standards of service provision actually underpin competitiveness (e.g. by securing an appropriately skilled workforce: Mohan and Lee, 1989). Nevertheless, debates about the role of the public sector in regional development still tend to use the basic/non-basic distinction. Mobile government-financed services are sometimes regarded as part of the export base, to the extent that their *output* is purchased by government demand from outside the region. This will have positive multiplier effects in a regional context. Hence the location of units which are in principle mobile has attracted much attention; this is particularly the case with the Civil Service.

Dispersal of government offices has been advocated since the early 1960s, both on cost grounds and because of benefits to recipient regions. Nevertheless, very few moves had taken place by 1979, and the incoming Conservative government lacked a strong commitment to an active regional policy. Relocation of the Civil Service returned to the agenda in the late 1980s, though primarily as a cost-saving measure rather than a part of a regional policy. The result was that the pace of decentralisation accelerated somewhat; whereas between 1979–88 only 17000 jobs moved from London and the South East, for the 1989–93 period the corresponding figure was 19000 (Marshall, 1996, 357, 359). However, the pattern of change was somewhat less beneficial to the peripheral regions than might be desirable. Major relocations, such as the Department of Health to Leeds, and the expansion of the Employment Department in Sheffield, have redressed some of the imbalance. Growth has, however, tended to be most rapid in the Midlands, Lancashire and Yorkshire rather than the North of England, Scotland and Wales. The poor performance of the peripheral areas is exemplified by Tyne and Wear, which lost 3765 civil service jobs between 1979–95, principally because *in situ* decline in established offices outweighed relocations (Marshall *et al.*, 1997, 611). Second, many relocations were of a classic 'branch-plant' or 'back-office' type; with exceptions such as the Department of Health's move to Leeds, it has proved difficult to move high-level jobs and it appears that the residential preferences of senior civil servants have played a part here (Winckler, 1990). While Civil Service dispersion has reduced the South East's dominance, the regions most in need of such jobs continue to suffer perennial problems of peripherality and a negative image, which disadvantages them in attracting such activities. A more strategic policy of decentralisation would have to be pursued if the neediest regions were to benefit from dispersal.

Public sector services are, then, a relatively stable source of employment and there have been calls for a controlled expansion of public sector employment to reduce unemployment. Lovering (1999) contends that, for all the hype about FDI in the Welsh economy, the public sector has been the main source of employment growth. However, tight limits on public spending, a result of Labour's self-denying ordinance, will limit scope for such action domestically. Internationally the need to meet convergence criteria for European Monetary Union (EMU) will dictate continued downward pressure on public expenditure. The most plausible result will be the generalised extension to the public sector of processes of workforce segmentation, through competitive tendering and commercialisation (Burrows and Loader, 1994; Pinch, 1994 and 1997).

5.5 CONCLUSIONS:
NEW INDUSTRIAL SPACES?

A key question raised by the studies considered in this chapter is the extent to which patterns of regional economic success can be replicated or generalised. Numerous studies of enterprise or high-technology industry do not, in fact, give very clear indications. For example, the key process in the development of high-technology industry is usually held to be spinoffs from research universities, but this process will not necessarily work for all innovations (e.g. those not dependent heavily on formal science). And Cambridge is not the only British research university with a strong high-technology (e.g. semiconductor) orientation, but others have not generated similar numbers of innovations (Storper, 1995, 202–7). As to small firms and entrepreneurship, Keeble *et al.*'s multivariate analyses point to numerous plausible connections, but these do not show precisely why certain places facilitate entrepreneurship. There is a tendency to rely on Marshallian explanations – something 'intangible' which permits inno-

vativeness in some places and not others. But there is a circularity in such arguments: innovation occurs because of the presence of a particular 'milieu'; a milieu exists in regions where there is innovation. This kind of argument cannot specify why localisation and territorial specificity should make technological and organisational dynamics work better in some places than others. Some economists such as Krugman refer to geographical 'accidents', which give regions initial advantages, leading to specialisation, but Pinch and Henry (1999) argue that such 'accidents' on their own are not enough to explain patterns of regional comparative advantage.

Given these criticisms, what insights are available from contemporary theorising on new industrial spaces (NISs)? There is much hagiographical writing in this literature, which concerns a disparate set of places and branches of production, raising doubts about its theoretical coherence. In his initial formulation, Scott (1988a) argued that vertical disintegration of firms produced transaction costs, which clearly have a spatial element. The greater the significance of these costs, the more there will be a tendency to reagglomeration in order to minimise transaction costs. The linkages likely to lead to agglomeration are those relating to flexible, non-standard, small-scale elements of production, such as those relating to shifts in demand or technology. There are several criticisms of this basic model (summarised by Henry (1992); see also Lovering (1990b), Scott (1988b) and Storper (1995)). However, in order to demonstrate that this process was responsible for concentrations of particular industries, one would need to demonstrate not only dense inter-firm linkages, but also that the transaction cost mechanism was operating *and* that this mechanism was more important than other factors in producing agglomeration.

Henry (1992) attempted to apply Scott's (1988a) thesis to hi-tech firms in Hertfordshire. Henry *was* able to demonstrate that local production linkages had played a role in the establishment and development of specific firms. However, he could not provide evidence for the operation of Scott's transaction costs mechanism. Only 22 per cent (34 of his respondents) indicated that their major production linkages were significant in their decision to locate in Hertfordshire and, of those, 12 did not have a major production linkage in Hertfordshire. Morgan and Sayer (1988, 215–16) argued that local inter-firm linkages were much more common in the M4 corridor than in South Wales, but these linkages did not compare in density with those in hi-tech firms in some American case studies. They attributed agglomeration to local labour-market conditions rather than to transaction costs and production linkages.

Henry's interviews with firms revealed some evidence of transaction cost mechanisms, but many inter-firm relationships were ones of dependence; suppliers or contractors relied heavily on connections with some major multinationals. Moreover, the identification of production complexes would require an elastic definition of the 'locality' since, for many firms, the most significant linkages were not with other firms in close proximity. Similarly, Massey *et al.*'s (1992) study suggested that hi-tech firms in Cambridge were linked into national and international networks much more than to local firms. For these reasons Henry was unable to conclude that agglomeration was necessarily due to the transaction costs mechanism. He felt that there were in fact several overlapping high-technology production complexes in Hertfordshire, and the locational logic of these did not follow Scott's precepts.

Later versions of the NIS theory criticise Scott's formulation for its (as they see it) excessive reliance on the transaction costs mechanism, which gives it a rather static cast. Instead the argument is that sustained competitiveness has more to do with capabilities leading to dynamic improvement than with achieving static efficiency. Industrial systems are made up not just of physical flows of inputs and outputs, but also of intense exchanges of business information, know-how, and technological expertise, in both traded and untraded forms. The key to understanding agglomeration is said to lie in the superior capacity of spatial configurations to enhance learning, creativity and innovation. Agglomeration is said to promote informal processes of learning and

exchange of tacit knowledge, not always codified, and it is these processes, involving a mixture of collaboration and competition, that lead to innovation.

The key argument is that technological excellence relies on knowledge or on practices which aren't fully codifiable, and therefore firms who master these become tied into networks with other firms, both through formal exchanges (trade) and *untraded interdependencies*. The latter include nationally derived rules of action, customs, understandings and values governing the development and dissemination of knowledge. These interdependencies can confer on regions absolute advantages which shelter them from competition.

Thus, the apparent 'resurgence of regional economies' is to be explained in terms of two sorts of analyses: (i) the tension between respecialisation and destandardisation of inputs and outputs to and from the production process which, *ceteris paribus*, raises transaction costs; and (ii) the untraded interdependencies which are attached to processes of economic and organisational learning and cooperation. Pinch and Henry (1999; see also Pinch *et al.*, 1997) illustrate these general remarks with a discussion of the British motor sport industry (BMSI) (figure 5.4). This displays a considerable degree of spatial concentration in South East England. While some of this may be due to 'geographical accidents' of the kind identified by Krugman (1996) (such as the availability of disused airfields, which encouraged club racing, providing a strong base of 'grassroots' knowledge), what now sustains this industry is quite different. Understanding these processes is important because the BMSI, employing over 50 000 people mainly in technologically advanced small firms, is precisely the kind of industrial success story governments are seeking to replicate. Henry and Pinch stress the key role of *knowledge* in the BMSI. This is an industry in which ideas and information are central to the permanent search for innovation and competitive advantage. Proximity permits a rapid process of diffusion of ideas about which technologies work best. More than that, because complex knowledge is difficult to

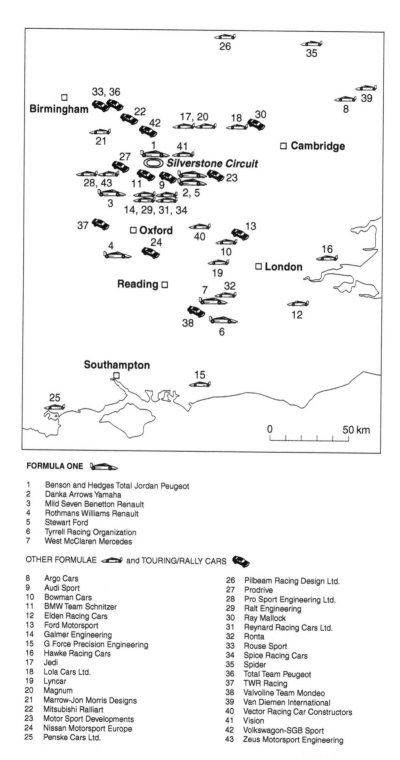

FORMULA ONE

1   Benson and Hedges Total Jordan Peugeot
2   Danka Arrows Yamaha
3   Mild Seven Benetton Renault
4   Rothmans Williams Renault
5   Stewart Ford
6   Tyrrell Racing Organization
7   West McClaren Mercedes

OTHER FORMULAE and **TOURING/RALLY CARS**

8	Argo Cars	26	Pilbeam Racing Design Ltd.
9	Audi Sport	27	Prodrive
10	Bowman Cars	28	Pro Sport Engineering Ltd.
11	BMW Team Schnitzer	29	Ralt Engineering
12	Elden Racing Cars	30	Ray Mallock
13	Ford Motorsport	31	Reynard Racing Cars Ltd.
14	Galmer Engineering	32	Ronta
15	G Force Precision Engineering	33	Rouse Sport
16	Hawke Racing Cars	34	Spice Racing Cars
17	Jedi	35	Spider
18	Lola Cars Ltd.	36	Total Team Peugeot
19	Lyncar	37	TWR Racing
20	Magnum	38	Valvoline Team Mondeo
21	Marrow-Jon Morris Designs	39	Van Diemen International
22	Mitsubishi Ralliart	40	Vector Racing Car Constructors
23	Motor Sport Developments	41	Vision
24	Nissan Motorsport Europe	42	Volkswagon-SGB Sport
25	Penske Cars Ltd.	43	Zeus Motorsport Engineering

FIGURE 5.4 Principal single-seater, touring and rally car constructors in the UK, 1997–1998 (Pinch and Henry, 1999)

disseminate over large distances, face-to-face contact is essential if ideas are to be communicated. One reason is that there is a large amount of 'tacit knowledge and built-in assumptions' that are unique to the ways in which things are undertaken in different places. Moreover the BMSI does depend, collectively, on information being shared, at least to some extent. If one firm developed a major competitive advantage, such that its cars were winning every race, interest in and support for motor racing would decline rapidly. Thus all firms have an interest in sharing ideas - though in practice there is little they can do to prevent others learning about and copying their innovations. It is this tacit knowledge, which is not formally traded but which is available to those firms and individuals 'plugged into' the spatially concentrated BMSI, which accounts for the concentration of small innovative firms, principally in central Southern England, in this sector. A transaction costs mechanism on its own could not be responsible for this concentration. Henry and Pinch have certainly identified a plausible industrial district, but how far this might be replicated must be questioned. The BMSI (and motor racing) is arguably unique in its reliance on knowledge (and the sharing of tacit knowledge) and it is not clear what other sectors operate in the same way. It is also (despite its international prominence) comparatively small, with employment at around 50000. If such regional success stories are to be replicated, this in turn poses important challenges for spatial policy (chapter 12).

Two further concluding points are in order. First, scepticism about the strength of the service economy is in order because, at least some service sector growth is fragile and unsustainable. The recession of the early 1990s which followed the 'Lawson boom' had the effect of eliminating much ephemeral activity in consumer services which had expanded with the easy availability of credit. This recession was led by the South East, which experienced increases in unemployment over a year before the effects were felt in Scotland and Wales (Townsend, 1993).

Second, how far are the forms of economic growth discussed here sufficiently 'embedded' as to withstand external competition? Clearly there are some – the City, consumer services and tourism – which are very much rooted in particular places but for other sectors this is less so. There is no obvious reason why certain routine activities in high-technology or services, that do not rely on face-to-face contact, could not be relocated overseas. This indicates the continuing need to build on and sustain existing clusters of activity where Britain has an edge. In this context it is increasingly clear that investment in the labour force is becoming a key source of competitive advantage and it is to labour-market regulation that we now turn.

NOTE

1 The statistical source does not actually differentiate between public and private providers of health care and some other services. This does not, however, affect comparisons between places substantially, and in the places where the public sector makes its greatest contribution to employment there will generally be few private providers of these services.

6

GEOGRAPHIES OF LABOUR MARKETS
FLEXIBILITY AND FRAGMENTATION

6.1 INTRODUCTION

Much conventional economic geography has, until recently, neglected labour or treated it simply as a passive factor of production. However, the work of Massey (1984) and subsequently Peck (1996) has been enormously influential in changing this situation. Fundamentally, labour markets are locally variable, resulting from the role played by places in spatial divisions of labour (Massey, 1984). Particular rounds of investment are associated with specific geographies and, as a consequence, different parts of the production process come to be located in different places. Massey argues that capital is acutely aware of spatial divisions of labour, and that capital consciously seeks out particular qualities in a labour force. Moreover, capital has considerable power over labour in the sense that labour continues to require the stability and support of community life (Peck, 1996, 15) whereas capital is in principle much more mobile. Attachments to place, barriers to mobility, and imperfect or incomplete information about employment opportunities are all ways in which geography complicates the workings of labour markets.

Peck (1996) argues for a distinctively geographical perspective on labour markets. In contrast to 'classical' markets, mediated solely by price, labour market segmentation means that firms will deliberately recruit from within *internal* labour markets (because

skills are firm-specific, and easier to develop from existing employees) rather than from outside the firm (Jones, 1996). Such barriers mean that the scope of price-regulating competition is constrained, and in turn this implies limitations on how far wages can be allowed to fall. More generally, there is no guarantee that the supply of labour will necessarily meet the demand for it. Against the background of high unemployment this might seem implausible, but it remains the case that labour supply is 'relatively autonomous' from demand, because the supply of workers entering the labour market is a result of 'intersecting social, demographic, educational and economic decisions', shaped by social forces such as 'state policies, ideological norms and family structures' (Peck, 1996, 26–7). State policies, in particular, substantially set and modify the parameters of labour supply. However, there are definite limits to policies predicated on removing a 'floor' from the labour market: if wages are cut to or below subsistence levels, ultimately market demand is undermined, leading to crises of overproduction. But there remains considerable scope for manoeuvre concerning the terms on which non-participation in the labour force is permitted, and concerning who bears the costs of reproducing labour (education, training, health care, etc.). This draws analytical attention, therefore, to the social regulation of labour markets, and to how this process varies both between and within nation states. Finally, from a more global perspective, the 'work of nations' (Reich, 1992) will be to a growing extent divided between those states whose workforce is educated and skilled to take advantage of and adapt to a knowledge-based, high-technology future. Active labour market policies, on this view, are crucial to economic success. It is no longer adequate to treat labour as the resultant of the economic decisions made by private capital.

The chapter therefore reviews key debates relating to geographies of labour in contemporary Britain. The key word here is *flexibility*: changes in the organisation of production have produced new, apparently more flexible forms of employment. Those changes have been both the precondition for and a consequence of the neoliberal labour market policies pursued with vigour since 1979. Consequently the chapter begins with a review of those policies (section 6.2). It then turns to a consideration of the evidence on flexibility, looking first at geographical variations in wage rates, and then at the emergence of the so-called 'flexible firm', drawing on evidence from studies in various contexts (section 6.3). This is followed by an examination of labour market segmentation, focusing on divisions within and between different groups and between places. Finally there is a discussion of recent and prospective policies designed to regulate the labour market (section 6.5).

6.2 PROMOTING OF FLEXIBILITY IN CONTEMPORARY BRITAIN

The Conservatives' diagnosis of the state of the nation emphasised that the balance of power between capital and labour had swung too much in favour of the latter. Restrictive practices, immunity from prosecution and liability, and the floor under the labour market provided by the welfare state, were all seen as integral and connected symbols of British economic decline. It was clearly necessary to recreate the labour market conditions that had obtained in the era of Britain's global economic hegemony. This was articulated rhetorically in various ways. First, unemployment was attributed to the

failures of workers to accept the wages employers were prepared to pay. Workers were to price themselves into employment, and to accept jobs on whatever terms were offered to them. Second, trade unions were identified as a crucial institutional barrier to the operation of market forces. There followed a number of pieces of legislation designed to weaken trade union power; even where the government did not use legislation, deflationary economic policies and cuts in public expenditure removed many jobs in highly unionised sectors. Third, geography was seen as part of the whole process of reworking work. Regional wage differentials were viewed as part of a 'natural' process of adjustment, an equilibrating mechanism which would ultimately be self-correcting. Workers were to respond to market signals either by lowering wage demands in areas of high unemployment, or by getting on their bikes, in Norman Tebbit's famous phrase, and looking for work elsewhere. In practice, however, because of cost-of-living differentials, inter-regional migration has not been a serious option, and the evidence for inter-regional wage differentials, other than the high wages paid in the South East, remains limited.

Given these diagnoses, a number of policy measures followed (Rubery, 1989). First some institutional barriers to the operation of market forces were weakened or removed, such as the abolition of the Fair Wages resolution, and the initial reform, then abolition, of Wages Councils. Second, industrial relations legislation was reformed. This greatly restricted the powers of trade unions and removed some of the immunities they had previously enjoyed. Companies were under no obligation to recognise unions and employers took full advantage, launching a vigorous campaign of union derecognition. Third, a degree of deregulation of the labour market was achieved through extending the length of service required before a worker could claim unfair dismissal (from six months to two years), and abolishing industry-wide collective bargaining. Aside from these legislative measures affecting employment regulation, the government indirectly used the public sector as a demonstration project for its wider programmes. Thus, despite disavowing incomes policies, the strict enforcement of cash control of public expenditure meant that, *de facto*, an incomes policy was pursued in the public sector as workers found that their incomes persistently lagged behind inflation. Moreover programmes of privatisation and decentralisation of pay bargaining were used to signal to private capital the kind of changes the government believed necessary. Finally, benefit regulations, which set a floor under the labour market, were made consistently tougher. Payments were restricted, eligibility criteria were tightened, and the unemployed were required to demonstrate that they were *actively seeking* work, not just that they were *available for* work (King, 1995).

The consequence was that 15 years after the 1979 election, the scope of labour market reforms exceeded the wildest dreams of the New Right: there was no regulation of working time (e.g. maximum working hours – and there is still resistance to this); no legal protection for labour hired on fixed-term contracts; no minimum wage legislation; minimal employment protection; and employees had no right to representation in the workplace. An OECD composite index of labour market protection ranked Britain lower than any country (other than the USA) in the industrialised world (Hutton, 1995, 94).

Of course, some of these changes resulted from secular trends in the organisation of production. Thus, new production processes, such as just-in-time systems, may require a wider range of skills of workers, as well as more flexible working patterns. Technological change is often associated with different working arrangements as workers are required to operate new and different machinery over a greater range of tasks. Finally there are long-term changes in the economy. The most important of these is the transition to a service-based economy which, especially in the personal services sector, is creating large numbers of part-time or temporary jobs. Trends towards employing larger numbers of women, towards outsourcing of elements of the production process, towards smaller plant, and towards employing workers with higher skill levels and

greater autonomy – all these may, in the right circumstances, tend towards a more flexible labour market, at least from the point of view of capital. In one sense, therefore, the Conservative government simply went with the flow of external changes. However, it is also indisputable that – by comparison with European states – they accelerated the speed of those changes, and as a consequence the degree of labour market polarisation in Britain exceeds that found elsewhere in Europe.

6.3 FORMS OF FLEXIBILITY AND RISK

I consider, first, whether there is evidence of greater geographical flexibility in wages, as neoliberals would predict. I then go on to discuss the model of the so-called 'flexible firm' before examining evidence from different places across the UK as to the extent of flexibility in the labour force.

Geographical flexibility

Martin and Tyler (1992) constructed indexes of real and nominal wages (the former allow for variations in costs of living, the latter do not) for the regions of the UK for 1979–90. In principle, the government argued, lower relative wages in regions where unemployment was high ought to stimulate job creation. However, no region-specific measures were introduced to deregulate local labour markets. Instead the government relied on the general policies outlined previously, in which the emphasis was firmly on decentralisation and localisation of labour. Martin and Tyler (1992) make three main points.

First, the increase in relative levels of unemployment in the North was associated with a decline in relative money wages, for both manual and non-manual workers, which is to say that some movement occurred in inter-regional money wage structure in response to differential movements in regional labour demand and unemployment. However, they question whether this was enough to sustain a claim that genuinely flexible labour markets had appeared. Second, while nominal wages were lower in the North than in the South, variations in real wages (after allowing for housing and other costs) were much smaller. This depressed real wages in the South relative to the North; in turn this led to higher-than-average wage claims; and these eventually led to the re-emergence of inflation from a base in the South East (Martin, 1989a). Third, there was some nominal adjustment of wages downwards in the North relative to the South, but it could be argued that in a flexible labour market very large movements in the cost of labour would have been required in order to achieve labour market clearing. For spontaneous processes of adjustment to operate properly, wage levels in the depressed regions would have had to have been pushed to near-subsistence levels. Furthermore, there is no evidence that the decline in relative wages of low-income workers has been associated with an expansion of demand for their services (Martin, 1998, 37), as neoliberal theorists would predict.

Other studies have broadly endorsed these conclusions, noting that there has been some dispersion of regional earnings, especially for non-manual workers. However, this is attributed largely to higher wages in the South East, especially in the services sector, where comparatively small numbers of very highly paid jobs distort averages. In addition, the higher wages in the region are partly an artefact of the cost of living (e.g. London allowances). Excluding the South East the evidence for variations in wages at the regional level is very limited (Beatson, 1995). In short, wage flexibility may have increased a little, but Martin and Tyler (1992) argue that if the claims of free-market theorists were to be supported, there would have to have been evidence of much greater dispersion in wages. In the absence of a much more vigorous attack on the welfare system, designed to force people to accept employment at almost any price, this is not very likely. The contrast with the USA is instructive: there is much greater evidence of wage flexibility, and associated higher levels of migration, but these result from institutional

factors (a minimalist welfare state, and a very flexible housing market) which do not apply to the UK. Longitudinal analyses show that the propensity of the unemployed to migrate between regions is actually below average, reflecting the objective constraints they face in housing markets (Fielding, 1995, table 10.3). Migration does not, therefore, exercise an equilibrating effect on unemployment differentials. Migration rates tend to rise during economic upswings and to fall during recessions. Migration should therefore be seen as a response to employment change, not to differentials in unemployment (Green *et al.*, 1998). Arguably the Conservatives preached greater labour market flexibility and simultaneously limited the scope for it, by permitting the sale of council housing and promoting owner-occupation to the point where the inter-regional house price gap widened substantially. Finally, the major changes in the income distribution in the past two decades have been *intra*-regional, as all regions have experienced a growing gap between the top and bottom deciles of the income distribution (see chapters 7 and 8). Given this absence of inter-regional variability, the likely impacts of economic convergence, following the Maastricht Treaty, cannot be absorbed by migration and will result in a widening of unemployment differentials.

The flexible firm?

Considerable attention was given in the 1980s and 1990s to the question of whether new forms of employment are emerging which offer a degree of flexibility in working arrangements which is said to be beneficial to both workers and employers. Optimists argued that this would permit a smoother dovetailing of paid employment and domestic life; critics suggested that advocates of the flexible firm generalise from a limited set of (usually high-skill service) occupations and people, whereas the reality for many is that flexibility permitted employers to exert an even greater degree of control over labour (Pollert, 1988 and 1991). Thus, flexibility may be a misnomer from the perspective of the employee.

The basic thesis is that the workforce is now divided into a core and periphery, as well as various subcategories according to the degree of economic security (this summary draws extensively on Allen, 1988). The core consists of secure, full-time permanent employees, who enjoy job security, and high levels of earnings and fringe benefits, in return for performing a wide range of tasks that cut across skill demarcation lines. These workers offer *functional flexibility*. As the nature of the business of a firm changes, and external conditions (such as markets for various products) alter, the core workers acquire new skills and/or use their existing skills for different tasks. The typical members of the core workforce are managerial or professional staff, and a range of multi-skilled individuals whose central characteristic is that their skills are not readily available in the wider labour market, so that the firm will attempt to retain them and separate them from the external labour market.

Around the core there is a series of outer layers, each representing a different group of peripheral workers. The first of these are full-time workers, but they enjoy less job security and poorer career prospects than those in the 'core'. They are hired to fill specific jobs, usually of a semi-skilled character. In the absence of career opportunities, the routine nature of the tasks encourages a high degree of labour turnover. This segment of the workforce grants the employer a degree of *numerical flexibility*. A second peripheral group comprises part-time workers and various temporary workers, ranging from agency staff, people on short-term contracts, and individuals on training schemes. These can be hired and laid off according to fluctuating levels of production. Finally, at the periphery there is a group of external workers not employed directly by the firm, who perform either very specialised (consultancy, information technology) or routine tasks (e.g. catering, security, cleaning). Numerical flexibility is achieved here through the use of subcontracting, outsourcing and self-employment.

A number of criticisms have been levelled at this model. One of the most important is the methodological point that it uses changes at the macro-level (e.g. in the structure of employment) as evidence of change at the micro-level in employer strategy. It thus confuses compositional and strategic effects, and presumes a degree of conscious strategy and agency which may not be apparent on the ground. Thus, while there is evidence of growth in non-standard employment forms (Dex and McCulloch, 1997), this does not necessarily mean that there are many 'flexible firms'. Moreover, compositional effects (the tendential shift from manufacturing to services) may be more important in creating new forms of work, such as the expansion in part-time employment; there is some evidence that the proportions of part-time workers employed in manufacturing has fallen both absolutely and relatively (Allen, 1988, 210; Rubery, 1989). What in fact is the evidence for the emergent flexibility?

Considering aggregate studies of firms, surveys for the early 1980s – which was when the IMS studies were published – indicated little evidence of considerable change. The 1984 Workplace Industrial Relations Survey (WIRS) covered 2000 establishments with more than 25 employees. It found little sign of substantial change in employment practices, with little use of the so-called peripheral sections of the workforce. The studies by those who developed the model of the flexible firm were based on small-scale samples which were virtually self-selecting: one case study selected its respondents precisely because they *had* introduced measures to promote more flexibility (Pollert, 1988). This could hardly be used as a basis on which to generalise about labour-market changes across the board. Similarly Pinch *et al.* (1991) quote a 1987 survey which indicated that only 5 per cent of employers were *consciously* adopting a 'core/periphery' strategy. O'Reilly and Rose (1998, 719) argue that specific changes are often concentrated within certain sectors; for instance, the use of short-term contracts is concentrated in the public sector.

There is apparently more evidence in support of claims that flexibility has expanded from wider, aggregate trends in the labour market. However, some of these do not lend themselves to one-dimensional interpretations. Certainly, there is evidence of growth in part-time work, but this reflects as much sectoral shifts as conscious decisions by firms. Figure 6.1 indicates the growth in part-time employment over the 1984–98 period, from 4.87 million to 6.38 million employees, an increase of 30.9 per cent; of these jobs, 869 000 were taken by women compared to 640 000 by men. The proportion of people working part time therefore grew from 20.6 per cent to 24.9 per cent (ONS, 1997, tables 1a, 2a). Considering the geography of part-time employment, in 1981 there were only 13 local authority districts in which part-time employment exceeded 30 per cent of the total. In 1996 this proportion exceeded 30 per cent in 248 districts and 40 per cent in 16, including Nairn and Penwith (44 per cent) and Rushcliffe, Skye/Lochalsh, Sutherland and Aberconwy (all 43 per cent). Many such districts, as might be expected, were in rural

areas where the main sources of employment were tourism and agriculture. Given secular trends towards greater use of part-time employment it may not be long before such jobs are in the majority in some localities.

There is clearly also growth in self-employment which expanded from 1.77 million in 1979 to a peak of 3.47 million in 1990, since when it has fluctuated at around 3.25 million. This growth is not easy to interpret. In some circumstances employers may indeed substitute self-employment for employment – staff may be asked to accept new working arrangements which are, *de facto*, a form of self-employment. This can be regarded as a form of flexibility. However, other elements of the growth of apparently self-employed individuals – the introduction of government schemes which help the unemployed into self-employment, the

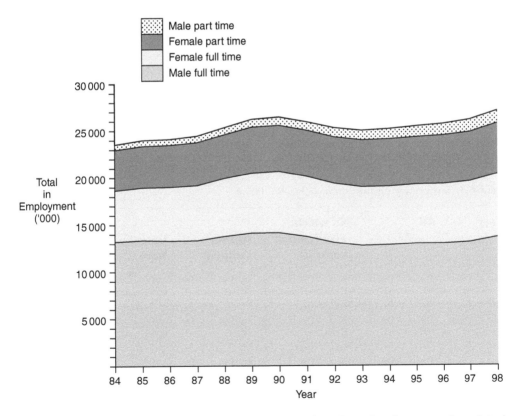

FIGURE 6.1 Employment trends, 1984–98 (ONS 1997; *Labour Force Supplement*, various dates)

growth of privatisation programmes – cannot so easily be assimilated into this vision of flexibility (or, for that matter, entrepreneurship). Self-employment is not always a voluntaristic, entrepreneurial response by individuals; it may instead be something on which people embark as a last resort. The stability and sustainability of self-employment need to be established if the claims of those who view it as a beneficial and progressive form of flexibility are to be sustained.

There is also evidence of growth in the numbers of temporary workers (including those on fixed-term contracts as well as agency staff, casual employees and seasonal workers). The total number of temporary workers rose from 1.19 million in 1992 to 1.67 million in 1997 (falling slightly in 1998). The growth of agency work, and of fixed-term contracts, were the main features of this (table 6.1). Some evidence implies that firms are increasing the use of such labour as they are increasingly concerned about the external environment (Dex and McCulloch, 1997). Finally there is anecdotal evidence of new strategies towards working practices: contracts offered on the basis of a total number of hours per year, to be worked at the employer's discretion; zero hours contracts (where people are in effect retained – though not paid by – a given employer, unless they are actually called on to work); and the 'Burger King' contract, in which employees in a fast food restaurant were only paid when there were actually customers in the shop. Set against these examples, however, those in secure employment are actually working longer hours, and the numbers holding second jobs doubled (to 903 000) between 1984 and 1996 (ONS, 1997, table 2b), which presumably indicates the pressures households are under in order to make ends meet. More than 30 per cent of British male employees work more than 45 hours per week, significantly higher than in Germany. If flexibility is a response to changing levels of – and higher levels of – demand, why are companies in other states, facing similar competitive pressures, responding in different ways? One inference might be that the British economy is not generating high-quality jobs in sufficient numbers, thus forcing individuals into circumstances where they have little choice but to work longer hours or take on second jobs. A second inference might be that the growth of second jobs has the effect of widening divisions between work-rich and work-poor households.

What of evidence from particular sectors? The public sector has been used for its demonstration effects of what was possible under a more flexible labour market. Indeed, it has been suggested that the core–periphery divide, which does so much work for theorists of flexible specialisation, is neither new nor largely a phenomenon of manufacturing industry; it is located within services and specifically within the state sector (Jessop *et al.*, 1988). Thus, considering the reshaping of

TABLE 6.1 Temporary employment, 1992–98 (000s)

Year	All	Of which: Fixed period contract	Agency temping	Casual work	Seasonal work/Other
1992	1195	577	81	263	274
1993	1251	621	91	273	265
1994	1386	744	114	302	226
1995	1512	818	161	305	227
1996	1557	801	201	319	235
1997	1673	861	227	332	254
1998	1661	845	254	305	257

Source: ONS (1997); 1997/1998 data from *Labour Force Supplements*. Figures refer to spring of each year.
Note: Row totals may not sum due to rounding.

work within the NHS, one could hardly account for privatisation and subcontracting in terms of changing external demand conditions. Arguably pay determination within the NHS has sought consciously to divide the core from the periphery of the workforce. Nurses, doctors and to some extent managers have been cast in an heroic role as the 'front-line troops' whereas support staff (principally ancillaries, but also lower-grade administrators and clerical personnel) are regarded as more dispensable. Ancillary staff have found their work subjected to competitive tendering both in this sector and local government. The effects on terms and conditions of service for large numbers of people have been described in numerous analyses (e.g. Pinch, 1994; Mohan, 1995a, chapter 6). For these reasons some now see the public sector as reflecting the 'flexible firm' *par excellence*, though questions have to be asked about the appropriateness of such forms of labour regulation in welfare services.

In similar vein, nationalised industry policies have been viewed as ways of reworking the nature of work in highly unionised environments. It may well be true that the productivity of some of these industries left much to be desired, but it is also undeniable that quite drastic reductions in employment, and a worsening of terms and conditions of employment for the remaining workforce, resulted from Conservative policies. Thus Sadler (1992) shows how, against a background of mass unemployment in North East England, the British Steel Corporation first sacked and then re-employed the same workers on a quasi-casual basis; employees were required to work less predictable hours, carry out a greater range of tasks, with fewer fringe benefits and no security of employment. (For other discussions of the effects of nationalised industry policies on employment, see Morris (1995), Turner (1995b).)

Flexibility in place

The two most developed attempts to consider the significance of new forms of flexible working were those of Hudson (1989c) drawing on considerable research on North East England, and Pinch *et al.* (1991) on the Southampton area. Hudson questioned whether labour market changes in Britain's (and Europe's) Old Industrial Regions (OIRs) in fact represent anything new. He argues that mass unemployment simply provided the opportunity for firms to engage in labour-market practices redolent of classical Fordism, if not *laissez-faire* capitalism in its Victorian variant. Thus, in the 'basic' industries of such regions (coal, steel, shipbuilding) he identifies efforts to increase productivity through classically Fordist methods – speeding up the pace of production, and an increase in automation within plants, sometimes though not always associated with new technology. He also identifies efforts to intensify work practices without introducing new technology: measures taken here include seeking greater flexibility, lowering demarcation barriers, and introducing competitive forms of wage bargaining. New firms moving into the peripheral regions have been able to exercise great care in selecting their workforces, choosing young, physically fit workforces, committed to their company, thus achieving enhanced labour productivity in what are arguably classically Fordist branch plants. Thus the Nissan plant at Washington reduces every operation to the lowest skill level possible, and minimises any scope for worker discretion. This is a lowest-common-denominator approach to skills and training (Garrahan and Stewart, 1992; see also Sadler, 1992). This is not the emergence of a flexible, multiskilled labour force, but the use of local labour market conditions to exert disciplinary power on those seeking employment. Further evidence that jobs generated by FDI are often low quality is provided by Munday *et al.* (1995) who comment on the low level of R and D activity and low proportions of graduate employees (see also Amin and Tomaney, 1991). Hudson also argued that the employment relationship was being reworked through the growth of subcontracting and of casualisation. This produced an extremely peripheral workforce, hired and fired almost at will, and is characteristic of pre-modern industrial relations. Hence Morris (1995) refers to a process of 'McJobbing' – generating routine jobs of a quality typically associated with fast-food outlets.

Hudson therefore questioned whether these labour market changes constituted a transition to a regime of flexible accumulation, and suggested that they were best viewed as strategies designed to preserve old modes of accumulation. A common linking theme – in both traditional (albeit declining) industries and the emergent new industries – was the reassertion of managerial control over labour and the labour process. There was little evidence of flexible production systems, defined by theorists of flexible specialisation as 'forms of production characterised by a well-developed ability to shift promptly from one process and/or product configuration to another, and to adjust output rapidly up or down without any strongly deleterious effects on levels of efficiency' (Storper and Scott, 1989 quoted by Hudson, 1989c, 20). Such production systems were conspicuously absent from old industrial regions. (For similar comments see Lovering (1996) and Morris (1995).)

Even in locations conventionally regarded as part of the UK's 'sunbelt' there was little evidence of labour market flexibility in the way described by analysts of the flexible firm. Pinch et al.'s (1991) work on the buoyant south Hampshire area showed that functional flexibility was important in half the manufacturing firms surveyed, but there was no evidence of an extension of flexible working (which might have been expected if flexible working was being adopted more widely). There was little evidence of numerical flexibility – even in the form of part-time work – in their area. Why was this? One reason related to the kind of firms found in the locality: numerical flexibility, for example, would be more appropriate to low-skill sectors, whereas Southampton was characterised to a greater extent by high-skill manufacturing, so that part-time work was less significant. Moreover, the concept was perhaps not appropriate for certain large, well-established and vertically integrated firms. The type of firms surveyed had not needed to use greater flexibility; they had responded to competitive pressures by an increase in the sophistication of their products and improved quality control. Thus they were under less pressure to restructure

labour practices than branch plants of manufacturing firms in peripheral areas. The threat of plant closure was less likely because the firms in their survey were further up the corporate hierarchy than peripheral plants in the North East. Pinch et al. (1989 and 1991) suggested that moves towards greater flexibility were more likely to be found when companies relocate production to greenfield sites. In a new location firms could exert much more control over the organisation of work – indeed the opportunity offered was often a key motive for relocation. In summary, they caution against eagerness to accept the idea of flexibility, or flexible accumulation. These studies suggest, taken together, that one should be sceptical about the extent to which flexible accumulation has become hegemonic; instead flexibility has been used mainly to boost efficiency from established production strategies.

Flexibility, insecurity and risk

Much of the debate around flexibility, especially in the more optimistic accounts, presents it as a positive development for both capital and labour. In reality it may be more realistic to acknowledge that, for labour, flexibility is about who bears the risks associated with economic change. Allen and Henry (1997, 182) argue that accounts of flexibility are invariably written from the vantage point of firms and how they use labour. This leads to neglect of the 'uncertainty and risks faced by men and women who are increasingly having to move into and out of different kinds of employment'. Second, flexibility subsumes a range of types of irregular, insecure and unstable work practices; we should be careful of attributing common characteristics across different groups. Thus we should not confuse issues such as job instability, job insecurity, or irregular employment. Job security is usually defined, legalistically, by statutory employment rights. Job stability is simply how long people stay with the same employer. In contrast to temporary work, therefore, part-time work should not necessarily be characterised as unstable. Moreover, people will move between jobs with different degrees of security and stability.

However, given the indisputable evidence about the changing distribution of work, several commentators have attempted to summarise emerging divisions according to the degree of economic security enjoyed by workers. The best-known interpretation is Hutton's (1995, 105–110) notion of the '30–30–40' society (but see also Coutts and Rowthorn, 1995). He provides a pessimistic view of a divided society, and one in which he anticipates a steady lowering of labour standards on the part of companies. While employers may prefer to retain valued and skilled workers, they have become much tougher with respect to the conditions on which they are prepared to employ unskilled workers at the bottom of the labour-market hierarchy. A combination of sustained high unemployment and the Conservatives' attack on employment regulation have exposed growing numbers of unskilled workers to brute employer power. While Hutton broadly concurs with Coutts and Rowthorn on the bottom 30 per cent (though he extends it to 40 per cent) he tends to characterise the intermediate 30 per cent as marginalised and insecure.

Although people in this stratum may have comparatively high incomes, they have a tenuous relationship to the world of full-time, tenured employment, and work at jobs which are insecure, poorly protected and with few fringe benefits. This is not just a semantic distinction, Hutton argues, because of the growing significance of part-time employment: over 70 per cent of new part-time jobs are for 16 hours per week or less, and are thereby not eligible for statutory employment protection. Since the majority of new jobs are part time, it follows that the workforce as a whole is becoming steadily less protected. Moreover, since the majority of women workers are part time, then it is women rather than men who are statistically more likely to be disadvantaged.

In opposition to Hutton, not all part-time workers are in insecure positions in the labour market. Many are from two-income households, and over 15 per cent have held their jobs for more than five years. However, Hutton counters this in a number of ways. Self-employment, for example, can of course provide greater autonomy for workers, but it

means that substantial numbers have only their own resources standing between them and the market. He believes that only those self-employed for more than two years can be regarded as in any sense economically secure. He also stresses the growth in temporary work in which individuals are wholly dependent on fluctuations in demand. Crucially, he highlights how the terms and conditions of those in full-time employment have been weakened. For example, even middle-class occupations are increasingly regulated by fixed-term contracts. There are some 3.5 million people who have held full-time jobs for under two years, and therefore do not qualify for employment protection. He adds to his category of the newly insecure the one million full-time employees who earn less than 50 per cent of the average wage and who are poor, by any definition. Hutton's arguments can be encapsulated in two statistics (drawn from Gregg and Wadsworth, 1995): in 1975 some 55 per cent of the adult population held full-time tenured jobs (i.e. jobs with employment protection) whereas the corresponding proportion for 1993 was 35 per cent. Moreover, labour market analysts such as Rubery (1989, 165–6) are pessimistic that workers can move from the unstable parts of the labour market to more stable and permanent jobs. Instead, workers are likely to become trapped in the unstable sectors, and this clearly has a spatial expression given what we know about the availability of jobs, networks of information about job opportunities, and the recruitment strategies of employers.

Thus we can argue, with Allen and Henry (1997, 183), that 'if flexibility is the language of the firm, perhaps risk is more appropriately the language of the workforce'. They draw on Beck's (1992) concept of a 'risk society'. Beck argues that until comparatively recently the risks and uncertainties associated with the workings of the marketplace were largely predictable. Unemployment could be managed through the institutions of the welfare state. The employment regime associated with such a golden age (if it ever actually existed) was characterised by standard contracts arrived at through collective bargaining, the concentration of work at large sites, and the expectation (for men) of regular, permanent employment. That employment regime has gone, to be replaced by a 'risk-fraught system of employment'. In this system people have to rely on their own resources to construct their own employment biographies. Of course, this 'risk-fraught environment' is partly a result of deregulatory measures, but these in turn reflect responses to wider trends in the organisation of production and work, so it remains to be seen whether regulatory systems can be devised to cope with such risks.

6.4 LABOUR MARKET SEGMENTATION AND EXCLUSION

Aggregate statistics on employment trends conceal important socio-spatial variations. In this section the emphasis is on the different experiences of men, women and those in minority ethnic groups.

Reviewing trends in male and female employment might lead one to celebrate the success of women in obtaining jobs. For example, in 1996 there were 264 local government districts in which women were the majority of those in employment. In 68 districts the proportion exceeded 55 per cent and in five districts (Rhondda, Penwith, Bearsden/Milngavie, Clydesdale and Nairn) it was over 60 per cent. Women are the majority in employment, by some distance in some cases, in several large towns and cities, such as Middlesbrough (55.3 per cent), Liverpool (54.5 per cent), Newcastle (54.2 per cent), Hull (54.8 per cent) and Bolton (59.8 per cent). There are varying reasons for this – in those areas mentioned the collapse of traditional sources of employment is important but proportions of over 55 per cent are also found in places where services (tourism, retailing and public administration) are important contributors to the local economy. By contrast in 1981 there were only 17 local government districts where women were the majority in employment.

We should be cautious about interpreting these figures for two reasons. First, much of the growth in jobs taken by women has been in part-time employment. The numbers and proportions of women employed part time have been rising steadily and part-time female employment now exceeds full-time female employment in 24 counties. There have been heated exchanges about this trend. Hakim (1995) suggested that this reflected women's preferences for forms of employment which could be combined with domestic responsibilities; thus she seemed to imply that part-time work was chosen voluntarily. Ginn et al. (1996) countered that Hakim had neglected constraints, such as the availability and affordability of childcare, which prevented women working full time even if they wished to do so. Moreover, employers deliberately construct jobs as part time; the structure of the benefits system makes this a rational response as employers do not have to pay National Insurance contributions for those employed under 16 hours per week. Part-time work may appear to be an expressed preference but this view can only be justified if one ignores constraints on labour-market participation.

Second, even full-time female employees typically earn rather less than their male counterparts and so there is a clear correspondence between locations with high female employment and low wage levels. This is especially noticeable in counties where the economy depends heavily on tourism and agriculture. Thus there are four counties/unitary authorities where over 50 per cent of full-time employees earned less than two-thirds of the male median wage in 1998 (Cornwall, Conway, Moray and Dumfries/Galloway). Also notable are the figures for North Yorkshire, Shropshire, East Lothian and Blaenau Gwent (all 49 per cent), Devon, Herefordshire and Lincolnshire (48 per cent) (source: Low Pay Unit).

Turning to the participation of minority ethnic groups in the labour force, there are signs of some improvement (at least for certain groups) although, in general, minorities occupy a worse position than whites. Figures given here are drawn from Owen (1996) who used the results of a question on ethnicity in the 1991 Census to analyse labour-market participation. He found that male unemployment rates for various ethnic groups were typically twice as high as for whites, reaching

21 per cent and 23 per cent for Pakistanis and Bangladeshis, and just over 20 per cent for those identifying themselves as black. For young black males unemployment rates were three to four times those of whites. However, the Indian unemployment rate was 11 per cent while that for Chinese men was 7.6 per cent (Owen, 1996, table 3.3).

Owen also identified regional variations. For minority ethnic groups unemployment rates were highest in conurbations in the North and Midlands, exceeding 40 per cent for Pakistanis in the West Midlands and South Yorkshire; a similar pattern was found for Bangladeshis. Thus, migrant groups who had been recruited to solve labour shortages in northern industries now found that they were the first to be shown the door after deindustrialisation (see also Phillips, 1998). For all minority ethnic groups together, unemployment rates in the principal northern cities and in inner east London were over 25 per cent (Owen, 1996, 59). Minority ethnic groups are characteristically employed at a lower level than whites with similar qualifications. However, the differentials are narrowing and minority groups are no longer disproportionately confined to low-skill, low-paid work. There is a 'glass ceiling' in the sense that men from all minorities are substantially under-represented in the most elite jobs (table 6.2). Beyond that, the position of minorities in employment relative to whites falls into three bands. The Chinese and African Asians experience disadvantage only with respect to top jobs in large organisations; Indians and Caribbeans experience relative disadvantage; and severe disadvantage is evident for Pakistanis and Bangladeshis on all dimensions (occupations, earnings and unemployment) (Modood, 1997).

Perhaps the most significant challenge facing the British labour market is the enormous decline in demand for unskilled and unqualified labour. One-fifth of men aged 16 to 64 are now without employment (Gregg and Wadsworth, 1995): they are classed either as actively seeking work (and therefore registered unemployed) or as having withdrawn from job search altogether (economically inactive). A total of 3.7 million people are in these categories. As an indication of the extent to which unskilled men have been expelled from the labour market, since 1986 the numbers of economically inactive men of working age in Britain has consistently exceeded the numbers classified as unemployed. The

TABLE 6.2 Employment disadvantage for ethnic minority men

	Chinese	African Asian	Indian	Caribbean	Pakistani	Bangladeshi
Employers and managers in large establishments	0.5	0.3	0.5	0.5	0.3	0.01
Professionals, managers and employers	1.5	1.0	0.8	0.5	0.6	0.6
Supervisors	0.9	0.8	0.6	0.7	0.4	0.4
Earnings	1.0	1.0	0.9	0.9	1.7	0.6
Unemployment rates	0.6	0.9	1.3	2.1	2.5	2.8
Long-term unemployed	–	1.6	3.1	5.9	7.7	7.7

Source: Modood, 1997.
Note: Scores compare the group in question with the national average. A score of 1 indicates parity.

differential experience of men and women is important here. Over the period 1977–92, the activity rate amongst the non-student male population of working age fell from 96 per cent to 88.5 per cent and by late 1998 it was 84.8 per cent; conversely the activity rate for women rose from 65.4 per cent to 73 per cent by 1998. The majority of the increase in female participation has been within households already containing employed men, and so the effect has been to enhance the unequal distribution of employment across households and between communities.

The decline in male activity rates is the result of a shift in demand away from low-skilled work, rather than any decline in labour supply. Male employment declined by 11.9 percentage points between 1977 and 1992; of this, 7.4 percentage points were due to the rise in inactivity and 4.6 percentage points to the rise in unemployment, so that 60 per cent of the decline was accounted for by a rise in non-participation and the rest attributed to an increase in (officially recognised) unemployment. For men aged over 55, the fall in activity rates accounts for 85 per cent of the reduction in employment. However, the largest employment reductions have been among workers lacking formal qualifications – 33 per cent of the unqualified labour force was without a job and 25 per cent of the stock of unqualified workers lost their job over this period. The greatest reductions in employment among the unqualified occurred between 1979 and 1986, indicating a temporal coincidence with the most intense phase of deindustrialisation, and a geographical coincidence with the former industrial heartlands. Consequently, the places affected earliest by this shakeout of male labour have for some 20 years now had to cope with enormously high levels of male unemployment, at levels which have generally been well above the national average.

This might be expected to lead to a widening of unemployment differentials but the reverse seems to have happened: regional unemployment differentials narrowed in the recession of the early 1990s. Green *et al.* (1998) and Beatty and Fothergill (1996) shed light on this. This apparent convergence is partly due to differential rates of withdrawal from labour-market participation. While the coalfield areas of the UK have experienced a collapse in male employment, official unemployment rates have not risen on a scale commensurate with the decline of mining. The main reason for this is a near-doubling of reported rates of ill health, because large numbers of men are classified as long-term sick (and are therefore not actively seeking work). This directly contradicts long-term trends towards better health. Moreover, the sickness rates obtaining in the coalfields are two to three times those evident in more prosperous parts of the UK. Unless one argues that redundancies were deliberately targeted on the least healthy members of the workforce, it is difficult to explain these differentials other than as the result of withdrawal from the workforce. As a result official unemployment rates substantially underestimate the extent of non-participation in paid work. The converse of this situation, in the more prosperous regions, was that the sustained economic growth of the 1980s provoked an expansion of the labour supply; as a consequence, during the recession of 1990–92, unemployment rates rose to levels approaching those of the northern industrial regions, thus contributing to inter-regional convergence (Green *et al.*, 1998, 78–9).

It is increasingly clear that spatially concentrated pockets of unemployment are the dominant feature of the unemployment map. Green *et al.*'s (1998) analysis is also helpful here. Comparing variation in unemployment rates at several spatial scales (TTWAs, local authority districts, and electoral wards), they found that while there was convergence at the TTWA level, at the local authority level and at ward level the total variation in unemployment rates was much greater in 1991 than 1981 (table 6.3). The proposition that local unemployment rates are converging only holds true at higher levels of aggregation. Proceeding to an investigation of the determinants of unemployment via regression analysis, Green *et al.* (1998, 91) argued that the presence of rented housing, the proportion of lower socio-economic groups, and the percentage of the workforce with higher qualifications are the driving factors, and that region is of secondary importance. This

TABLE **6.3** Between- and within-region variation in unemployment, 1981 and 1991

Aggregation level	280 Travel-to-work areas (TTWAs)		459 Local Authority Districts		10 000 Census wards	
	1981	1991	1981	1991	1981	1991
Total variation	2520.5	1944.9	4719.7	4992.5	220.6	283.6
Between region variation	775.2 (30.7%)	373.8 (19.2%)	1191.5 (25.2%)	436.7 (8.8%)	24.6 (11.2%)	12.7 (4.5%)
Within region variation	1744.9 (69.2%)	1571.1 (80.8%)	3528.2 (74.8%)	4555.8 (91.3%)	195.9 (88.8%)	270.8 (95.5%)

Source: Green *et al.*, 1998.

is consistent with analyses suggesting that the economic position of the less educated is deteriorating (Schmitt and Wadsworth, 1994); it also suggests causal linkages with processes of residualisation and polarisation in the housing market (chapters 8, 9). As a consequence 'the North–South divide is no longer relevant as a description of the geography of unemployment' (Green *et al.*, 1998, 91).

6.5 REGULATING LABOUR MARKETS

The extent of exclusion from the world of paid employment raises questions about whether the labour market can be regulated in such a way as to promote greater social inclusion and produce a better-trained workforce. Here, a comparison of the Conservatives' policies (the Training and Enterprise Councils (TECs)) and Labour's programmes (the Welfare to Work initiative) is instructive.

Until the 1960s training in the UK had been almost solely the preserve of the private sector, but concern about relative growth rates led to state intervention in the form of Industrial Training Boards (ITBs). These raised levies on firms to fund training, and planned their activities to meet anticipated skill shortages. The levy system was an attempt to deal with problems of underinvestment; skills are collective goods which if left to markets, are undersupplied because firms find it cheaper to poach workers than provide their own training. Subsequently the Manpower Services Commission (MSC) was established as a national tripartite body, charged with manpower planning, industrial training, and the operation of government training schemes and employment services. Over time the MSC became increasingly focused on the management of unemployment. This was partly because of the perceived weaknesses of industrial training schemes which relied upon levies. Employers effectively held a veto over this system, since by demonstrating their unwillingness to pay – or even by seeking exemptions from – levies, they could significantly impede progress (Keep and Mayhew, 1994, 316).

The Conservatives were hostile to the MSC and ITBs: the former was a quintessentially corporatist body; both were excessively interventionist, and anathema to a government which made clear its preference for voluntarism and an employer-led approach to training. Although mass unemployment in the early 1980s necessitated a huge expansion of government-funded training programmes, the subsequent reduction of unemployment in the mid-1980s and the continued electoral success of the Conservatives provided the opportunity for a radical break from corporatism.

The government handed over responsibility for training to a national network of employer-led bodies, the TECs. They were limited companies, charged with administering government training schemes, identifying and addressing local skill shortages, and stimulating training and enterprise. They were funded by government for training schemes aimed at the unemployed. Consistent with the government's voluntaristic ideology, in-company training was to be funded by the private sector. In certain respects they exemplified the Thatcherite restructuring of the state, seen by some as emblematic of a transition to a post-Fordist state form. They were private-sector led, rather than corporatist; pro-active and entrepreneurial, not bureaucratic; established by local negotiation and agreement rather than in conformity with a national blueprint (Peck and Jones, 1995; see also chapters 3 and 11). But the development of TECs was hampered from the start, and a key reason (Peck, 1991, 1996; Peck and Jones, 1995; Jones, 1997) was the uneven landscape on which they had to operate.

Though conceived in an expansionary economic climate, most TECs became operational in the recession which began in mid-1990; as a result, like the MSC before them, TECs found that their main role was the management of unemployment. As with other market-led reforms in the public sector, they were subject to a financial regime which rewarded them both for increased 'outputs' (placing people in jobs; obtaining qualifications) and increased value-for-money (lower unit training costs). Peck and Jones (1995, 1385) characterise this as a 'geographic version of institutional Darwinism' as an uneven cost terrain was used to force down costs and stimulate competition between TECs. This 'pseudomarket' was beset with pressures inimical to best practice: it encouraged short-termism, the neglect of those with special training needs, and the avoidance of areas of the labour market where skill formation is costly. This was a 'low road' to flexibility, compounding skill shortages, and operating to the disadvantage of those people and places in greatest need – as the collapse of the South London TEC showed (Jones, 1997).

Furthermore, this process of regulatory undercutting and competition was associated with local experiments in workfare, in which entitlement to benefits became contingent on participation in training. The TEC operating regime created space for such experiments to take place, without the opprobrium that would have ensued had the government implemented workfare nationally. Such initiatives subject the unemployed to labour-market discipline through inducements (such as 'benefit-plus' schemes) to participate in low-grade schemes (Peck, 1996, 253). This promotes competition between places to lower the floor of labour standards. So far, under the Labour government, little change has taken place in the TEC structure or operating regime. However, their relationship with regional development agencies seems unclear and they have had little involvement (so far) in Labour's welfare-to-work programme, suggesting that their future role is somewhat uncertain.

It is against the background of the failures of TECs to improve the skill base that we now consider the Labour government's major innovation in labour market policy. In broad terms Labour accepts the Conservative's nostrums about flexibility, while insisting that a minimum wage (at some level) is necessary. But Labour has signalled its determination to attack long-term unemployment via its welfare-to-work initiative, the 'New Deal', under which the 18–24 age group will be offered a range of options (subsidised employment; full-time education/training; voluntary sector or environmental task force placements), refusal of which will trigger benefit cuts. Labour insists that these are not cheap options and that they will be of high quality; these arguments are essential against critics of the element of compulsion in the scheme.

The New Deal is based on a view of long-term unemployment which prioritises labour supply and which asserts that a temporary programme to enhance employability will lead to a permanent reduction in unemployment. There are potentially two theories here, which could provide a justification for the programme. One is that long-term unemployment depletes skills and damages motivation, adversely affecting future employ-ability. This creates a pool of people who are effectively unemployable and disconnected from the labour market. If reconnected in some way they would put downward pressure on wage inflation and permit the economy to operate at a higher level of employment. The second theory is that people remain unemployed for longer than necessary because of the level of benefit payments; the policy implication is that benefits should either be reduced, or sanctions increased, to make the long-term unemployed accept whatever jobs are available (Turok and Webster, 1998).

Both these theories imply that the problem of long-term employment is quite distinct from the overall unemployment problem, and that the relationship between the two varies according to the overall level of unemployment. In fact, there is a clear and consistent relationship: long-term unemployment is a simple function of overall unemployment, and the main problem therefore arises from deficient demand for labour. This deficiency varies substantially from place to place. Locations which need the New Deal most have experienced a collapse of traditional industries and in these places the weak private sector will generate few New Deal opportunities, which may well be of poor quality. Moreover, the distributions of the target groups for the New Deal (the long-term unemployed; unemployed youth; lone parents; the long-term sick) are all highly correlated with one another. The New Deal will therefore be 'trying to push all four groups into jobs in local labour markets already suffering from an acute oversupply of labour' (Turok and Webster, 1998, 313). In such locations the problem is not 'local deficiency in the work ethic; it is a straightforward reflection of job availability' (Peck, 1999, 17). Unless Labour can stimulate demand for labour (and the party has rejected demand management as a macroeconomic tool), little can be done about this. In particular, job subsidies simply will not work if there are no jobs; consequently, in such locations, welfare-to-work may lose credibility because it will be seen as low-quality and because it will raise employability but not employment. Not

unlike TECs, therefore, the pre-existing geographies of labour markets will mean that those in greatest need will gain little from welfare-to-work. Finally, this is a workfarist scheme in the sense that it is primarily about recreating a work ethic, rather than creating the conditions for an expansion of work or, more ambitiously, producing a much more highly skilled workforce so as to compete on the basis of high skill standards rather than low wage standards.

6.6 CONCLUDING COMMENTS

On the evidence presented in this chapter the UK labour market is plainly heading in a much more flexible and fragmented direction, even if the ideal-typical model of the flexible firm is far less common than its proponents imagine. Part-time work, casual and temporary employment, and second jobs are all growing in absolute terms and as a proportion of the labour force. Allen and Henry (1996) characterise this as a shift towards 'precarious employment', as a result of an expansion of various forms of flexible and insecure contracts. In this view, because of the limited security associated with such work, risk is increasingly displaced onto employees (because of the erosion of employment rights). The contract service industries they describe are exemplars but the evidence of polarisation in the labour force, depending on the degree of security of employment, seems indisputable. Moreover, the scale of exclusion from the formal economy constitutes an even greater problem which, given the decline in demand for unskilled labour, can only be tackled through initiatives such as intermediate labour markets. These would focus on socially useful work (environmental projects, for instance) and would be targeted at areas of greatest social exclusion (CSJ, 1994; CCBI, 1997).

There are, of course, interpretations of these labour-market trends which are much more optimistic. These suggest that stable jobs for life have disappeared and individuals should welcome this as freedom from drudgery. They should respond, instead, by developing portfolios of skills which enable them to obtain contracts, and hence incomes, from a range of sources. The flexibility and fulfilment of such working patterns are said to be empowering. However, such accounts (e.g. Handy, 1995; RSA, 1998) risk generalising from a limited set of cases (primarily, individuals with the property and cultural assets necessary to survive such working patterns – though see Ekinsmyth, forthcoming, on the ambiguities of flexibility for freelancers in publishing). They offer little scope for those on the margins of, or excluded from, the labour force. For this group, instead, Labour's New Deal may indicate a future in which benefits are contingent on participation in official government schemes, in a British version of workfare. There are variations on this theme: whereas the Conservatives saw workfare as a deterrent to fraudulent claimants, Labour suggest that the New Deal is a more active policy to reintegrate the unemployed into the labour market. Debates will continue regarding whether this is a British workfare state in operation (Peck, 1999).

The particular characteristics and geography of British labour markets raise three issues of broader significance. First, there are the consequences for social policy of an increasingly polarised society; will it be possible to maintain support for a comprehensive welfare state across social strata with such divergent economic prospects? (See chapter 9.) Second, in a European context, disparities in unemployment levels hinder monetary and economic integration, because imbalances in demand for labour make it difficult to expand economies without provoking inflation in low unemployment regions well before surplus labour is fully utilised (for a UK example, see Martin, 1989b). Economic and monetary union is likely to exert adverse demand and supply pressures on high-unemployment regions. This will exacerbate regional unemployment disparities (Martin, 1998, 17) which have displayed a high degree of persistence. Finally, it is implausible that the problems of concentrated long-term unemployment can be solved without substantial investment in the UK's human capital. For as Szreter (1997, 80) remarks, 'the history of modern economic growth ... shows that the necessity of the accumulation of a nation's social and cultural capital (the quality of its human resources) is every bit as fundamental ... [to] economic growth as the accumulation of its material capital' (cf. Wilkinson, 1996). Drawing on comparative evidence concerning literacy levels and expenditure on education and training, Szreter argues (p. 101) that the UK has failed to invest in human resources on a scale comparable with the world's long-term economic high performers. This relative underinvestment needs to be corrected; high-quality 'human capital' is a cause, as well as an effect, of a competitive national economy.

7

GEOGRAPHIES OF MONEY
AND FINANCE

There is little work on geographies of money in most standard texts on the UK's human geography. The exceptions concentrate on personal incomes: Coates and Rawstron (1971) used Inland Revenue statistics to demonstrate the disparities in pre-tax personal incomes between regions. Several authors have subsequently commented on the extent of inter-regional and intra-regional disparities in personal incomes (Martin, 1995). Authoritative recent reviews of the income distribution are available elsewhere (Hills, 1996) so here I simply summarise the most salient points.

Income inequality in the UK has widened steadily since 1979. Whereas during the 1960s and 1970s dispersion of household incomes was practically unchanged, and indeed declined slightly in the early 1970s, since 1979 the trend has consistently been upwards (figure 7.1). The Conservatives' response to such evidence was to assert that a rising tide would ultimately lift all boats, but such claims were undermined by analyses of post-1979 income changes which showed conclusively that in real terms the bottom 10 per cent of the income distribution had in fact fallen behind the rest of the population (DSS, 1994), its incomes declining by 14 per cent between 1979 and 1992. The South East clearly leads the income distribution;

whereas in the late 1970s average household income in this region was about 10 per cent above the national average, by the early 1990s the South East's lead had risen to almost 25 per cent (Martin, 1995, 27). Elsewhere, East Anglia, the South West and the East Midlands experienced an improvement in their relative position, while the remaining regions experienced a deterioration (figure 7.2). These regional variations in part reflect the cost of living, but if allowance is made for such variations, the respective positions of the regions do not change greatly (Borooah *et al.*, 1996).

The gap between high and low earners has

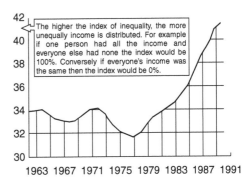

The higher the index of inequality, the more unequally income is distributed. For example if one person had all the income and everyone else had none the index would be 100%. Conversely if everyone's income was the same then the index would be 0%.

Note:
From 1949 to 1964 inequality changed little. From then until 1977 incomes gradually became more equal. From 1977 income inequality has grown rapidly, wider now than at any point in the past 50 years.

FIGURE 7.1 Income inequality in the UK, 1963–91 (Rowntree Commission, 1995)

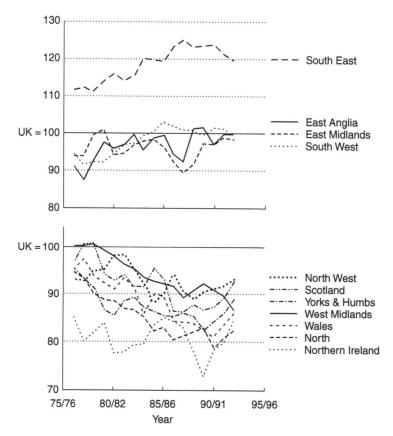

FIGURE 7.2 Regional income inequalities, 1976–93 (Martin, 1995, 28)

widened in all regions, but particularly the South East. The ratio of incomes of those in the top deciles of the income distribution to those in the bottom 50 per cent is shown in figure 7.3 for both 1980–81 and 1991–92. For the South East, this ratio increased from 4.85 in 1980–81 to 6.71 in 1992. For all other regions in England, the ratio in 1991–92 lay between 4.78 (Northern region) and 5.36 (North West). This is a comment on the particular type of economic growth experienced in the South East: because of the boom in financial services in the region, there emerged a highly polarised income distribution, with those on high incomes pulling even further ahead of the rest.

There are sharp regional contrasts in the income distribution. First, the poorest households are disproportionately found in the northern regions: the Northern, Yorkshire–Humberside and North Western regions, plus Wales, Scotland and N. Ireland, all had a share of the households in the bottom 10 per

cent of the income distribution that exceeded their population share (table 7.1). Similarly, following the EU's definition of poverty (less than half the national average income) all these regions had above-average proportions in this category, exceeding 40 per cent in Wales, N. Ireland and the Northern region. Correspondingly, in those regions, social security benefits make a much larger contribution to household incomes. Finally, the converse of this is the disproportionate representation in the South East of those on high incomes. For 1995–96, 30.7 per cent of the working population in South East England had incomes of at least £20 000; for the UK the corresponding figure was 24 per cent. In most regions the proportions were typically around 20–21 per cent with the exception of the Northern region (18.5 per cent) which had fallen behind Wales, Scotland and Northern Ireland.

No data on income are available from the census but local studies demonstrate that

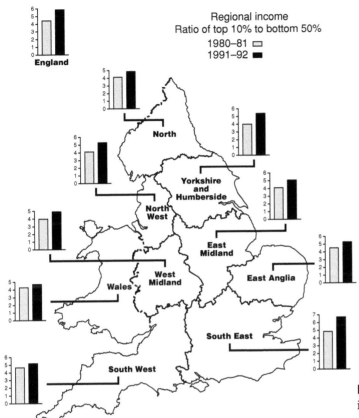

Regional income
Ratio of top 10% to bottom 50%
1980–81 ☐
1991–92 ■

FIGURE 7.3 Intra-regional income disparities, 1980–81 and 1991–92

micro-scale variations are consistent with what is known about emerging patterns of social polarisation (chapter 8). For example, Cossey (1996) demonstrates stark disparities even within Manchester, a comparatively poor city: in some wards (Didsbury, Chorlton) average incomes per head of working age population were more than twice those in wards such as Moss Side or Hulme. She also reported variations of a factor of four in the ratio of ward income to the number of children resident in the ward. At an even finer scale, Noble and Smith (1996) use time series data on housing benefit to analyse the distribution of those on low incomes in Oxford and Oldham. Both towns experienced increases in the proportions of their population in receipt of means-tested benefits. However, in Oldham particularly, the result had been a greater concentration of low-income households into particular locations; indeed 'many areas of Oldham failed

to improve or even fell back over the 1980s' (Noble and Smith, 1996, 317). This work clearly showed that, at a fine spatial scale, the interaction of economic trends with housing markets produces locations in which there are very high concentrations of low-income households. Public-sector renting now being an option primarily for those with incomes at around or below the Supplementary Benefit level, it is not surprising to find Enumeration Districts (EDs) in Oldham where around 50 per cent of households rely on state benefits. Oldham is not untypical of deindustrialised northern cities (Byrne, 1995b; Campbell, 1993). Peripheral council estates in such places experience a 'fiscal haemorrhage': retail outlets charge higher prices; insurance may be priced out of the reach of households; credit may be available only on ruinous terms. Economic change combines with the housing system to produce what are virtually moneyless communities (CSJ, 1994). Such

TABLE 7.1 Regional statistics on incomes and income sources

	Regional shares of:			%of	%of	%of
	Households in bottom 10% of income distribution (average, 1990–93)	Population (1991)	Location quotient (col.1/col.2)	Households with gross normal weekly income below £170, 1992 (*)	Household income from Social Security benefits, 1995–96	Households with employment incomes exceeding £20 000 1995–96
Greater London	11.6	11.8	0.98	N/A	10.5	N/A
Rest of South East	11.5	18.6	0.62	N/A	10.7	N/A
All South East	N/A	N/A	N/A	26.6	10.6	30.7
East Anglia	3.0	3.6	0.83	26.8	13.9	23.9
South West	6.7	9.0	0.74	30.9	12.0	21.0
W Midlands	8.8	9.1	0.96	33.2	15.7	20.8
E Midlands	6.1	6.8	0.89	27.9	11.5	20.4
Yorks/Humbs	10.1	8.5	1.18	37.1	16.1	20.2
North West	12.3	11.1	1.11	38.0	16.3	21.0
Northern	9.0	5.3	1.69	40.9	19.0	18.5
Wales	6.4	5.0	1.28	40.7	16.5	20.0
Scotland	11.9	8.8	1.35	36.8	16.5	22.2
N Ireland	2.6	2.4	1.08	40.9	23.6	20.6
UK	100.0	100.0	1.00	33.0	13.7	24.0

Source: cols 1–4: Martin (1995), tables 2.4, 2.5; col. 5: ONS (1996), table 8.7; col. 6: Inland Revenue (1997), table 3.11.
Note: * In 1992 this was approximately half the average household income.

exclusion from the material benefits of every-day life is an inevitable corollary of the oper-ations of the British financial system; it is also a corollary of the political system, as greater social segregation undermines the willing-ness of taxpayers to support such marginal locations (chapter 9).

However, the geography of money cannot be confined to the geography of incomes and the remainder of this chapter summarises work on the ways institutional features of the distribution of public and private finance shape patterns of uneven development. The spatial organisation of the financial system may have profound implications for patterns of uneven development. For example, inves-tigations into the problems of the 'Special Areas' in the 1930s revealed that, even at that time, the financial system was displaying the centralism and anti-industrial biases for which it has subsequently been censured by several writers. Taylor and Thrift (1983) describe how the decline of regional banks left Northern entrepreneurs bereft of any source of long-term venture capital, thereby inhibiting new enterprise formation. It is therefore surprising that standard works on regional problems and policy ignore the spa-tial structure of the financial system. At the micro-scale, too, the withdrawal of financial infrastructure (banks, post offices) from Britain's poorest communities amply demon-strates the power of financial institutions, in the absence of countervailing regulations, to exacerbate the socio-spatial marginalisation of deprived communities (Leyshon and Thrift, 1995).

Geographies of money are important in a further sense: ostensibly non-spatial public expenditure policies may underpin uneven development and outweigh substantially any expenditures on formal regional policies. Despite the significant proportion of GDP taken by public expenditure, there is no con-sistent accounting framework for the spatial incidence of such expenditure (for one excep-tion, see Short, 1981). Most relevant research focuses on individual expenditure pro-grammes. A full consideration of such issues would raise questions about how far the programmes of government departments should be (or in practice *can* be) coordinated.

It would also raise questions about whether, and to what extent, the allocation of public expenditure should precede, rather than follow, economic development.

Finally, we should note the implications of government decisions on monetary policy. The UK's experience in the 1980s provides a salutary warning of the impacts monetary policy can have on regional development. First, the severe 1980 and 1981 budgets exac-erbated recessionary impacts and, because of the impact on interest rates, gave firms little alternative but to close capacity and make staff redundant (chapter 5). The economic pendulum soon swung to the other extreme, as relaxations of credit control, reduced inter-est rates, and tax cuts all combined to fan the flames of an economic boom. The resulting corrective action (confined to large increases in interest rates) had effects throughout the UK, however, dragging the whole country into recession. Commentaries on European Monetary Union (EMU) have also drawn attention to the likely regional implications of adjusting to this process (chapter 12; Armstrong, 1998).

These comments serve to justify a consid-eration of geographies of money and finance. The chapter is divided into three sections. First, there is a discussion of the role of pub-lic expenditure and taxation policies as they affect regional development. This is followed by a discussion of the spatial organisation of the financial system, including such issues as the role of venture capital in fostering regional enterprise, and the operations of the pension funds. The concluding section con-siders proposals for redressing the spatial biases evident in both private and public financial systems.

7.2 GETTING AND SPENDING: GEOGRAPHIES OF TAXATION AND PUBLIC EXPENDITURE

Public expenditure accounts for some 40 per cent of Britain's GDP, a proportion which has remained substantially unchanged for three decades. However, both the sources of public expenditure, and its distribution, do change over time according to demographic trends, economic circumstances and political priorities, and so public expenditure is not always either raised, or distributed, in the most equitable ways. Moreover, hidden institutional geographies serve further to modify the distribution of public expenditure. Given the scale of public expenditure and the substantial impacts of even minor changes in its distribution, it far outweighs the scale of assistance available under the rubric of formal spatial policies (see also chapter 12).

Spatial impacts of taxation policies

The taxation system is a major influence on disposable incomes. Decisions on taxation policy are taken in a national context but because they are applied to a profoundly uneven surface (the pre-existing distribution of incomes) their effects are far from spatially neutral.

Hamnett (1997) shows that because of the economic changes of the 1980s, and particularly those of the 'Lawson boom' of the 1983–88 period, there was a marked upward shift in the distribution of pre-tax incomes, particularly in income bands above £25 000 p.a. Given the nature of the services-led economic boom taking place, a boom concentrated in the South East, one would expect large numbers of those on high incomes to be located in that region. Hence the benefits of reductions in the highest rates of taxation would be expected to display a marked regional bias.

This is precisely what happened. The 1988 budget cut the basic rate to 25 per cent for the first £19 300 of taxable income, and abolished all higher tax rates and tax bands, replacing them with a single rate of 40 per cent for taxable incomes over £19 300. As Edgell and Duke (1991, 7) observed, this meant that, in less than a decade, the 'top rate of taxation has been reduced drastically to the point where it is now nearer the 1979 basic rate (33 per cent) than is the current basic rate'. For most households paying income tax, the net result of the 1988 budget was savings of between 7.2 and 7.5 per cent of their annual incomes, but for those with taxable incomes of above £20 000, £30 000 and £50 000, respec-

tively, the savings were 8.1 per cent, 12.6 per cent and 29.3 per cent (Hamnett, 1997, table 7). Such income groups were disproportionately concentrated in South East England, which had 40.6 per cent of those earning £20–30 000, 49.6 per cent of those in the £30–50 000 band, and 53.6 per cent of those earning over £50 000. Consequently, a regional analysis of the 1988 tax cuts demonstrated that abolishing the higher rates had reduced tax revenues by £2.6 billion for 1988–8, of which £1.6 billion (61 per cent) was accounted for by the South East; that region contained only 33 per cent of taxpayers. The consequence was a further upward twist in the consumer boom taking place there, as personal disposable incomes rose by 17.3 per cent, compared to 11.2 per cent nationally.

These tax cuts clearly had a sharply differentiated impact. Less obvious in its influence on the geography of disposable incomes is mortgage interest tax relief (MITR). This was originally introduced to facilitate entry into owner-occupation; it involved granting tax relief, at an individual's highest marginal tax rate, on interest paid on the first £30 000 of a loan. Its regional impact is therefore a function of the distribution of house prices and of the incomes of those buying properties with mortgages. Given the comparatively smaller proportions of properties in the peripheral regions costing over £30 000 (at least until the 1980s) and the fact that fewer residents of such regions have been in higher-rate tax bands, it is clear that the northern regions of England have benefited from this subsidy much less than the South. MITR peaked at £7.7 billion in 1990–91 (mainly because of high interest rates), of which £3.04 billion accrued to Greater London and the South East.

Typically MITR was worth around £1.9 billion in Greater London and the South East during the late 1980s, whereas most other regions benefited to the tune of £3–400 million. This was both a consequence and a partial cause of the regionally focused house price boom at that time; levels of inflation in house prices in turn persuaded more individuals to embark on owner-occupation in the anticipation of capital gain. Even in the mid-1990s MITR was worth nearly £1 billion to the South East, though it subsequently

TABLE 7.2 Regional impacts of the 1988 tax cuts, mortgage interest relief, defence expenditures and venture capital

	1988 Tax cuts			Mortgage interest tax relief (1995–96)		Defence procurement expenditure (1995–96)		Venture capital investment 1993–96		
	Tax revenue foregone (£m)	Regional share of UK total	No. of beneficiaries ('000)	(£m)	Regional share of total	(£m)	Regional share of total	Amount invested (£m)	Regional share of total	Location quotient
South East	1600	61.5	490	980	36.3	2600	38.5	3853	42.4	1.21
East Anglia	100	3.8	30	110	4.1	100	1.5	247	8.7	0.68
South West	200	7.7	70	270	10.0	1250	18.5	516	5.7	0.61
E Midlands	150	5.8	40	180	6.7	250	3.5	752	8.3	1.20
W Midlands	100	3.8	60	220	8.1	400	5.8	966	10.6	1.25
Yorks/ Humbs	100	3.8	50	220	8.1	350	5.1	501	5.5	0.73
North West	150	5.8	60	260	9.6	600	8.5	1034	11.4	1.26
Northern	50	1.9	20	120	4.4	500	7.5	331	3.6	0.96
Wales	50	1.9	20	100	3.7	50	1.0	192	2.1	0.42
Scotland	100	5.8	50	200	7.4	600	9.0	601	6.6	0.87
N Ireland	0	0	10	40	1.5	100	1.0	103	1.1	0.32
UK	2600	100	930	2700	100	6800		9096		

Sources: cols 1–3: Hamnett (1997), tables 12, 13; cols 4, 5: Inland Revenue (1997), table 5.3; cols 6, 7: Ministry of Defence (1998), table 1.9; cols 8–10: Mason and Harrison, (1999) table 1.

declined as interest rates fell. Its abolition in the 1999 Budget was possible because by then it was worth much less than in the early 1990s.

Considering the amounts involved, then, a plausible case can be made for regarding MITR as a major counter-regional subsidy. Its precise impact on wider economic changes is disputed (see Hamnett, 1994) but what cannot be gainsaid is that the regional impact of tax cuts and of MITR has been substantially to the advantage of the South East's economy. Such tax cuts and reliefs have combined with other public programmes to underwrite the region substantially.

7.3 HIDDEN REGIONAL POLICIES: THE SPATIAL IMPACT OF NON-SPATIAL PROGRAMMES

An immediate problem is the absence of aggregate figures on the overall impacts of public spending by standard region. Consequently, attention has focused on individual programmes, such as defence procurement.

Research into the origins and expansion of Britain's high-technology 'M4 corridor' (see also chapter 5) has emphasised the key role of defence procurement expenditures. There are several key industrial sectors in this nexus – computers, electronics, aerospace, military hardware – all of which rely heavily upon the MoD. Spending on such areas by the Ministry has grown by accident rather than design and reflects the presence of key institutions to the South and West of London, including military bases, training establishments, and research laboratories. Given this concentration and the proximity of Whitehall and government, it appears that the expansion of the M4 corridor was almost a *fait accompli*. However, it received a major stimulus during the 1980s when, as Boddy and Lovering (1988) note, the 'remilitarisation of the British economy' produced a widening gap between the regions of the UK. Of the annual MoD procurement budget of £10 billion, some £5 billion was spent in the 'south' of England. More recent figures show that while the total sums spent have been reduced, regional disparities remain; the South East still receives some 42 per cent of procurement expenditure (table 7.2). Plainly, these sums dwarf expenditures on what is formally and conventionally viewed as regional policy. In employment terms, defence expenditures supported 11 per cent of manufacturing employment in the South East and South West standard regions in 1985, compared to as little as 1.3 and 2.0 per cent in Yorkshire and the West Midlands, respectively. However, whether it can therefore be regarded as a 'hidden' regional policy may be more controversial. This is because there is some dispute over whether defence industries can be regarded as autonomous sources of regional change or whether, instead, this pattern of development merely reflects (rather than leads) wider changes in the relative fortunes of regions.

The gradual thawing of the Cold War has seen a reduction in defence spending. Lovering (1995a) argues that this principally led to the withdrawal of contracts from major manufacturers, contributing to deindustrialisation (chapter 4). The R and D activities west and south of London were substantially preserved. However, this process has been market-led and there has been no systematic effort to investigate how public expenditure might have been used to support diversification into non-military production. Just as public expenditure has cushioned localities dependent on defence contracts, its sudden withdrawal has caused major dislocations.

A strategic, regional approach to public expenditure policy has also been absent from important infrastructural policy decisions. Adequate infrastructure – particularly communications and transport – is essential given contemporary trends in the organisation of production, which emphasise transport as a key element in links between producers and their supply chains. However, public expenditure policy emphasises reacting to the present distribution of economic activity rather than anticipating future developments and attempting to promote a more balanced pattern of development. For example, high levels of subsidy to public transport in South East England have been criticised for helping preserve existing patterns of overdevelopment in that region. Other examples are decisions to invest in new airport facilities (e.g. Stansted) in the South East, rather than strengthening under-used regional airports, and the apparent lack of commitment to high-speed rail links to ensure that northern and western parts of Britain benefit from the Channel Tunnel. Growing private sector involvement in infrastructure planning is inimical to such strategic thinking. The Private Finance Initiative (PFI) exemplifies the problems. The PFI was introduced by the Conservatives as a means of expanding capital expenditure without either raising taxation or increasing the Public Sector Borrowing Requirement (PSBR). The commercial criteria used mean that the benefits of PFI will be confined to projects which will generate sufficient returns to attract financial backing. Investment in public facilities will go not to where it is most needed but to locations which are most commercially viable. This will probably force organisations such as hospitals to operate in ways inimical to community needs (e.g. accepting more private patients; discharging patients even more quickly). The likelihood is that PFI-financed expenditure will become concentrated in the core regions of the economy while elsewhere desirable infrastructural developments await their turn in the political queue.

The territorial dimension of public expenditure has received renewed attention because of the devolution debates. Anti-devolutionists contend that, in order to sustain present levels of spending, newly-elected assemblies would have to raise taxes substantially (Heald and Gaughan, 1996). For example, Scotland's general government expenditure was around 19 per cent above the UK average in 1995–96 while expenditure in Wales was 12 per cent above that average. Spending in England was 4 per cent below the UK average. Some of the difference may reflect particular conditions – for example, the costs of delivering various public services in remote areas. However, such differentials have provoked a debate about the formula used by central government for allocating public expenditure which according to its critics, effectively guarantees that Scotland's

advantage should be preserved. This is the Barnett formula, devised in 1978, which is a population-based method of sharing out *changes* in public expenditure plans between the countries of the UK. It does not determine *overall* expenditure levels. Critics, for example from the peripheral English regions, argue that if devolution proceeds which preserves the advantages given to Scotland and Wales, English regions will be put at a double disadvantage (not least because of the greater institutional capacities available to Scotland and Wales) in stimulating economic development. They argue that Scotland's economic performance relative to the rest of the UK is much stronger than when the formula was devised. Supporters of the formula point to Scotland's poverty, poor health and low population density, all of which are said to justify higher expenditure. Even if the formula were abandoned, because it only applies to *changes* in public expenditure, the benefits for the English regions would be minimal. Moreover, the political problem for the government is that it cannot renege on its commitment, in framing devolution legislation, to preserving the formula. Some updating of the formula may take place, therefore, but change will be marginal (see House of Commons Treasury Committee, 1997).

The general point raised by these examples is the ways public expenditure programmes operate to counteract or to support existing patterns of uneven development. MacKay (1994 and 1995) finds consistency in the degree of territorial fiscal redistribution within European states, though much of his evidence relates to 'an economic environment more sympathetic to non-market forces' (1995, 220) – in other words, the era of Keynesian economic management in the 1960s and 1970s. By contrast, the Conservative budgets of 1980 and 1981 reduced public expenditure, thereby weakening key countercyclical elements of the economy. The consequence was a widening of unemployment and income differentials. MacKay correctly observes that demand-led social security expenditure mitigated the worst impacts of recession on disadvantaged areas (see also Walker and Huby, 1989), but this does not justify his presentation of public expenditure as an *automatic* stabiliser for regional economies. Some public expenditure decisions (e.g. reductions in state support for nationalised industries) contributed much to the widening of regional divisions in the 1980s. If public expenditure is a stabilising influence, the extent of that stabilisation is a matter for political decision, and growing social polarisation may well undermine the electoral base for such fiscal transfers. Finally, public expenditure decisions may lead to divergence rather than convergence between regions.

7.4 GEOGRAPHIES OF PRIVATE FINANCIAL INSTITUTIONS

Two issues are considered here. First, because of their significance in absolute terms, the spatial structure of the UK's pension funds, and the characteristics of their investments, are analysed. Second, the geography of venture capital is considered because of its influence on the pattern of new enterprise formation. A third issue is also relevant but is not discussed at length. This is the effects on low-income communities of a 'flight to quality' by financial institutions. The principal evidence relates to the withdrawal of banks from poor locations through rationalisation of branch networks (Leyshon and Thrift, 1995), and the growing difficulties posed for low-income households in obtaining insurance, especially in high-risk areas (Whyley *et al.*, 1998). These trends reflect financial institutions' accurate analysis of the socio-spatial patterning of risk and, of course, the power of these institutions to shape those patterns.

These issues have to be seen against the background of wider developments in the organisation of finance capital in Britain, notably the decline of regional banking and the centralisation of financial control and power. Martin and Minns (1995) suggest that the British financial system has passed through three phases: a bank-oriented phase, followed by a market-oriented phase, leading to the contemporary securitised system. In the first of these, from the nineteenth cen-

tury and lasting until the early 1970s, banks played a key role in channelling savings into capital stock and other forms of investment. This was followed by a market-oriented phase, in which capital and equity markets became the dominant destinations for savings and funds. Most recently there has been the development of a securitised system characterised by the emergence of a range of financial instruments, and associated with regulatory liberalisation and a degree of global integration. This phase is dominated by institutional investors, such as pension funds and life insurance companies. These developments have been paralleled by a 'progressive loss of regional financial independence to London' (p. 128): local and regional banks were steadily absorbed by the national clearing banks, while the activities of regional stock exchanges became absorbed into the London Exchange. Gamble *et al.* (1995, 258–60) argue that the financial system is overcentralised and that this inhibits the development of new enterprises in peripheral regions.

The financial system exerts influence over patterns of regional development through control mechanisms whereby funds are invested in – or denied to – enterprises. According to the assumptions of neoclassical economics, capital flows should respond to – and in the long run eliminate – regional differences in the rate of return. Martin and Minns (1995) question this, and argue that the process is not unlike that of cumulative causation. Local investors and financial institutions evaluate the prospects of particular places; as institutions centralise geographically, investors lose faith in, and outlets for, local investment; external investors become reluctant to invest as the cumulative effect of decline becomes apparent. This exacerbates initial pessimistic evaluations of particular economies, and the consequence is that the systematic vulnerability of regional economies will encourage liquidity preference, thereby discouraging investment in capital assets and possibly leading to investment in other financial assets. As a result there will not only be a 'core' and 'periphery' in the financial system itself, there will be a net flow of funds from the latter to the former, fanning the flames of uneven development and indeed being an integral part of that process. These arguments are relevant to both pension funds and venture capital.

The pension fund system

The resources assembled by the UK's pension funds system are substantial: on some estimates the total amount invested is equivalent to 60 per cent of the UK's GDP. These funds are invested in a range of financial instruments – primarily equities, government securities, and property, with a proportion retained in cash. Although there is some evidence of growth in overseas investments, some 70 per cent of pension fund assets are invested in the UK. The pension fund sector is heavily spatially centralised: 46 per cent of the largest pension funds (those with over £10 million in assets) and 63 per cent of total accumulated assets are owned by companies or organisations headquartered in South East England. Furthermore, many pension funds of organisations not located in the region are managed from within it; some 95 per cent of pension fund assets are controlled from the South East of England (Martin and Minns, 1995, 135).

The pattern of investment of these funds further supports the economy of the South East, as restrictions on the proportions of company shares which can be held by pension funds create a bias towards investment in large companies, most of which are headquartered in the region. Martin and Minns suggest that two-thirds of all pension fund equity investment is held in the shares of the top 90 companies located or headquartered in the South East (p. 136). The top 100 companies – in the Financial Times 100–share index – are biased towards financial services themselves (so that there is a degree of self-investment), utilities, and the extraction industries. However, as many of these companies will have substantial overseas activities, it is plausible that pension fund investment could fund expansion abroad and contraction at home. Significantly absent are the production industries.

In principle this pattern of investment need not be a cause for concern; the companies benefiting from the pension funds can use new capital for further investment and expansion, which would have macro-economic benefits as well as region-specific ones. In practice, only a small proportion of pension fund activity is related to primary investment; the bulk is concerned with buying and selling of existing shares from other shareholders (the secondary market). Of share dealing that is not concerned with the secondary market, much goes into the funding of takeovers, repayment of debt, or payment of dividends, which can have profoundly negative effects in the form of asset stripping and plant closures.

Hallsworth (1996) develops these arguments in an analysis of the ways in which the short-termism of the British financial system favours relatively unproductive forms of investment, particularly in land and property. For reasons explained above (and in Hutton, 1995) liquidity preference and short-term investment are endemic. Moreover, because British company law prioritises the rights of creditors to recompense from tangible corporate assets, the land on which companies sit is a key asset. Thus, when financial difficulties loom, institutional investors will be less motivated to develop long-term strategies to assist companies who are losing money; rather, the system favours short-term profit-taking, and cashing-in of land values. This may be one explanation for the extent of redundancy and capital shake out in British industry in, for example, the early 1980s. Conversely, in terms of patterns of new investment, the price of land in Britain encourages speculative construction of retail and office developments on undeveloped plots of land whose values then escalate dramatically. This 'meshing of the economic system and property markets' (Hallsworth, 1996, p. 35) was ultimately an unproductive use of assets, though rational from the point of view of the companies concerned. What is neglected on this view is investment in the productive capacity of the economy. It also has quite distinctive consequences for urban form, creating swathes of low-density development which can only be accessed by road (chapter 13).

The geography of venture capital in the UK

Post-1979 economic policies have undeniably focused on the promotion of an 'enterprise culture'. However, that culture has spread unevenly throughout the UK (chapter 5). One reason for this is that there is a general problem for new enterprises in obtaining finance for start-ups or expansions, and a specific problem facing particular regions. There has been expansion in the availability of 'venture capital', but such funds exhibit their own spatial biases.

Martin (1989b, 390–1) defines venture capitalism as an activity in which 'investors support entrepreneurial ventures with finance and business skills to exploit market opportunities'. It may embrace a range of forms of support, but defining features of venture capital funds include long-term commitments of assistance, with a view to capital gains, and a degree of active 'hands-on' involvement in the activities of the enterprise receiving the finance. By comparison with the USA, growth in the British venture capital industry is a recent phenomenon. Nevertheless, the total funds committed in the 1992–96 period totalled £9 billion (Mason and Harrison, 1999, table 1).

However, the UK venture capital sector exhibits a high degree of uneven development, whether one considers the distribution of firms or of financial commitments. In a recent survey, 63 (of 101) head offices of venture capital firms were located in London. Other regions had very few head offices (apart from Scotland), although some larger London-based companies had regional branches. The distribution of investments showed a similar pattern. For 1992–6, the South East accounted for 36 per cent of investments and 42 per cent of the amount invested. The next largest regional shares were in Scotland, the North West and the West Midlands. However, it is probably more meaningful to relate these distributions to existing levels of business activity; Mason and Harrison (1999) do this by comparing the pattern of venture capital with stocks of VAT-registered businesses. In terms of sums invested, the North West, West Midlands, East Midlands and

the South East attracted more than their 'expected' share of venture capital investments. Conversely, Northern Ireland, Wales, the South West and East Anglia received substantially less than their 'expected' share (table 7.2). There are signs that geographical disparities have reduced over time; analyses of the late 1980s had demonstrated a much greater concentration in the South East. This improvement reflects an expansion in venture capital investment activity in the Midlands and North West, but the position of the regions most deficient in venture capital has not improved since the 1980s.

It might be argued that these spatial patterns are a demand-led phenomenon (Martin, 1989b, 397), reflecting the ways in which the South East (and some adjoining localities) led the process of economic recovery, in contrast to the comparatively slow pace of recovery elsewhere. This might help explain variations in demand for venture capital. However, this is not the whole story. London's historically pre-eminent position as a national and international financial centre made it almost inevitable that a comparatively specialist branch of financial services, such as venture capital would concentrate there. This in turn affects the support given to projects. Evaluation and monitoring of, and involvement in venture capital projects, are all facilitated by spatial proximity. Martin implies this creates images of the southern regions as being rich in investment opportunities and low risk (because proximity permits closer monitoring of projects). This is not so for more peripheral regions, where lack of contacts and the absence of previously successful projects combine with limited local knowledge to create a perception of such locations as lacking in enterprise and therefore 'high risk' (Martin, 1989b, 398).

If there is any foundation to such remarks, there is a need to establish local venture capital markets, through which enterprises can obtain the support they need. Some public sector bodies, such as the Scottish and Welsh Development Agencies, or local authority enterprise boards, have responded by providing some venture capital funds. In addition there is limited evidence of a northward shift in the activities of private-sector venture capitalists. This process will take time, and there are further reasons for believing that in any case more is needed than an expansion of conventional venture capitalism. Motivated by the prospect of capital gains, venture capital firms focus on fast-growing businesses, and relatively large businesses at that – the costs of evaluating projects make investments of under £250 000 uneconomical. Harrison and Mason (1996, 9) therefore argue that alternative sources of venture capital must be sought to stimulate enterprise more widely. They describe the impacts of the Business Expansion Scheme (BES), whereby individuals obtained substantial tax reliefs on investments in unquoted companies. However, the BES seems merely to have reinforced existing tendencies in British finance, namely a South Eastern, anti-technology, and pro-property and services bias. To counter these tendencies, some authors argue that what is needed is a reinvigoration of regional financial networks, the prospects for which are considered in the concluding section.

7.5 ALTERNATIVE GEOGRAPHIES OF MONEY

Four points can be made about possible policies to deal with the issues discussed here. These relate to: the need for a 'regional audit' of all forms of public expenditure; the possibility of creating community-based institutions to deal with financial exclusion; the development of a regional financial infrastructure; and proposals for more fundamental reforms of the financial system to promote a 'stakeholder' economy.

Numerous calls have been made for the planning of public expenditure in a coordinated way, which would ensure that the distribution of expenditure more accurately reflected patterns of economic disadvantage. However, given that most public expenditure is allocated departmentally on needs-based criteria, it is difficult to see what alternatives there are. Moreover, efficiency criteria relating to capital expenditures, and the growing use of private finance in government pro-

grammes, would appear to militate against expenditure being allocated to areas of greatest need. The reorganisation of the machinery of government into regional offices (see chapter 11) may improve coordination but is not guaranteed to be associated with a redistribution of expenditure. At most there may be some reworking of the formulae used to allocate funds to the countries of the UK (the Barnett formula) as a response to the devolution lobby.

In response to the apparent 'flight to quality' of the financial institutions, there have been proposals for community-based financial institutions, such as credit unions or local exchange trading systems (LETS). The former encourage small-scale savings and make loans to individuals with a regular savings record; the latter facilitate the non-monetary exchange of services (e.g. Lee, 1996). Despite the current fascination of geographers with LETS, they are at present small-scale initiatives and are not concentrated in areas of greatest need. There is no equivalent of the American community development banks, which have a very specific social role in underpinning enterprise in disadvantaged communities. Such community-based initiatives may well contribute to creating 'social capital' in some communities, since they are essentially participatory and democratic organisations, but as initiatives against economic marginalisation their impact will be limited.

There have been proposals from both left and right for creating a more decentralised regional financial infrastructure. One suggestion has been the re-creation of a regional banking system analogous to that which existed in the nineteenth century, which would stimulate regional enterprise because banks would develop a better understanding of those who were applying for credit and of the needs of regional economies. Leyshon and Thrift (1995, 331) draw attention to the parallels between this argument and arguments for the development of a richer 'institutional tissue' as a basis for regional economic success (see also chapter 12). However, tendencies towards centralisation at a supra-national level in banking are such that regional banks would be swimming against a strong tide. The decline of building societies, which often had their roots in particular regions, suggests that simply 'unbundling' the clearing banks will not create a responsive and viable regional financial infrastructure.

An alternative is Gamble *et al.*'s (1995) proposal for the formation of regional development banks (RDBs). They argue that a decentralised banking system would avoid decision-making according to rigid national criteria, and would therefore promote more sensitive lending practices. They also emphasise the ways in which RDBs could form part of a collaborative regional strategy in technology and research and development, by demonstrating the compatibility of firm and regional economic interests (p. 260). Regional institutions could facilitate cross-sectoral collaboration between unrelated firms who could mutually benefit from technological change. Furthermore, novel financial instruments could be used to spread risk, especially in the SME sector, thereby facilitating enterprise formation and expansion. These would have to be accompanied by requirements to engage in long-term investments, which raises wider issues.

The implications of Hutton's (1995) concept of a stakeholder society are that merely effecting administrative reforms – e.g. the creation of new tiers of a financial infrastructure – will not have the desired impacts unless accompanied by the creation of a more inclusive system of corporate governance. The privatisation of the Trustee Savings Bank (TSB) exemplifies this (Plender, 1997, 175–9). The TSB had strong roots in disadvantaged communities and regions, and demonstrated considerable success in providing financial services for poorer social groups, but privatisation forced it to operate in much the same way as other financial institutions. Plender argues that the privatisation of the TSB therefore represented a substantial loss of social capital of a kind not generated by – indeed antithetical to – financial institutions driven by shareholders' needs. Unless community, employee and supplier interests are allowed to influence the policy of financial institutions – as well as those of shareholders – then the realities of regional economic power (Minns and Tomaney, 1995) will remain unchallenged. Hutton's work does demonstrate that alternatives to short-termism exist; he shows how financial institutions in Germany and Japan have worked collaboratively with industry and local government to secure the future of key businesses, driven by a recognition of mutual interest in survival. It remains to be seen whether a more supportive regional financial infrastructure can be developed in the UK.

CLASS, GEOGRAPHY AND SOCIAL POLARISATION

8.1 INTRODUCTION

In certain respects this chapter is pivotal to the arguments of this book. The patterns of, and debates on, social stratification are the result of and responses to the economic changes discussed above (chapters 4, 5 and 7). The resulting fragmented socio-economic landscape is something to which the welfare state must respond (chapter 9), but those responses are constrained according to what is deemed politically acceptable (chapter 10). Likewise the extent to which effective policies can be devised in response to uneven development or sustainability (chapters 12 and 13) is constrained by what is acceptable to the 'comfortable majority' (Galbraith, 1992).

Much contemporary academic and political attention has been given to the question of whether Britain is a classless society. Thatcher's government vigorously denied that class divisions existed, while Major's government emphasised that Britain was classless. Both did so against the background of growing social polarisation, which was arguably a deliberate and intended effect of their policies. Recognising the scale of divisions generated by this 'two nations' project, Blair has spoken frequently of his desire for a return to a 'one nation' politics, and for the elimination (or at least reduction) of social exclusion.

To achieve this will be no small task. A succession of reports have amply documented the extent and growth of social divisions (CSJ, 1994; Rowntree Commission, 1995; Hills, 1996). These divisions pose questions about the future of class analysis, and about whether to speak of social polarisation, social segregation or social exclusion (sections 8.2 and 8.4). There have been various theoretical responses which have focused on two key issues: the character of the so-called 'service class', and the evidence for the existence of a socially and spatially segregated 'underclass' (section 8.3). Some of the consequences of social division receive attention, notably those relating to social cohesion (section 8.5).

Necessarily there are omissions in this chapter. Empirically there is little demonstration of inequalities in, for instance, education and health status. Conceptually, there is no attempt to disentangle the complexities of the interrelationships between gender, race and class. I agree with Westergaard (1995, 140) that it is 'analytically myopic' to treat these as conceptually parallel and separate dimensions of inequality, but to achieve the converse of this would require a book-length treatment in itself. As debates on inequalities in health demonstrate clearly (e.g. Popay *et al.*, 1998), separating the effects on health inequalities of class, gender, age, ethnicity (and, as those authors argue, place) is a task of considerable complexity. The core argument advanced is that important class-based social divisions are having visible adverse consequences with respect to social cohesion, rendering problematic the construction of a programme to deal with them.

8.2 DEBATES IN CLASS THEORY

There are long-running debates about the validity of Marxist and Weberian concepts of class. The former stress position in relation to the means of production whereas the latter emphasise status-based conceptions of class (Crompton, 1993; Westergaard, 1995; Marshall, 1997). As originally formulated by Marx, class implied a specific relationship to the means of production: in a bipolar model, society was divided into the bourgeoisie, or capitalist class, and the working class. Between the two there existed a fundamentally antagonistic relationship: the concern of the bourgeoisie was to minimise wages paid to workers in order to maximise the extraction of surplus value (the source of profit); the working class sought to maximise the returns on their own labour power. Marx's key insights were that this relationship was an *exploitative* one, that it was a *necessary* one (capital and labour could not exist without one another), and that there was a constant *class struggle* over the process of capital accumulation. Marx's propositions about social class extended to arguing that the working class would ultimately recognise the nature of its exploitation, develop a collective consciousness, and ultimately transform society through revolutionary action. Implicit in this view was the argument that class interests and class action followed directly from class position.

Conceived in this simplistic – albeit analytically powerful – way, Marxist ideas perhaps have limited contemporary applicability. Pahl (1989) criticises Marxism's deterministic model of political mobilisation, arguing that links between class structure, class consciousness and class action have not been demonstrated. He suggests that, if class as a concept is ceasing to do any useful work, it should be abandoned. Other authors reject the historicism of Marxist analysis, but argue that the basic purposes of class analysis remain valid: these are the mapping of the

'character and sources of power embedded in economic organisation; the disparities of human life experience that arise in consequence; and the ways in which, politically and otherwise, people act out their varieties of class-shaped experience' (Westergaard, 1995, 2; compare Goldthorpe and Marshall, 1992, 382–5).

If so, Marxist categories need to be revised substantially, but there are social strata which Marx would certainly have recognised. Hutton (1995), for example, writes of exploitation of casual and temporary workers in the 'most brutish corner of the labour market' in terms which would not be unfamiliar to Marx. Although deindustralisation and the decline of large firms have largely eliminated Marx's industrial proletariat, the growth of small firms, and the emergence of ostensibly 'new' forms of industrial relations, have not been associated with a diminution of exploitation in the workplace. Moreover, contemporary debates about the 'underclass' are redolent of Marx's concept of a reserve army of labour. Runciman (1990) notes the parallels between the terminology used to describe the 'underclass' in both the 1890s and 1980s.

At the top of the wealth pyramid, likewise, little has changed: income and wealth are still substantially concentrated in the top 1 per cent of the population, while Scott's (1991) work on the upper classes shows, in addition, the centrality of the Home Counties to the networks of power through which this small stratum of the population exercise influence (see also section 2.4). Westergaard (1995, 123–5) suggests that proponents of 'classlessness' in contemporary Britain have 'lowered the ceiling for [their] concerns' and ignored the substantial evidence for continued concentration of wealth. Furthermore, because of the close connections between this group and large corporations, their decisions involve the deployment of vast corporate resources. Westergaard proceeds to argue that the 'taken-for-granted assumptions which any society has for the ongoing conduct of its general affairs' are generally set in favour of this social elite (notably, that resources be deployed to the long-term benefit of private capital, and that public intervention should not offset this presumption), thereby circumscribing what is politically possible. Runciman (1990, 383–4) argues that although this upper class is too small to figure as a separate entity in national social surveys (1–2 persons per 1000 population), it merits separate treatment in class analysis because of its disproportionate influence.

Accepting these points of broad continuity, what is the most fruitful way to proceed with a discussion of class? Three key approaches are identified here. First, there have been attempts to develop and extend Marxist categories which rely on concepts of exploitation in the workplace. Second, and associated especially with discussion of the service class, there have been efforts to incorporate questions of status and control of employment assets. Third, some would abandon Marxist ideas altogether, in favour of an argument that consumption sector divides are now the most significant in contemporary societies. These developments can be read as responses to major socio-economic changes since Marx's time. Thus, the growth of the state has produced substantial numbers of workers whose employment is not purely a function of capitalist rationality. The very substantial growth of managerial occupations blurs distinctions in the workplace, as managers in large organisations control substantial resources without owning them. The scale of the service sector and the nature of the activities carried on within it also pose challenges for a simple bipolar model.

E. Wright (1985) argued that while these changes meant that class divisions were by no means transparent, Marx's basic arguments about exploitation still held true. He therefore developed an analysis of relationships of exploitation and control in the workplace; exploitation could be revealed by investigating the degree of autonomy exercised by individuals, their ownership of the means of production, and their involvement in decision-making and the supervision of employees. One strength of this analysis was that it offered some insights into class positions and relationships *within* the state sector. Managers in large public enterprises clearly have substantial autonomy and control, which may be exercised in an authoritarian,

exploitative way. Wright also pointed to the important concept of *contradictory* class relations. Ownership of the means of production now tends to be separated both from investment and production decisions and from control of the labour process. Ownership is in the hands of shareholders but the use of assets is delegated to people who may not own any part of the company for whom they work even though their decisions affect many people. Hence many people, especially in managerial occupations, are in positions which Wright saw as *objectively contradictory*; managers may control labour without ownership; the petit bourgeoisie may own capital, without controlling labour; some craft employees may own their tools and have substantial autonomy, while remaining employees. Wright subsequently revised his analysis, arguing that the key divide (cutting across both capitalist and state socialist societies) was given by the degree to which individuals occupied positions which carry powers to direct investment and/or the labour process.

Wright's analysis sparked considerable debate and, perhaps most importantly, it led to comparative cross-national research which revealed the limitations of his scheme. For Marshall (1988, 144), Wright substantially overestimated the size of the proletariat, or traditional working class. Approximately half the British population were assigned by Wright to this category, a figure considerably at variance with the British census, in which the categories of skilled, semi-skilled and unskilled employees together comprised only 36 per cent of the economically active population. This discrepancy arose because Wright regarded the large majority of white-collar employees as having such limited autonomy that they should be placed within the working class. Marshall, by contrast, suggests that one-quarter of Wright's 'proletariat' ought to be reassigned to the category of skilled non-manual workers, because those so reallocated had more in common with the middle classes than the proletariat. They were more likely to regard themselves as middle class, and their voting patterns were closer to those of the middle class than to any other group (Marshall, 1988,

144–6). Part of the problem arises from what Marshall saw as Wright's static and structuralist conception of class. If, on the other hand, we see class as a process as well as a structure, then for many people routine clerical work (to give one example) represents only a stage in a career trajectory, and a temporary one at that. Marshall also criticises Wright for the imprecision of other categories, especially that of the marginal working class, because this category (some 11 per cent of the population) is extremely heterogeneous. It included many people who are non-owners of capital and non-managers, but who claim to have high autonomy in carrying out their work. As Marshall shows (p. 151) this includes chartered accountants as well as school cleaners because of divergent perceptions of job autonomy.

These (and other) problems have led to attempts to differentiate positions within labour markets in terms of employment relations. In the work of Goldthorpe and colleagues, distinctions are made 'between individuals having relatively advantaged and disadvantaged conditions of employment; between those involved in agricultural and non-agricultural employment; and between those having different employment statuses' (Marshall, 1997, 2). A key stratum identified is the service class, or salariat. People in this class 'exercise delegated authority or specific knowledge, in return for which they enjoy relatively high incomes, security of employment . . . and . . . autonomy at work' as well as fringe benefits resulting from employment. The skilled and unskilled wage-labourers, by contrast, 'supply discrete amounts of labour in [an] exchange of money for effort'. Such workers are also subject to more intensive supervision or control from above. The 'routine clerical' class is defined by 'employment relationships that take a mixed form between the extremes of the service relationship and the pure labour contract'. Small proprietors, farmers and agricultural workers are differentiated from other proprietors and employees because of distinctive elements of the production process and because of property ownership in small firms and agriculture.

The Goldthorpe scheme may lack the overt political force and prescriptive intent of Marxist analyses but it arguably has more purchase on patterns of socio-economic stratification than, say, Wright's analyses, especially through its discussion of the service class. The recent revisions of official socio-economic classifications have adopted a variant of the Goldthorpe classification, updated by reference to recent surveys of terms and conditions of employment. These showed that the distinction between a service relationship and a labour contract was still a valid way of differentiating between a service class and a working class (O'Reilly and Rose, 1998, 728). However, an occupation-based classification also has to deal with the growing proportion of the population who are more-or-less permanently excluded from the world of paid work. One response has been to categorise this group, with some exceptions, as an 'underclass', but this is problematic for several reasons (see section 8.3, below).

In opposition to those whose categories are based on employment relationships, arguments from a Weberian perspective have suggested that the key contemporary socio-economic divide is in the sphere of consumption, between those reliant on the state for collective provision, and those able to satisfy consumption needs through private provision. Housing is the main empirical arena for the exploration of these arguments (Saunders, 1990; see also Busfield, 1990, on health care). Saunders draws on Weber's arguments that while class (conceived in terms of property ownership) is important to social stratification, it must be complemented by analysis of political party and social status. The implication is that membership of a status group not only conveys benefits but also predisposes people to act in particular ways. In particular, it is argued that those who satisfy consumption needs through the private sector will tend to align themselves with political parties which emphasise self-reliance, independence and the market, rather than the 'dependency culture' associated with state provision.

The problem is whether consumption cleavages can be regarded as an *independent* dimension of social stratification. Hamnett

(1989) suggests that they cannot: there is a strong association between occupational class and variations in consumption, though it is true that at the individual level the relationship is indeterminate (see also Warde, 1990). A further step in the argument is that consumption divides affect political behaviour. For example, it was widely believed that the 'right-to-buy' legislation of the Conservative governments caused a switch in voting behaviour among those purchasing their council houses. However, the evidence appears to support the view that occupational class remains the best predictor of different aspects of behaviour and attitudes (Crompton, 1993, 170). This is not surprising, Goldthorpe and Marshall (1992, 393) contend, because many individuals will be extensively engaged in both the public and private sectors simultaneously (most obviously with respect to health care, where even those with a comprehensive private package plainly have an interest in the NHS's ability to provide emergency care). It is thus highly unlikely that collective identities would form around consumption issues. Nevertheless, the continued expansion of private provision could produce substantial minorities which have little direct interest in public services; there is also evidence that support for public welfare services is by no means uniform (Curtice, 1996; see chapter 9).

An extension of arguments about consumption divides is provided by postmodernist social theorists who argue that classes have dissolved and fragmented. Social identities are multiple and cut across one another; they are situated in 'imagined communities, membership [of which] ... is a function of taste, choice and commitment'. As a consequence rigid social categorisations are *passé*. For Pakulski and Waters (1996, 4) various processes are 'decomposing' class boundaries: the redistribution of property; the segmentation and globalisation of markets: an increasing role for consumption; the decline of traditional careers, and so on. Identities and lifestyles, and therefore political commitments, are based less on employment status than on a range of other sources of difference (age, gender, ethnicity, nationality). Acknowledgement of such social divi-

sions has provided a welcome recognition of diversity but it has also had disabling political consequences. Geography's focus on identity and consumption has produced some interesting studies of consumption patterns, but neglected the basis of these patterns in a profoundly unequal restructuring of society (Byrne, 1995a; Leyshon, 1995). Marshall (1997, 16–18) berates postmodernist announcements of the death of class for their 'data-free sociology' and for being 'detached ... from empirical reality'. He accuses them of either ignoring substantial bodies of evidence (on health and voting patterns, for example) or citing evidence which largely confirms the continuing salience of class. This is not to deny the need for attempts to document the contemporary moves towards 'fragmentation of identities and pluralisation of lifestyles' but these processes need to be set against the background of deepening patterns of socio-economic inequality. Consumption and lifestyle may be increasingly significant, but the 'economic factors identified by the nineteenth- and early twentieth-century theorists of social class still play ... *the* major role in the structuring and persistence of systems of social inequality' (Crompton, 1993, 185).

Two further criticisms may be advanced of class analyses based on employment relationships. First, Pahl (1988 and 1989) suggested that the most appropriate unit for analysis was the household, taking account of the situation of all its members. He was pointing to two developments in particular. First, the likelihood of a woman participating in the workforce was partly a function of whether or not her partner was in employment. There was an emergent distinction between work-rich and work-poor households (see section 8.4). Second, employment-based classifications ignored various informal forms of work, such as self-provisioning, services performed on an exchange basis for neighbours, or DIY, which were important to household strategies. Pahl originally hypothesised that an expansion of these activities would allow households to mitigate their deteriorating position in the formal labour market. However, his analysis subsequently revealed that self-provisioning required skill, income and contacts which for most people were obtained through participation in the formal economy. Hence these activities served to widen, not to reduce, inter-household inequalities.

A second qualification concerns the respective priority to be accorded to gender and race *vis-à-vis* class. Although the entry of women into the labour force has substantially benefited many households, broadly speaking, women continue to occupy the most marginal positions in the labour market, due to the gendered construction of particular occupations and gendered assumptions about responsibilities for caring (McDowell, 1991; see also chapter 6). Furthermore, Pahl's analysis implicitly presumes a two-adult household, but for large numbers of households headed by women the reality is that class combines with gendered constraints to confine them to marginal positions in housing and labour markets. In the case of ethnic minorities processes of institutional racism combine with class inequalities to produce high levels of exclusion from paid work and spatial concentration in disadvantaged locations (e.g. Peach, 1996; Phillips, 1998). But gender and race would appear to be best theorised as compounding class divides, not as analytically separate and parallel domains of inquiry.

The approach taken here, then, focuses on the consequences for social stratification of major changes in the UK's economic structure in the past two decades. First, there is a discussion of two key developments in the class structure, namely the emergence of the 'service class' and the 'underclass'. This is followed by a review of evidence concerning the nature and extent of social polarisation, segregation and exclusion.

8.3 CLASS AND SPACE: THE SERVICE CLASS AND THE 'UNDERCLASS'

Contemporary debates about social stratification in Britain focus on two central issues: the nature of the service class, and whether or not there is an emergent underclass. The nature of the middle class has been a persistent concern for class theory and the service class debate is an extension of that. The underclass debate, on the other hand, has come to renewed prominence because of heightened levels of social segregation, and also because of the problems posed for class analysis by the neglect of large numbers of economically inactive people.

The service class

Much recent controversy in class analysis focuses on the middle class, and specifically on whether a distinctive service class is emerging. Two trends have led to an expansion of the numbers in professional and managerial occupations. First, there is the growth in public services, staffed largely by individuals with professional qualifications. Second, the growing complexity of businesses has necessitated the recruitment of many professional workers to perform specialised tasks (most notably because of developments in IT and in financial services). In both the public and private sectors, furthermore, there is now a substantial cadre of managerial staff. Goldthorpe's formulation (1995, 314) included the class of professional, managerial and administrative *employees*. The emphasis is important: these are people who do not own the enterprises for whom they work, but offer important specialised services to them. Goldthorpe distinguishes the 'service relationship' from the 'labour contract': service-class employees render service to their employing organisation, in return not only for their salary and fringe benefits, but also job security, pension rights and well-defined career opportunities. This contrasts with contracts on fixed (hourly, weekly) rates, with no fringe benefits and limited prospects of advancement. The exercise of professional autonomy or managerial authority is not, for Goldthorpe, relevant to categorisation (compare E. Wright's (1985) extension of Marxist categories).

Against this view it has been argued that Goldthorpe's definition encompasses very different forms of service: professionals typically perform specialist business services or organise non-household forms of welfare provision; managers typically engage in the control of labour power; still others (design professionals; research and development staff) are engaged in essentially conceptual tasks. Savage *et al.* (1992) therefore emphasise the internal differentiation of the service class. They stress, in certain respects drawing on E. Wright's (1985) efforts to extend basic Marxist categories, that the service class is fragmented into three groupings: those formed around property (the petit bourgeoisie), bureaucracy or organisation (managers), and cultural capital (professionals). This is a somewhat eclectic approach which blends strands from quite different analytical traditions. However, it does raise the issue of the value of allocating all professional and managerial employees to the service class, which can be pursued by exploring the political and other attitudinal attributes of this group.

Goldthorpe initially regarded the service class as a conservative force, since they held rewarding or desirable positions within the division of labour, and were therefore likely to support the status quo. Heath and Savage (1995) show that identification with the Conservative Party is at least twice as likely as with the Labour Party. Nevertheless there are differences, in that professionals tend to be more likely to support Labour than the Conservatives, but this may be due to occupational choice (in other words, people chose jobs in line with their political inclinations). It is not surprising that public-sector professionals would support parties committed to more egalitarian programmes, but this is not because they are members of the service class. In this respect we must recognise the diversity of this grouping.

In other respects there are some signs of common ground between members of the service class. Several commentators have noted the involvement of service-class members in various anti-development protests in South East England (e.g. Short *et al.*, 1986). Barlow and Savage (1986) suggest that this is an attempt to achieve a form of exclusionary closure, designed to consolidate the market situation of those with scarce skills, by restricting new in-migration. Others note the impact of the service class at a more local scale in the gentrification of certain inner city neighbourhoods (Butler, 1995). Studies of social change in rural areas have also argued for commonalities of tastes and lifestyles among the middle class, to the extent that migration is producing a tailoring of rural environments to the interests of middle-class newcomers. This is evident in the style of new housing developments, the middle-class character of leisure activities, and so on (Cloke *et al.*, 1995b; Urry, 1995). A corollary may, of course, be the marginalisation of those unable to buy into this vision of a rural 'idyll'. In these senses, the service class can have an important influence on the character of places and on local politics.

This raises the question of the relationship between class and space. It is common to regard the South East as a middle-class region: the geographical concentration of the control functions of the UK economy defines a geography of class, as professionals and managers concentrate in the region. The growth of specialist producer services, a key area of expansion in the 1980s and 1990s, has re-emphasised the South East's key role in the service economy (Marshall *et al.*, 1988; Savage *et al.*, 1992, 160–2). However, we should not read off the geography of the service class from the distribution of employment: in certain respects causality works the other way, as employment follows the residential preferences of the service class. Of course, the South East *does* have an over-representation of professionals and managers – the proportion of these groups is 16 per cent and 24 per cent above the national average – but Savage *et al.* (1992, 162) argue that the key role of the South East in class terms is in promoting upward social mobility rather than merely in terms of the density of middle-class employment in the region. Fielding (1992 and 1995) shows that: the South East attracts a disproportionate share of those moving from education into service-class occupations; people tend to achieve more rapid upward mobility in the South East than elsewhere; and managers and professionals show a high propensity to migrate away from the South East at, or before, retirement age, and while many stay in their own social class, some transfer into small business ownership. Thus the 'geography of opportunity in Britain is deep-rooted and relatively unchanging' (Fielding, 1995, 186).

What also emerges from this analysis is that, given this geography of opportunity, workers with skill assets 'are more likely to stay within specific regions, where markets for their skills are well-developed' (Savage et al., 1988, 472). This is increasingly likely to be within the South East: relative economic buoyancy allows considerable scope for moves between firms without relocation. The effect may be that the South East drains other regions of talent, leaving them with less of the skilled labour on which growth might be based (Fielding, 1992; Allen et al., 1998, 119). Finally, as the South East's social structure has become dominated by service-class groups, this constrains what is politically possible: 'the combination of the social and spatial power of middle class groups is the most intractable problem standing in the way of addressing the gross inequalities of geographically uneven development in the UK' (Massey 1995, 342). This power of veto contrasts sharply with the powerlessness and dispossession of the so-called 'underclass'.

The underclass

Concern about the underclass derives from two sources. Historically, in Victorian times, there was a middle-class fear of the 'dangerous classes' – an unemployable stratum of the population, with tenuous involvement in the labour market, and surviving through occasional work, reliance on poor relief, and criminal activity (Morris, 1994). Second, the US experience has generated concern about the deleterious effects on work incentives of the welfare state. The issue is also racialised, with much attention being devoted to the breakdown of black family structures and the rise of illegitimacy. But because the American economy consistently generates new jobs, to a much greater degree than European economies, moves in and out of work are more frequent. In Britain, by contrast, the problem is one of persistent long-term unemployment rather than one of semi-permanent attachments to the labour force. In both

Britain and the USA there are strongly moralistic overtones to underclass debates, with conservatives being predisposed to attribute causality to the failings and moral degeneracy of those in poverty.

The idea of an underclass has developed as a counterpart to the notion of social classes and acquires meaning within that framework. Those working-age adults who are not in employment (on some counts, up to 40 per cent of the working-age population) clearly fall outside conventional Marxist categories as they neither own capital nor earn a living as wage-labour. Runciman (1990) allocates people to classes based on the nature and degree of economic power attaching to their roles through their relation to the processes of production, distribution and exchange. He would therefore define the underclass as including those whose role places them 'more or less permanently at the economic level where benefits are paid by the state to those unable to participate in the labour market at all'. Critics of such a view argue that strictly the concept applies only to those who are conclusively *outside* the labour market (the aged, chronically sick and disabled) since in principle others of working age *could* participate in the labour market (Morris, 1993). However, defined in these terms the concept embraces people of such widely divergent prospects (especially the elderly) that it is debatable how helpful it is. Furthermore, if classes are alleged to exhibit common behavioural patterns and attitudes, it is hard to see that such a heterogeneous group has much in common (compare Runciman, 1990, with Marshall et al., 1996).

There are, then, important difficulties in determining the position of the underclass in relation to conventional class categories. Here, I consider definitions which attempt to capture different elements of the concept of a 'class apart'. First, Buck (1992) analysed the underclass in relation to its degree of *attachment to the labour force*. Using the 1979 and 1986 Labour Force Surveys, he estimated the total numbers of economically inactive and long-term unemployed heads of household, excluding the retired, long-term sick and students. On these figures the numbers in these categories had grown from 0.7 million households (1.96 million people) in 1979 to 1.6 million households (4.58 million people) in 1986. Crucially, Buck pointed out that while the growth in single parent households over that time period was 38.8 per cent, rates of growth were far higher for single person and childless couple households. Buck's study showed that even after controlling for other factors the percentage of economically inactive council tenant households was 15 per cent above the average for the country as a whole, highlighting the extent to which council housing had, even by the mid-1980s, become an essentially residual tenure. On the basis of labour market participation this is clearly a very heterogeneous group. Buck's broad conclusions were endorsed by Marshall *et al.* (1996), who also showed that there was little to distinguish the so-called underclass in terms of behavioural patterns and attitudes.

An alternative definition of the underclass has been in terms of *social segregation* and *social networks*. Morris's (1993) study of Hartlepool shows that even within the population whose attachment to the labour market is at best insecure, there are divisions. Those whose social networks show concentrations of unemployed people are more likely to remain out of work than those who have a pattern of insecure, temporary employment. The latter group will at least have some contact with a workplace and with people within it, which may help them obtain information about forthcoming vacancies. Morris, however, questioned some of the more gross stereotypes about the underclass. Some discussions of the term argue

that the underclass adopt attitudes whereby they (whether by accident or design) avoid contact with employed people, and therefore do not develop informal social support networks that are common among the employed. Morris showed that this was not the case, arguing that there remained channels through which employed and unemployed still socialised. For her the crucial problem was the lack of access to paid work.

A third definition of the underclass is that of Byrne (1995b) who emphasises *spatial exclusion*. Byrne is considering arguments which stress an ecology of dispossession in particular places, and some of the behavioural and cultural patterns which may or may not follow. Byrne produces a cluster analysis of census and other socio-economic data to classify the enumeration districts (EDs) in Middlesborough. In one cluster – east Middlesborough – containing 26 per cent of the population, male unemployment was 40 per cent and over half of all children were living in a household containing no working adult (table 8.1). The EDs containing the most deprived population were spatially concentrated, especially on the peripheral estates. The schools attended by the children living on these estates are those where very low proportions of children obtain at least 5 A–C grades at GCSE level. His emphasis is on global economic changes leading to deindustrialisation, but he also shows how – locally – this is combined with policies of modernisation (especially the construction of peripheral estates) and the local impacts of national policies. Crucially, spatial separation

TABLE 8.1 Contrasts between east and suburban Middlesborough, 1991

	East	Suburban
% Economically active males unemployed	37	8
% Households in owner-occupation	30	93
% Households in public tenancies	68	3
% Households with no car	73	19
% Households with children, without workers	55	10
% Gaining 5 GCSEs at A–C grade	6	38
% No GCSEs	23	7

Source: Byrne, 1995b.

occurs; residents of such peripheral locations cannot access social life and social goods – especially education and, therefore, work – on the same terms as the rest of society. The consequence is that those with any form of stable employment move out into owner-occupation. The residualisation of public housing, plus the emphasis in government policy on school choice, combine to reinforce one another, so that – whether intended or unintended – social segregation has increased as an effect of government strategy. The concentrations of deprivation he describes are common to peripheral public sector housing estates in most large British cities (Blackman, 1995; Byrne and Rogers, 1996). Byrne argues, sailing close to the wind of underclass theorists, that this spatial segregation may produce concentrations of deprivation which are so intense as to reproduce a new cultural order.

These are concentrations of substantial disadvantage, but do they support Byrne's assertion that a new cultural order is reproduced? The criticisms of such analyses take two forms. First, there is the ecological fallacy – that it is illegitimate to make inferences about individual behaviour from area-level statistics. Byrne acknowledges this but argues that it misses the point. He is not making statements about individual behaviour. For him the key issue is the existence of spatial locales where the concentration of deprivation is so intense that cultural effects may operate at the neighbourhood level. These effects may influence norms regarding the legitimacy (or otherwise) of particular forms of behaviour. A second criticism of his analysis might be that more poor people live outside such locations than within them. He suggests that this is also largely irrelevant to debates about a spatially segregated and dispossessed working class. It is the existence of the concentrations themselves which is the issue, not the absolute numbers of people they contain.

Byrne emphasises that initially at least such places were a success, offering a greatly improved environment. Under conditions of employment differentiation, however, it was inevitable that places like east Middlesborough, containing socially homo-geneous populations drawn largely from the unskilled or semi-skilled socio-economic strata, would become marginalised. Promoting market forces in both housing and education then reinforced this process of segregation. However, Byrne still insists that despite the existence of spatial concentrations of deprivation, the key issue is not the behaviour of individuals but the generalised problem of poverty in an economy which is not generating enough well-paid jobs.

We are now in a position to consider the various explanations advanced for the concept of an underclass. Three key issues are worthy of comment, and a distinction between 'liberal' and 'conservative' perspectives is used for analytical purposes.

First, liberals and conservatives differ in the emphasis placed on *structural and cultural processes*. Conservatives emphasise the breakdown of traditional beliefs, mores and behavioural patterns. This is indexed principally through a rise in anti-social and criminal behaviour, and perhaps most of all in the rise of illegitimacy and the concentration of large numbers of female-headed households on public sector estates. Added to this is a refusal to accept paid work, though some conservatives see this as less a matter of conformity to social norms and more a rational response to incentive structures, notably the welfare system (Silver, 1994, 555). Opponents of such explanations stress the impact of structural forces. Deindustrialisation has removed jobs for the least-educated members of society and their labour market options are limited by a poor infrastructure – notably the lack of availability of child care and transport. In turn, housing options are limited mainly to peripheral estates or to poor-quality private rented accommodation. The children of such disadvantaged groups are largely constrained to enter schools where attainment is low; the likelihood is that they too will leave without the qualifications necessary for success in a changed job market. From these perspectives then, the stress is upon the impacts of economic restructuring as opposed to cultural/behavioural dispositions.

A second contrast can be drawn between

liberals and conservatives in terms of their attitude to the *role of state intervention*. For conservatives the key is the dependency culture created by the welfare state, which not only fails to penalise those who do not seek work, it even rewards births out-of-wedlock by making housing easily available. Conservatives also implicate state control and planning of housing in the construction of poor-quality environments which promote anti-social behaviour (Coleman, 1985). These processes are likewise mutually reinforcing, producing concentrations of single-parent households, unable to socialise their children, living in poorly designed environments. Liberals have a somewhat different perspective. They would criticise Coleman's design determinism and the risible notion that young women consciously choose to have children in order to obtain a flat in undesirable council estates. Instead, liberals focus on the role of state policies in producing large concentrations of poor households in public sector housing. Such estates may be distant from emerging patterns of economic activity. More importantly the financial regime of public sector housing has forced rents up to historically high levels in order to cover costs. In general, therefore only those eligible for maximum housing benefit can afford public sector rents, which means that *de facto* public-sector estates are populated to a large degree by those with limited access to paid work, who are then deterred from seeking paid work by benefit traps.

A third contrast is evident in the significance attached to *gender and family structure*. For conservatives, great emphasis is placed on the decline of the traditional family and its replacement by allegedly feckless single mothers. In the absence of strong father figures, acting as role models for young boys, there is a lack of discipline and authority which then translates into high levels of anti-social behaviour and crime. The explanation is cast in terms of the failure of mothers to manage both men and boys, and is linked economically to discourses of welfare dependency, since the generous provisions of the welfare state make it possible for mothers to do without stable male earners in their households. Liberal critics question whether absent fathers really make that much difference, and argue that such views devalue the enormous contribution women make in sustaining a sense of community solidarity. Campbell (1993) contrasts the different responses of men and women to the 'economic emergency' she sees as characterising Britain's peripheral estates. Women create and sustain networks of informal social solidarity and self-help; men have been unable to redefine their roles in relation to economic change and have responded by dominating communities through criminality, violence and coercion. She shows how efforts at community regeneration were invariably sparked off by women and how challenges to such efforts inevitably came from young men, turning inwards on their own community in a despairing response to their own social exclusion. She also questions the somewhat romanticised versions of community offered by conservative accounts.

It could, of course, be argued that adopting the terminology of 'underclass' debates serves to reproduce a conservative stereotype of a major social problem, by labelling people and confirming their marginality. However, my argument is that the issue needs to be confronted because of its spatiality – the emergence of substantial areas cut off from the mainstream of economic and social life. It is therefore important to highlight the dangers of such crude generalisation, the multifaceted character of the concept of an underclass, and the range of possible explanations.

8.4 SOCIAL POLARISATION SOCIAL SEGREGATION AND SOCIAL EXCLUSION

Social polarisation, segregation and exclusion all have wide currency but different connotations. Whereas polarisation and segregation are visible and (perhaps) readily quantifiable, exclusion refers to processes which are sometimes much less tangible. All three terms have implications for political strategy and public policy, particularly the latter because of the quite specific way it has been defined by New Labour. Complexity in the discussion of these terms arises for several reasons. First, there are the units of analysis: individuals, households or places? Second, there are methodological questions concerning the definition and measurement of polarisation/exclusion and the variables to be deployed in analysis. Third, the underlying processes are complex. They may include deindustrialisation and the growth of services, with a consequent bifurcation of the employment structure; the growth of part-time and temporary work; changes in household structure; the operation of (quasi-)market processes in service provision, notably in housing and education; and the effects of institutional racism.

Social polarisation

Dorling and Woodward (1996) defined polarisation as a widening gap between individuals, households or groups of people in terms of their economic and social circumstances and opportunities. At the individual level, economic change has had markedly different impacts on individuals, depending on their location and occupation; the key feature has been the expulsion from the workforce of men of working age (chapter 6). At the household level, the proportion of households supported solely by the income of a male breadwinner fell from 40 per cent to 23 per cent of all households between 1979–81 and 1989–91. Williams and Windebank (1995) contrast work-poor and work-rich locations. In the latter a majority of households have at least two earners, but this is not always the case, as witness inner London, Wales and Northern England where nearly 30 per cent of households have no earners and as few as 43 per cent of households have two or more earners (table 8.2) (the figure for inner London probably relates to the higher proportion of single-person households in the capital). These figures do not tell us whether these jobs are full time or part time and what the balance is between them, but they do draw attention to the contrasts even at an aggregate, regional scale. More detailed studies reveal the locality-specific character of inter-household polarisation. Thus Pinch's

(1993) study of Southampton indicates that, in this fairly buoyant labour market (the survey was conducted in the late 1980s) the character of social polarisation was that of multi-earner households pulling ahead of the rest in relative terms. By contrast Morris's (1993) analysis of Hartlepool, where large-scale redundancies had removed many jobs for unskilled men, indicated (as would be expected) polarisation between households in work and those not in work. But she also described polarisation between those households with access to paid work: the contrast was between households with a relatively stable attachment to paid employment, on the one hand, and those with discontinuous attachments to employment, on the other. She recorded an increase in employment histories broken by periods of unemployment, reflecting the general collapse in demand for unskilled labour.

Focusing on the spatial dimensions of polarisation, several authors have presented empirical evidence of the nature and extent of social divisions within the UK (e.g. Rowntree Commission, 1995; Hudson and

TABLE 8.2 Regional distribution of household types by number of wage earners in household (%)

	Number of earners		
	0	**I**	**2+**
RoSE	20.98	28.84	50.18
Outer London	21.83	30.46	47.70
East Anglia	22.95	28.55	48.50
E Midlands	23.33	27.46	49.20
W Midlands	24.14	27.55	48.31
South West	24.84	27.73	47.42
Scotland	26.22	28.55	45.23
Yorks/Humbs	26.73	27.31	45.96
North West	27.21	26.83	45.97
Wales	28.58	28.17	43.25
North	28.88	27.21	43.91
Inner London	29.96	32.79	37.24
Great Britain	24.82	28.29	47.89

Source: Williams and Windebank (1995), table 3 (in turn 1991, Census, Sample of Anonymised Records).

Williams, 1995, 75–124; and various contributions to Philo, 1995). These can be one-dimensional – for instance, drawing attention to broad regional contrasts in quantitative indicators, without stressing the more qualitative dimensions of the experience of poverty in particular places (Woodward, 1995). For example, much contemporary literature refers as much to economic insecurity as to absolute or relative poverty (Hutton, 1995). During the early 1990s new dimensions of insecurity, such as negative equity, mortgage arrears, indebtedness and even repossession, were experienced in unexpected numbers in unexpected places (Mohan, 1995b; Dorling and Cornford, 1995).

Furthermore, popular perceptions of polarisation are urban ones, relating to inner cities or peripheral council estates. However, we should not ignore the existence of rural poverty (Cloke et al., 1995a). Substantial minorities (up to 40 per cent in some localities) were existing on incomes close to income support levels. Their problems were compounded by isolation, inaccessibility and the need (often) to run a car in order to access such opportunities as were available. Furthermore, they report, the existence of poverty and relative deprivation in rural areas is routinely denied by the majority of residents, including those on low incomes. There tended, instead, to be a fatalistic acceptance of some negative elements of rural life as being a necessary price to pay for living in a desirable environment. Against this the notion of rural areas as 'idyllic', problem-free spaces was used to play down the divisions that exist there, and the experience of poverty was therefore problematic and difficult, when set against the general trappings of affluence in rural areas.

Much analysis of polarisation is actually static because of reliance on census data and there is a need for a more dynamic perspective. Green (1994) analysed the extent to which concentrated poverty exhibited a greater degree of spatial concentration in 1991 compared to 1981. She ranked wards on five indicators – lack of car ownership; unemployment; households in rented accommodation; economic inactivity; and social class. A ward was defined as experiencing

concentrated poverty if it was in the top 10 per cent of wards nationally on at least three of these five indicators. On this definition 8.9 per cent of wards experienced concentrated poverty in 1991 compared to 7.5 per cent in 1981. There was more continuity than change: 6.3 per cent of wards featured in both of these years, with 1.2 per cent featuring in 1981 only and 2.6 per cent in 1991 only. This relies on a relative definition of poverty, and has few theoretical underpinnings, but appears to support the view that poverty is becoming more concentrated spatially. Does this mean, however, that social *polarisation* is taking place?

One problem is that definitions of census variables change over time, and the absolute levels and proportions of the population falling into particular categories will also change, so that comparisons of absolute levels on a particular indicator may mislead. Dorling and Woodward (1995) tackle this by comparing, over time, the proportion of the population in a given category for a particular place with the national average for any given indicator. Thus, to investigate the proposition that polarisation was occurring in terms of the distribution of people in paid work, they show that in 1981 80 per cent of the population lived in wards with between 90 per cent and 110 per cent of the national proportion of people in employment. By 1991, that proportion was down to 69 per cent. They showed that the number of people living in areas with 83 per cent or fewer of the average proportion of workers increased three-fold (and covered 6 per cent of the working population). This was a reversal of trends during the 1970s, and was not due to a redistribution of people of working age: children and pensioners had, if anything, become more evenly spread. Although the unemployed had not polarised in the same way, the distribution of people of working age not in work – which is broader than a conventional definition of unemployment – exhibited marked polarisation. In 1981, 43 per cent of the population lived in wards where the proportion not working varied by less than 10 per cent from the national mean. That proportion had fallen to 30 per cent by 1991. In practical terms this means that people are now more likely to live either in a ward where many other people of working age work, or in a ward where the large majority of adults do not work. Their analysis is represented in figure 8.1, in which the shaded areas demonstrate those wards which had moved further away from the national average between 1981 and 1991.

Social segregation

A qualitatively different dimension of socio-spatial stratification is evident in the segregation of whites and non-whites and in the different geographies of the two groups. The degree of concentration or dispersal of minority groups has exercised social geographers since the mid-1960s, though recently there has been a welcome move away from 'mapping minorities' to charting the geographies of racism (Jackson, 1987). None the less the degree of segregation of minority groups remains an important issue. Peach (1996) attempts an answer to the provocative question of whether Britain has ghettos, arguing that the UK does not have the extreme degree of segregation of non-white minorities characteristic of North American cities. Even at the level of individual local authorities, there is no district in which minority ethnic groups constitute a majority, the highest proportions

being found in Brent (44 per cent), Newham (42 per cent) and Tower Hamlets (36 per cent). In certain wards, taking all minorities together, the proportions may reach 90 per cent, but the number of wards is small and this is still far from the situation of US cities where figures of 100 per cent for African-Americans alone are not unusual. Peach deployed the index of dissimilarity to measure the degree of segregation of minority groups from one another. He showed that, in cities with at least 1000 Bangladeshis, the Bangladeshi population were by far the most segregated group, having an index of dissimilarity of 73 at ward level against whites, indicating that 73 per cent of Bangladeshis would have to move to achieve an identical distribution to that of the white population. For Pakistanis and Indians, the indices of dissimilarity were some 10 and 16 points lower than for Bangladeshis. Black groups (whether Black Caribbean or Black African) exhibit segregation levels which are generally lower than for the foregoing, and Peach (1996, 227) notes a decline in segregation levels for Black Caribbeans over the 1961–91 period. He also shows that, if class alone controlled their distribution, minority ethnic groups would display a much lower degree of segregation. Thus he concludes that there is segregation but that it does not

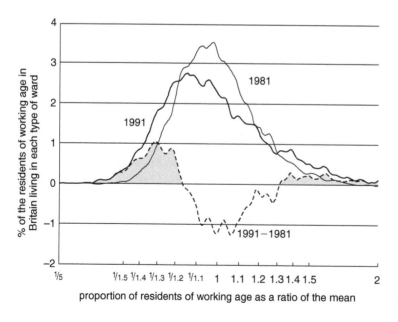

FIGURE 8.1 Residents of working age not at work in Britain, 1981 and 1991 (Dorling and Woodward, 1995)

approximate to the American model of 'hyper segregation'.

Precisely how one interprets this, of course, remains contentious. Peach is inclined to prioritise cultural factors such as religion, diet and family networks. He downplays the evidence of institutional racism which has confined minorities to particular estates within the stock of council housing, on the grounds that such allocative procedures cannot explain the prior degree of clustering of minorities in particular locations. He therefore emphasises 'recency of arrival, poverty, defensive clustering and a positive desire for clustering' (pp. 233–4).

Critics would argue that this neglects a long history in which black newcomers largely found work in marginal niches in the labour market, producing a racialised division of labour.[1] Similar segmentation emerged in the housing market, as immigrants were forced into poor private rented accommodation and the worst owner-occupied housing in inner city areas (Smith, 1989a; Phillips, 1998, 1682). This early pattern of settlement has produced a distinctive and enduring geography. Cultural factors shaped the character of particular minority communities but the distribution of these communities largely reflects the social, economic and political conditions prevailing at the time of migration. Thus was established a 'pattern of black concentration, segregation *and* deprivation', with two long-term repercussions. First, the poor living conditions of the immigrant population 'reinforced the perceived marginality of black minority ethnic groups'; the association between 'race' and deprivation was readily made. Second, the spatial clustering of the immigrant population meant that they subsequently became 'trapped in marginal urban areas in regions of industrial decline' (Phillips, 1998, 1683). This was particularly the case in the recessions of the 1970s and early 1980s (chapter 4): the massive shake out of labour from declining sectors had disproportionate effects on groups which had once sustained industries such as textiles with their cheap labour, and minorities experienced systematic discrimination in competing for such jobs as were available.

Having said that, Phillips suggests there are real signs of advancement for some minority ethnic group members, albeit against a background of continuing disadvantage in the aggregate. This reflects the differential success of minorities in the labour market (table 6.2 above). The occupational profile of Indian men is very similar to white men. By contrast, Black Caribbean, Pakistani and Bangladeshi males are significantly underrepresented in the upper echelons of the occupational hierarchy. These differentials are reflected in housing status, which clearly follows class lines. Thus, Indians are most likely to have moved into suburban owner-occupation (albeit in much lower proportions than their white counterparts) while the Pakistani and Bangladeshi populations remain relatively spatially concentrated. There are, then, some signs of (constrained) mobility. Nevertheless whether by design or default, 'certain geographical spaces ... the suburbs, the high-status residential areas, and the rural' are still seen as out of bounds (Phillips, 1998, 1688). Moreover, on almost all socio-economic indicators, black people are more at risk of poverty and unemployment than other groups. Such economic constraints combine with institutional forces in the housing market to maintain high levels of segregation. Moreover, the quality of housing available to minority groups is generally inferior. Black tenants have borne the brunt of the residualisation of public housing, in part because of continued evidence of segregationist policies by some local authorities. Although such policies are less evident than hitherto, in practice racial harassment constrains the range of realistic options open to black tenants, and predisposes local authorities to minimise racial conflict through their allocation decisions. There is also evidence of institutional discrimination and selectivity in private rented and owner-occupied sectors.

It is these constraints which should be given priority in explaining patterns of segregation. Peach may well be correct in his argument that the UK does not have the degree of hyper-segregation characteristic of cities in the USA, but this should not blind us to the realities of continuing inequality. Peach is also apt to prioritise the positive role of clustering in sustaining family and community ties, since this can play a role as a 'resource and a refuge in a potentially hostile society' (Phillips, 1998, 1698). However, the more powerful white population has the balance of control over institutional resources and these, combined with racial harassment, serve to 'maintain social and spatial distance' between black minorities and the white majority.

Social exclusion

A relative terminological newcomer to debates about inequality is the new government's choice, 'social exclusion'. This is in certain respects defined more broadly than either social polarisation or social segregation. Early French usage of the term specifically focused on groups excluded from participation in society for non-economic reasons (mental illness or physical incapacity) (Silver, 1994, 532–5). Exclusion has been used to refer to non-participation in various forms of associational activity, and also being unable to access goods and services on the same terms as the majority population. The use of the term exclusion signals something of a paradigm shift on Labour's part, away from a concern for equality and towards a focus on social inclusion and equality of opportunity. The emphasis is on education and training to prepare 'excluded' individuals for re-entry to paid work, and on social responsibilities and obligations rather than unconditional rights (e.g. increasingly stringent conditions attached to welfare payments). The terminology of exclusion has also been used in a more explicitly geographical sense. The main focus of the Social Exclusion Unit (SEU) thus far has been localised problem areas, particularly 'estates whose condition is critical or soon could be', and which have become 'no go areas for some, no exit zones for others' (SEU, 1998). The implied threat is of sores in the body politic from which moral decay could infect the whole of society.

The SEU's definition of social exclusion is multifaceted. First, drawing on economic evidence of social polarisation, attention is drawn to the numbers of poor neighbour-

hoods (on various definitions, between 1370 and 4000 localities of varying size) with concentrated socio-economic problems. Characteristically these areas had more than double the national unemployment rate and on average over 40 per cent of children in these neighbourhoods were living in low income families. Perhaps more significantly, gaps between affluent and poor areas have widened (cf. Dorling and Woodward, 1995).

However, the problems go beyond narrowly economic indicators. In a number of areas surplus housing leads to near abandonment. This is not just a function of declining demand; it reflects the growing reluctance of people to accept tenancies in unpopular neighbourhoods even where there are lengthy waiting lists, not least because of high crime levels. The affected areas also experience a combination of interrelated problems in terms of access to education and welfare services (Byrne and Rogers, 1996; see also chapter 9). Access to basic goods and services is problematic: the financial infrastructure is skeletal or non-existent (see also chapter 7); social facilities are minimal; and (in part reflecting the location of peripheral council estates) public transport is limited, severely circumscribing opportunities for those without cars. Retail facilities are of poor quality and expensive, so that maintaining a healthy diet consumes a much greater proportion of household income (Cummins and MacIntyre, 1997; Piachaud and Webb, 1996). There are suspicions – to put it no more strongly – that this process is producing subnormal nutrition and therefore perpetuating health inequalities (James et al., 1997). Given these related problems of marginality it is not surprising, finally, that census enumeration and civic participation (expressed in voter registration and turnout) are low (Hoggett, 1997; Dorling et al., 1996), since many residents see little point in engagement with formal political processes and structures.

The diagnoses of the causes of and solutions to these problems are revealing. The multicausal nature of exclusion is conceded but the relative impact of the various causal processes is not analysed. The decline of traditional sources of employment has had disproportionate, spatially concentrated effects.

Many poor communities also show the 'scars of weakening family structures, with more divorce and more children born outside marriage'. Government policies are said to have 'contributed'; blame is pinned on 'poor housing design', which undermined community cohesion, and policies on housing finance, creating circumstances in which only those most excluded from the labour market could afford social rented housing (see chapter 9). There are valid diagnoses of the contradictory and overlapping effects of past public policies (SEU, 1998, 34–40), and useful exemplars of small-scale local initiatives which work (pp. 41–51). However, concessions to a moral agenda and to fashionable criticisms of design determinism distract attention from the economic and political processes which marginalise and concentrate the disadvantaged. Concerns have been expressed that regardless of rhetoric about the multifaceted nature of exclusion, the use of the term signals an important departure from 'old' Labour's strategy of equality. Lister (1998, 219–21) argues that Labour's view of social exclusion is that participation in paid work represents the primary obligation for those of working age (see also Levitas, 1996). Thus policies will be targeted primarily at reequipping the excluded for participation in the labour market. Absent from policies on exclusion will be two key elements: a commitment to a more redistributive taxation system, and an economic strategy containing demand-side as well as supply-side measures to stimulate employment (see chapter 6). Against a background of entrenched social inequalities a key question is whether genuine social inclusion is possible without greater economic equality (Lister, 1998, 224).[2]

8.5 CONSEQUENCES AND COSTS OF SOCIAL POLARISATION

Poverty clearly has direct impacts on health and educational attainment, and there is growing evidence of divergence in the prospects of rich and poor areas. Perhaps more controversial are arguments about the impacts of social polarisation on social cohe-

sion. The focus here is principally on health and on perceptions of crime, the former as an index of the material impact of polarisation, the latter as an indication of the decline in social cohesion. These two themes are linked through Wilkinson's (1996) work which contends that health inequalities are greatest in societies where social cohesion is weakest.

Geographical variations in mortality in Britain remain substantial; for example, at district level, male all-cause standardised mortality ratios (SMRs) in England vary by a factor of 2.5, between Manchester (154, or 54 per cent *above* national rates allowing for age structure) to Elmbridge, Surrey (63.5). There are clear associations between mortality and levels of social deprivation. Moreover, mortality continues to remain relatively high in places where it has always been high. More worryingly, the experience of certain places is deteriorating relative to the national average. Trends in all-cause SMRs revealed that in 27 places SMRs increased between 1981–85 and 1990–92, with Oldham (SMR rising from 121 to 131), Greenock (123 to 130) and Birkenhead (112 to 121) deteriorating notably (Dorling, 1997, table 5). There are also several places (e.g. Blackburn, Halifax, Preston and Southwark) where infant mortality rates (IMRs) were around the national average four decades ago but are now in some cases twice the current national average. Dorling also compares mortality for each decile of the population grouped by SMRs for 1950–3 and 1990–2. In 1990–2, the 10 per cent of people living in those areas with the highest death rates recorded an SMR of 142.3 while those in the best decile recorded an SMR of 76.2. For 1950–3 the corresponding gap was between 131.0 and 81.8. Thus the range in SMRs between the top and bottom deciles has increased from 49.2 to 66.1 (Dorling, 1997, table 14).

This analysis was conducted for local authorities, and given the evidence presented above concerning social segregation one would reasonably expect smaller-scale variations to be greater still. Studies of Scotland, Wales and the Northern region have confirmed this (McLoone and Boddy, 1994; Phillimore *et al.*, 1994; Higgs *et al.*, 1998). In the Northern region the all-cause under-65

SMR for the most deprived 10 per cent of wards increased from 145 to 158 between 1981–91 and the ratio of the SMRs for the most-deprived and least-deprived 10 per cent of wards increased from 1.73 to 1.95 over this period (Phillimore *et al.*, 1994, table II). There was an absolute increase in mortality for males in the 15–44 age group in the most deprived 10 per cent of wards. This provided striking evidence of the ways in which the poorest areas were 'coming adrift' from the rest of the population, but even without the experience of the most deprived fifth of wards there has been no narrowing of inequality in mortality in the rest of the population. Finally, in statistical terms the association between mortality and deprivation appeared to have strengthened (at least as measured by correlation coefficients) during the 1980s (Phillimore *et al.*, 1994, 1127). The poorest areas of the Northern region had levels of mortality roughly equivalent to those obtaining nationally around 1950. In other words they were lagging some 40 years behind the rest of the UK, in terms of mortality experience.

Precisely how these differences are to be interpreted remains contentious. A straightforwardly materialist account would link these widening inequalities to the divergent economic prospects of the places in question. This would undoubtedly oversimplify a strong case. We need to think about precisely how the characteristics of places impact on individuals resident there. Recent analyses of the influence of place on health inequalities have suggested that 'health disadvantage is exacerbated in socially and economically impoverished settings' (Curtis and Rees Jones, 1998, 662), indicating that relationships between material circumstances and health are by no means always linear ones. Several recent studies have argued that health inequalities are greatest in societies where social cohesion is weakest. It is therefore appropriate to review evidence on the impacts of crime on social cohesion.

Key issues are whether social polarisation generates greater levels of crime, and whether and in what ways social cohesion is threatened by rising crime levels. Criminologists debate the connections between rising levels of crime and the economy, but it is clear, regionally, that the highest incidence of crime in the UK is found in the major Northern and Midland conurbations which experienced the greatest deindustrialisation in the 1980s (Fyfe, 1997, table 10.3). Locally, crime is spatially concentrated: 40 per cent of crime occurs in 10 per cent of areas; 10 per cent of residents in inner city areas experience at least one burglary a year, double the rate found elsewhere; 25 per cent of ethnic minority residents in low-income, multi-ethnic areas say that racially motivated attacks are a problem; and drug use is significantly higher in poor neighbourhoods than elsewhere (SEU, 1998, 28).

What may be more important than the actual levels of crime is the way citizens respond to it. Taylor (1996, 320) argues that crime, and fear of crime, may be part of a 'central problem' of our times: 'how citizens can deal with difference and still live in an organised civil society'. He discusses suburban crime prevention initiatives (Neighbourhood Watch, private policing etc.) in south Manchester not in the dismissive tones of writers who suggest that fear of crime is an exaggerated response to 'moral panic', but as part of a search for 'personal economic safety and neighbourhood order'. Participation in these initiatives is 'a more or less active, socially-situated response to social anxieties' (p. 321).

Of course, perceptions of crime levels probably are exaggerated, but several themes can be identified about the geography of urban crime. Reportage typically portrays local 'communities' (which begs the question) as being 'under siege' from external threats. A powerful image concerns the 'encroachment and contamination' of the suburbs by the inner city, taking material form in 'ram-raids' (smash-and-grab raids which involve driving cars into shop windows and ransacking the shop's goods). The responses to this by suburban communities may not take the form of North American cities (gated, guarded residential estates) but these responses deserve serious attention; Taylor even refers to them as an emergent 'social movement'.

The primary response of Taylor's suburban residents was a defensive, exclusionary one, but he also argues that different socio-economic groups still have to share a number of public spaces (principally the streets and public transport). He focuses on public responses to begging and street homelessness, and (in the context of a widely perceived decline in the character and quality of public space) shows how people devise strategies to avoid harassment (real or imagined) from those living on the streets. His work indicates a growing awareness of incivility, manifest in the 'presence on a predictable basis of unpleasant or threatening groups in particular urban-centre sites' (Taylor et al., 1996, 178). Fyfe (1997, 251–2) shows how perceptions of risk (of crime or other threats) have an impact on behaviour, serving in particular to restrict options available to women. In addition, he argues, there are defensive responses to the perceived threat of anti-social behaviour in the public sphere, including increased surveillance of streets and other spaces such as shopping centres, including the use of closed circuit TV systems. Such measures may not have gone as far as the aggressive exclusionary initiatives described in Los Angeles by Davis (1992), but the result is a 'subtle privatisation of public space as commercial imperatives define acceptable behaviour' (Fyfe and Bannister, 1998, 263). This will particularly affect those unlikely to participate in the consumption experience through what Sibley (1995) describes as a 'purification of space'. It may have more disabling consequences, denying society of the experience of disorder and difference which is central to producing tolerance, social cohesion, and a democratic political culture (Fyfe and Bannister, 1998, 263–5).

These geographies of crime, and fear of crime, combine with images of social polarisation to produce symbolic geographies of urban space, in which particular neighbourhoods acquire fearsome reputations as 'no go areas', zones of lawless masculinity and anti-social behaviour (Campbell, 1993). Such places may be perceived as having the potential for a breakdown of public order which, on occasion, is actualised (Taylor et al., 1996, 175–6). The causal connections between pub-

lic disorder and deprivation are not direct and obvious: considering the urban 'riots' of the 1980s, there are contradictory definitions of what counted as a riot, numerous locally specific reasons why certain events occurred, and large numbers of locations which did not riot (Keith, 1993; Power and Tunstall, 1997). However, a constant (if not always explicitly stated) theme has been that of race: the presence of a distinctive and stigmatisable community was undoubtedly influential in the identification of 'symbolic locations' by Sir Kenneth Newman, Commissioner of the Metropolitan Police in the early 1980s. These were places in London perceived as volatile flashpoints from which disorder might spread. Keith (1993, 232–52) argues that the 'variables of race, crime and public order did not just interact, they came in part to define each other'. The resulting discourse – 'lore and disorder' – had the effect of reinforcing existing stereotypes of minority groups, thus lending weight to calls for more punitive measures of social control while delegitimising claims from black communities for additional resources. That this process of racialised criminalisation was not just a response to the disorders of the 1980s is evident in recent responses by the Metropolitan Police to street crime, which have involved targeting black youths particularly. Conversely, the failure to obtain a conviction in the Stephen Lawrence case was yet another indication of the failings of the justice system to respond equally to the needs of minority groups in disadvantaged areas.

In short, the argument of this section has been that there is evidence that social polarisation is affecting the health prospects of communities; there is increasing evidence of divergence in mortality experience. At the same time, there are concerns that social cohesion is declining: the responses to crime discussed here are exclusionary and defensive ones, which indicate that diverse social groups occupy, to a growing extent, separate spaces rather than a common public realm. Thus greater social fragmentation is mirrored (albeit imperfectly) in diverging patterns of health inequality and (as indexed through fear of crime and perceptions of danger) declining social cohesion. This is the terrain of Wilkinson's (1996) arguments about health inequalities. He is seeking to explain the apparent paradox that health inequalities have widened in countries such as Britain, despite substantial improvements in living standards. If health inequality had simply been a residual problem of absolute poverty we would expect it to have been eliminated by growth (Wilkinson, 1997, 593). His answer to this paradox is that more egalitarian and cohesive societies are more healthy than less egalitarian ones. Possible reasons for this might include the psychological effects of relative deprivation and the impacts of under-investment in human capital. However, Wilkinson suggests, in some measure following Putnam's (1993) work, that people living in places with high levels of social capital (dense networks of community and associational participation) are more likely to trust their fellow citizens and value solidarity and toleration. Kawachi and Kennedy (1997) argue that measures of social trust provide a statistical link between income distribution and mortality in the USA. The UK does not have the gross social fragmentation of the USA – at one end of the spectrum, an 'over-class' in private, guarded estates; at the other, the ghetto underclass, with mortality increasing from causes of death such as AIDS, suicide and murder. However, the evidence in this chapter suggests that it is moving slowly in the same direction: increased social polarisation, widening health inequalities, and a greater awareness of the differential levels of risk and crime, all suggest a less egalitarian and cohesive society. This raises the question of how far the institutions of the welfare state can reverse these developments.

NOTES

1   In what follows I draw heavily on Deborah Phillips' (1998) recent comprehensive review. For more details see Karn *et al.*'s four-volume study of *Ethnicity in the 1991 Census.*

2   The policies proposed by the SEU to tackle neighbourhood decline are reviewed in chapter 12.

9

SPATIAL DIVISIONS OF WELFARE

9.1 INTRODUCTION

Welfare states are now confronted by circumstances not envisaged at their inception. The Beveridge Report of 1942 assumed, for example, that unemployment would be generally short term and affect a comparatively small section of the population, and that households would typically be supported by a male breadwinner, whose female partner would remain in the home. A contributory social insurance fund, in which the number of contributors greatly exceeded that of dependents, would meet contingencies arising from illness, unemployment and old age. These assumptions have been undermined by increased life expectancy, changing patterns of labour-market participation, growing fragmentation in the labour market, and mass unemployment. Moreover, the hypermobility of capital imposes severe constraints on what sort of welfare policies are thought to be politically possible (chapter 2; see also Taylor-Gooby, 1997; Hay, 1998; Pierson, 1998).

The responses to such changed circumstances are not pre-given and automatic, as Esping-Andersen's (1990) analysis of the diversity of welfare regimes makes clear. Commentators on social policy used to argue that the 'logic of industrialism' would lead, inexorably, to convergence between states (usually in the direction of greater state finance and provision). However, such broad generalisations should yield to explanations which are sensitive to the particular institutional conditions pertaining in an individual state (Melling, 1991). Convergence alone is not satisfactory; national policy developments cannot simply be 'read off' from international economic changes.

In particular, attention must be paid to the terrain on which national welfare policies are constructed and implemented. My argument here is that growing social polarisation – not least its geographical expression – has undermined the so-called 'consensus' about the welfare state and prepared the ground for the pursuit of divisive, competitive policies. Furthermore, these are policies which will be difficult to reverse. A related point is that there has been an intellectual critique of the welfare state, from both left and right of the political spectrum. For the right, the welfare state created a culture of dependency, undermining community-based efforts to develop welfare services. Furthermore, centralised bureaucracies could not absorb the information necessary adequately to respond to the needs of individuals and communities. Market signals therefore were seen as a superior way of coordinating service delivery. Commentators on the left had also emphasised the repressive and remote nature of welfare bureaucracies, though the preferred solution was greater democratisation and decentralisation. Given these critiques a reappraisal of the welfare state would have occurred at some stage. My point here is that growing social polarisation and individualism created a political landscape in which pro-market advocates of reform found a receptive audience.

The chapter begins by discussing the ways in which welfare policies were formulated against the background of an uneven socio-economic landscape (section 9.2). This is followed by examination of three key trends in recent social policy: the shifting boundary between public and private provision, the trends towards decentralisation and localism, and the move towards 'quasi-markets' in the public sector (section 9.3). Some alternative possible scenarios for welfare provision are then considered in section 9.4 but the conclusion reached is that current tendencies towards greater competition are not likely to be reversed except at the margins.

9.2 SPACE AND SOCIAL POLICY: RESPONDING TO UNEVEN SOCIO-ECONOMIC LANDSCAPES

Geographically uneven development was a key rationale for the establishment of the post-war welfare state, but geographical analyses were initially limited to the identification of continued regional disparities (Coates and Rawstron, 1971). A subsequent focus on territorial justice was welcome, insofar as it helped identify those unequal distributions which were inequitable. In broad terms there is evidence of territorial justice in service provision, and of considerable improvements since the pre-war era (Boyne and Powell, 1991 and 1993). However, these improvements were achieved in an era of consensus politics (chapter 2), which did not last, for several reasons. Mutterings of discontent became strident during the 1970s; state welfare provision was discredited by association with the failures of Keynesianism and an emergent fiscal crisis, and by its perceived inflexibility and propensity for prioritising producer interests. The New Right critique advocated monetarism, privatisation and competition rather than bureaucratic planning. The Conservatives conjured up images of the burden of public expenditure, the civil disruption associated

with union power, and the disincentive effects of the dependency culture. They promised an era of prosperity and self-reliance instead (Ginsburg, 1992).

Why did this platform attract support in 1979, given the rising tide of unemployment? One reason is a central paradox of the welfare state: its successes undermine the conditions for its continued well-being. Through the security it offered against certain risks, and the chance for advancement provided by broadening access to educational opportunities, the welfare state played a key role in raising living standards for a substantial proportion of the population. This generated new social divisions in two senses. First, divisions arise between those primarily dependent on public services, and those whose welfare is more immediately dependent on private consumption. Thus, sectoral divisions – consumption cleavages – are said to outweigh and override class alignments, especially those associated with housing tenure (Saunders, 1990; see chapter 8). Second, growing affluence generates growing consumerism (and hence some dissatisfaction with standardised public services) and also an incentive for the securely employed middle classes to defect from reliance on state welfare. This opens up the possibility of a system of much more residualised provision for the poor and dependent. These changed material circumstances affected the articulation of interests around welfare policy; the Thatcher governments' policies were based on an acute awareness that the political landscape was shifting and that the potential was there for a somewhat less inclusive political project. The defection of skilled manual workers to the Conservatives in the 1979 election, allegedly seduced by the promised sale of council housing, is cited as crucial evidence for this changing landscape.

Crucially, moreover, the landscape was shifting in a geographical sense too: the leading edge sectors of the economy were heavily concentrated in the South East, while the 'sunset' industries dominated peripheral, Labour-supporting Britain. It may seem simplistic to infer social attitudes from these changes and draw a contrast between a collectivist 'north' and an individualist 'south', but the British Social Attitudes surveys display interesting variations in attitudes to welfare, although only six regional divisions are available (table 9.1). Despite the small sample size, there appears to be a gradation from Scotland downwards in the extent of support for collective service provision, and in attitudes towards those who rely on the welfare state, with attitudes being least sympathetic in South East England. The main contrasts are between Scotland/the North and the South of England; differences between the Midlands and the North of England are not statistically significant. These data show consistency with previous studies and the regional contrasts seem to persist despite a narrowing of regional economic differentials (see also Fieldhouse, 1995).

The argument developed here is that, in the light of evidence on spatial and social variations in attitudes to welfare, welfare policies are played out on a much more fragmented terrain than has been the case for many years. Consequently there is a real problem of reconciling the interests of groups whose economic prospects are increasingly

TABLE **9.1** Regional variations in attitudes to the economy and the welfare state, 1993–95 combined

% Agreeing that:	Scotland	North	Wales	Midlands	London	South
Unemployment benefit too low and causes hardship	69	57	50	55	55	47
Government should redistribute income from better-off to the less well-off	60	61	48	48	51	43
Ordinary people do not get their fair share of the nation's wealth	70	71	65	67	62	59
Big business benefits owners at the expense of workers	66	61	58	63	59	56
There is one law for the rich and one law for the poor	75	73	73	74	68	66
Unemployment should be higher priority than inflation	77	70	63	72	72	65

Source: Curtice (1996), 9.

divergent: in Galbraith's (1992) terms the 'comfortable majority' may be increasingly reluctant to support a residual (if substantial) minority. I argue, therefore, that the Conservatives produced welfare policies which both drew upon emerging social divisions and sought actively to exploit and widen those divisions, and that there is little prospect of such divisions being substantially reduced.

9.3 RESTRUCTURING THE WELFARE STATE, 1979–1998

Offe (1984) identified the 'embarrassing secret' of the welfare state: it was both indispensable for, and yet incompatible with, capital accumulation, so no state had been able successfully to demolish it. Of course this has not prevented states from pursuing the (perhaps illusory) goal of reform, but there are electoral constraints on this. Change has been piecemeal, and has involved regulatory experimentation and tactical advances and retreats. The priority accorded to specific services has varied, reflecting a long-standing distinction between the deserving and undeserving poor, or between services enjoying mass support and those intended for a minority. Consider recent arguments about the relative priority given to housing or social security in contrast to the NHS: there are arguments about the extent of growth in NHS spending, but at least they are arguments about growth. By contrast, social security entitlements and housing have seen substantial cutbacks, as have services for socially marginal groups such as the mentally ill. Similarly, the unemployed face higher hurdles: they must now demonstrate not just availability for work but that they are actively seeking it. For the young unemployed, benefits are now contingent on accepting a place on Labour's 'New Deal', which some see as an emergent system of workfare. Benefits are indexed to prices not wages, opening up wider gaps between those in and out of work. Social security is a more discretionary, arbitrary and disciplinary system, through the operation of the Social Fund, a *de facto* cash-limited system of loans for those in urgent need, replacing grants and special payments. In housing, those unable to participate in owner-occupation find their options more restricted as the more desirable local authority estates are disposed of. Local authorities no longer have the statutory obligation to rehouse the homeless, and in any case are heavily constrained financially in what they can offer. For many, hostel accommodation or even homelessness is becoming a fact of life (Burrows *et al.*, 1997). Against this background three sets of developments are considered: efforts to shift the boundaries between public and private provision; efforts to place more of the burden onto community-based, informal care; and the introduction of market principles in public services.

The public–private mix for welfare

The inauguration of the welfare state did not spell the end of private provision: in education and health care there has always been a private sector. Moreover, recent growth antedates the 1979 election; the biggest stimulus to private health care was the 1974–79 Labour government's efforts to remove private beds from NHS hospitals (Mohan 1995a, chapter 7). In addition, while growth in private health insurance owes much to labour market trends (companies offering perks to valued employees), individual choice is also important – an illustration of what Klein and Millar (1995) term 'do-it-yourself' social policy.

But individual choice is not the whole story; state policies have done much to facilitate the expansion of private welfare. Private health care provision was encouraged through various small-scale changes in regulations, though tax relief on insurance premiums was only granted in the 1989 reforms, and then only to the retired (Mohan, 1995b, chapter 7). This concession was reversed by Labour. Housing has seen the vigorous pursuit of privatisation via the right-to-buy scheme, which offered large discounts to council tenants to encourage purchase of their home. Private education received support as well through the assisted places scheme. The scale of such measures is not always appreciated. For example a 1983 relaxation of regulations, permitting the costs of long-stay care to be met from the social security budget, led to a dramatic expansion of such care in unplanned locations and unplanned quantities. Within five years provision of private long-stay beds exceeded NHS provision in a number of locations. The geography of this – predominantly concentrated in coastal locations in southern England – posed thorny problems of regulation and planning.

There is, as a result of these developments, an emerging *spatial division of welfare* between places with comparatively high levels of private provision and those without (see Burchardt *et al.* (1998) for a wide-ranging review of private welfare). Private health insurance is a good illustration. There has been a clear social gradient in coverage since 1982 with the top two SEGs (professionals; employers/managers) consistently recording at least double the rates of all other groups. Developments between 1987 and 1995 indicate very little change in these patterns. Geographically, in 1987 the Outer Metropolitan Area (16 per cent) and the rest of the South East (15 per cent) had much greater insurance coverage than any other region, but the *General Household Survey*, on which this commentary draws, does not provide a geographical breakdown for 1995. However, given their social composition it would not be unreasonable to expect coverage of 30–35 per cent in parts of Berkshire, Buckinghamshire and Surrey[1] (table 9.2).

These patterns are reflected in the distribution of private hospitals and in the proportions of surgery carried out privately. Even in the late 1980s, approximately 30 per cent of all operations carried out in the NW and SW Thames RHAs took place in private hospitals or in NHS paybeds. The post-reform years have almost certainly witnessed an expansion of the contribution of the private sector, not least because of a relaxation of regulations on NHS authorities placing contracts for treatment with private hospitals (Mohan, 1995a, 158–73).

Private education exhibits similar social and spatial gradients although the proportion of the population involved is smaller

(around 5 per cent of households with children educated them privately in 1995). However, the income gradient is steeper; about 80 per cent of households using private schools were in the top 40 per cent of the income distribution (Burchardt *et al.*, 1998, 23). There are some overlaps between the use of different services, with approximately 5 per cent of households being owner-occupied and using both private health care and education, but Burchardt *et al.* (1998, 28) do not believe that the users of private welfare constitute 'a separate and distinct part of the population'.

The expansion of private health care and education undoubtedly helps insulate the middle classes from dependence on a cash-strapped public sector, although the significance of these services in consumption-sector divides is questioned (Busfield, 1990). However, the housing market has seen a much greater – and irreversible – shift in the balance between public and private sectors. Two key statistics highlight this: the number of owner-occupied dwellings has increased by some six million since 1979, while approximately two-thirds of remaining council tenants are dependent on income support.

The extension of home ownership was a key plank in the Conservatives' election strategies. This goal was pursued through the right-to-buy legislation and the deregulation of housing finance. The 1980 Housing Act gave local authority and housing association tenants the right to buy their homes at substantial discounts. Generating receipts of over £26 billion this was the largest single privatisation programme; if account is taken of the value of discounts the total rises to some £50 billion (Wilcox, 1996). The deregulation of housing finance facilitated access to mortgage finance to those moving into owner-occupation for the first time. These developments have fundamentally altered the 'status, quality and geography of public housing' (Forrest and Murie, 1988, 12). This is evident in both the geography of house sales and in the related process of residualisation of the remaining public sector housing stock.

The disposal of council houses under the right-to-buy legislation peaked in the early

TABLE 9.2 Private medical insurance coverage, 1982, 1987 and 1995: (a) private medical insurance by socio-economic group, Great Britain; (b) private medical insurance cover by standard region (percentages)

		Insured		
		1982	**1987**	**1995**
(a)	Professionals	23	27	22
	Employers/managers	19	23	23
	Intermediate and junior non-manual	9	9	9
	Skilled manual/own account professional	3	3	4
	Semi-skilled manual and personal service	2	2	2
	Unskilled manual	2	1	1
	All persons	7	9	10
(b)	North	3	3	
	Yorks/Humbs	7	6	
	North West	7	7	
	E Midlands	5	7	
	W Midlands	7	8	
	East Anglia	6	10	
	Greater London	10	13	
	Outer Metropolitan	13	16	
	Outer South East	9	15	
	South West	8	8	
	Wales	3	4	
	Scotland	3	4	
	Great Britain	7	9	

Source: OPCS *General Household Survey,* various dates.
Note: Subsequent *General Household Surveys* have not sought data on private health insurance.

1980s with sales exceeding 200 000 in 1982. Forrest and Murie's (1988) analysis covers 1979–85 in most detail; in this period over 750 000 council houses were sold. Regionally, the proportion of stock sold ranged from under 10 per cent in Greater London, to over 20 per cent in over 30 local government districts. None of these were in the Yorkshire/Humberside or Northern regions, nor were any London boroughs included. High sales were particularly evident in southern and rural areas. Of the bottom 30 districts, half were inner London boroughs but other low sellers included Hull, Leeds, Manchester and Sheffield. Sales tended to be highest in places with high existing levels of owner-occupation (Dunn *et al.,* 1987). Thus, 'rather

than sales "evening up" tenure structures, they tended to accentuate local variations' (Forrest and Murie, 1988, 120). In addition, sales tended to increase with the relative affluence of tenants and localities, and tended to be highest in areas where the council stock was mainly houses and lowest where there were large proportions of flats and one-bedroomed properties.

Council house sales are also associated with social polarisation within cities. This is partly a function of the historical geography of council housing: better-quality houses (semi-detached or terraced properties) in suburban estates have proved much more popular than flats, which are the characteristic form of inner city council housing. The result is a split within the public sector, between the 'mixed tenure areas of more popular housing and the one-class estates of less popular dwellings' (Forrest and Murie, 1988, 167). The result of sales and the decline in new council-house construction has been a net decline in the available stock: sales have exceeded new construction since 1980. Those dependent on public sector renting are therefore restricted in their options towards a more limited number of estates. In practice, those at the margins of the housing market now dominate access to council housing; such tenancies as are available are allocated to those in the most severe need. Thus *households* with problems have become concentrated in *estates* with problems (Forrest and Murie, 1988, 168). Tenants then have little chance of moving (if all new tenancies are taken up by the homeless, there is no scope for transfers) and little chance of improving their circumstances (living on stigmatised estates makes access to paid work more difficult).

Social polarisation within council housing therefore results from deliberate political decisions to sell council houses and to restrict new building, which combine to widen disparities between estates and to restrict the options open to those in the social rented sector. A third development has further exacerbated segregation. Since 1987 government policy towards rented housing has been dominated by raising rents to market levels. The logic was that this would attract private investment in rented housing: those on low incomes would be protected by housing benefit. Housing associations – to all intents and purposes now the sole providers of new social rented housing – were forced to rely to an increasing degree on private finance for capital developments, and the corollary has been that rents have been raised well beyond the rate of inflation. Consequently rents are only affordable by households likely to qualify for housing benefit. Public sector renting is thus a classically residualised tenure: the quantity of housing has been reduced by sales; the remaining stock is by definition less desirable; and it is open only to those on very low incomes. Two-thirds of public sector tenants are now in receipt of housing benefit

(Green *et al.*, 1997; note, however, Murie's (1997, 449) comment that a gradual residualisation has been evident for some 30 years). This means that households face severe penalties if a household member accepts employment. Interestingly, for reasons explored by Burrows (1999) there is evidence of increased mobility within the social rented sector. This contributes to and is an effect of problems of community instability. High levels of housing turnover are often a prelude to housing abandonment, which is a growing problem, especially in deindustrialised cities of northern England, despite high levels of demand.

Finally, restrictions on the availability of social rented housing have been associated with a steady rise in homelessness and in the numbers of households in temporary accommodation. Government estimates refer to those officially accepted as homeless by local authorities; these peaked at 157 000 in 1991 and there are now around 125 000 homelessness acceptances each year. A less restrictive definition, including concealed families (e.g. those living with parents), shared households, and those living in hostels, among others, produces estimates nearer 450 000 (Ginsburg, 1997).

The social rented sector illustrates some of the inabilities of market-led solutions to respond to social need. Problems of market failure, compounded by political mismanagement, also beset owner-occupation. The deregulation of housing finance raised house prices rapidly and led to the entry into ownership of some comparatively marginal households. Realising the inflationary consequences of their policies, the Conservatives announced a withdrawal of some support for ownership, notably the restriction of MITR to one allowance per dwelling, thus ending double tax relief for unmarried couples. However, giving five months' notice of this change further fuelled inflationary pressures in 1988. The house-price boom collapsed as a result of the interest-rate hikes required to choke off inflation. The immediate effects were twofold. First, up to 10 per cent of households were left with negative equity (houses worth less than the outstanding mortgage) although subsequent price recovery has largely eliminated this problem. Second, increased interest rates in the early 1990s, combined with recession (itself partly an effect of rising interest rates) produced an 'explosion in mortgage arrears' (Ginsburg, 1997, 146) and ultimately substantial numbers of house repossessions, peaking at 75 000 in 1991 (Mohan, 1995b). Nearly half a million households lost their homes through repossession in the 1990s. The impact of unchecked market forces in housing had other effects: regional house price differentials limited labour mobility especially during the boom years, a problem compounded by lack of choice of housing tenure in parts of the South East where levels of owner-occupation approached 90 per cent (Mohan, 1995a). The housing market must cope with an increasingly flexible labour market and for this a range of housing options is required. This will require, at a minimum, a sustained programme of public sector house building, and determined efforts to counter the residualisation of social rented housing.

Blurring the boundaries: community and voluntary forms of welfare provision

The previous section illustrated some changes in the boundaries between state and market in welfare provision. Perhaps of equal significance, though rather less visible, are the ways in which responsibilities for caring have been displaced onto informal sources – particularly unpaid care by relatives or neighbours. In addition, the role of charitable and voluntary provision has increased. These developments are exemplified with reference to health and social care.

The emphasis on community care is not just a post-1979 development; during the 1970s Labour proposed a shift of resources in favour of community services. The Conservatives seem to have assumed that community-based provision would be available to take up the slack left by the informal sector. In practice, the pressures on informal care have been considerable. First, the substantial increases in acute hospital activity, and associated reductions in lengths of stay, have resulted in patients being dis-

charged quicker and sicker. Second, long-stay institutions (hospitals for the elderly, the mentally ill and those with learning difficulties) were run down swiftly and this meant that residents were discharged often before adequate alternative provision became available. A very uneven patchwork of services resulted. The pace and timing of hospital closure (necessary to release funds for community-based services) were partly determined by the prospects of achieving land sales, and partly by the resource position of individual health authorities. Provision of community-based services was determined by the price and availability of accommodation, not all of which was appropriate to the needs of those discharged. The resultant clustering of community care services could cause local opposition as well as pressure on available public services, expansion of which was unlikely in a climate of severe restrictions on local government finance. The development of community care was also constrained by the reduced availability of social rented housing. For all these reasons policies of deinstitutionalisation have been controversial, never more so than when violent crimes are committed by former patients of psychiatric hospitals. That disproportionate numbers of the street homeless are former long-stay patients is one index of the failures of community care (Mohan, 1995a, 112–18).

One reason community care has not been successful is because of the unrealistic assumptions made about the availability of informal care. This assumption is that the burden of caring will be borne largely by female relatives of those needing the care, an assumption which on occasion has been explicitly enunciated (Mohan, 1995b, 104). Recent economic changes serve to restrict the pool of potential carers. There is a contradiction between an 'economy that increasingly depends on women's labour and a restructured welfare state making the same demands' (McDowell, 1991, 412). The vast majority of carers are female relatives and the proportion of non-relatives involved is small (Twigg, 1992). There are also variations in the availability of informal carers due to differential female participation in paid work (Duncan, 1991; McDowell, 1991). The growing flexibility and insecurity of employment militate against combining paid work with caring, which means that (in the absence of substantial expansion of statutory services) many women will continue to be trapped in the home with little prospect of respite.

The boundary between public and private provision has also been blurred by the encouragement of voluntary and charitable service provision. One element of this is charitable fundraising, ranging from high-profile campaigns to preserve world-famous institutions such as the Great Ormond Street Children's Hospital, to (rather less glamorous) local appeals for equipment, or community campaigns to retain hospitals through a mix of public finance, charitable support and voluntary labour. There are considerable variations in this and the extent to which voluntary provision can fill gaps in statutory provision is controversial (Lattimer and Holly, 1992; Mohan, 1995a, 176–80). More generally the non-profit sector seems set to play a growing role in welfare delivery; recent studies reveal considerable expansion in this sector (Kendall and Knapp, 1996), though this is partly an artefact (e.g. transfer of housing from councils to non-profits). One implication of this concerns the relations between the voluntary sector and the state: Wolch (1990) argues that there has developed a 'shadow state' in that non-profit organisations are dependent on the state for revenues and carry out service responsibilities previously undertaken by the state. This, she argues, constrains their autonomy and the extent to which they can pursue the voluntary sector's traditional role of pressurising for social change. A second issue is whether the voluntary sector can shoulder all the burdens of state retrenchment, and specifically whether it can do so in all localities. One assessment of community care suggested that 'where in Britain the patient lives is probably the most significant factor in determining quality of care' (Murphy, 1991, 61) – a judgement which could be extended to include many other welfare services. A third implication is the extent of decentralisation in the welfare state. One could argue that variations in service provision by local authorities are a healthy sign of local democ-

racy, but this argument cannot easily be extended to the NHS. The changing balance between public and private service provision could also be celebrated as a sign of enhanced consumerism and choice, consequent on rising living standards. But it could equally signal an abdication of responsibility on the part of central government and a move towards a position in which the availability of services depends on the economic success of individual localities. The Conservative response to this problem was to argue for local discretion and blame local managers for poor-quality provision. Although Labour appears alive to the dangers of this, there is still a tendency to blame local management when things go wrong (not least because a corollary, higher spending, would entail higher taxation). In the absence of clear national guidance for the difficult and unpopular rationing decisions which will ensue, local authorities are placed in an invidious position (Blackman, 1998).

Geography and quasi-markets

The difficulties attendant upon rolling back the frontiers of the state in welfare provision led to a search for ways of releasing resources through greater efficiency within public services. Ultimately the preferred solution was the introduction of quasi-markets throughout much of the public sector (Bartlett and Le Grand, 1993); legislation was introduced to force schools, hospitals and other agencies to compete to attract business. The resultant policies have generated much criticism. Here I discuss two key issues: the tension between planning and markets, and the relationship between quasi-markets and social segregation.

The tension between planning and markets is evident *par excellence* in the NHS, where the reforms of 1991 were designed to eliminate the rigidities and perverse incentives engendered by some forty years of planning. For most of the NHS's history, hospitals had served defined catchments, albeit with limited referral of patients from other health authorities. Successive tiers of specialisation of services meant that for all practical purposes comprehensive hospital treatment was available within each of the Regional

Health Authorities (RHAs) in England. Funding was allocated to health authorities in part on a population basis (modified for local age structures and social conditions) and in part to reflect the costs of running existing services. Thus, inner city health authorities characteristically received a disproportionate share of resources because (for historic reasons) that was where most hospitals happened to be. Crucially, health authorities were charged with a dual responsibility: running services within their jurisdictions, and ensuring the provision of care for their local population. Critics argued, with some justification, that this was unsatisfactory as a health authority had no incentive to consider placing contracts for treatment elsewhere: to do so it would have had to risk the political opprobrium of hospital closures (Enthoven, 1985; Paton, 1992). The proposed solution involved a separation of the purchasing and providing functions of health authorities. Provider units would compete to attract funds from purchasing authorities; the latter would, on the basis of assessments of the needs of their population and of the services offered by provider units, determine where to place contracts. These would not necessarily be with hospitals that had, through proximity and inertia, always provided care for particular localities.

It is possible to present these reforms as a narrowly technical measure, but that would ignore the political and geographical context. Despite warnings as to the difficulties of such a reform (notably, inadequate cost information and the bureaucracy involved in negotiating contracts) the Conservative government was clearly determined to demonstrate its ability to effect change in the core of the welfare state: it could not leave the NHS alone especially against the background of annual funding crises. Geographically, the reforms seemed to offer the chance to kill two political birds with one policy stone. First, post-reform resource allocation was based purely on demography and not on the pattern of hospital activity. Thus a hypothetical health authority in the South East, with few local hospital beds and a tradition of referral links to London's hospitals, would previously have seen part of its resources swallowed up by hospitals in London. After the reforms, such a health authority now had all the money its population would justify, and could choose where to spend it. This effected a major redistribution of financial power. It may be no accident that the years prior to the reforms had been marked by a chorus of complaints from South Eastern constituencies about the failure to redistribute funds from London (Mohan, 1998). For, second, the London hospitals were perceived as high-cost yet politically untouchable: closing one of the major teaching hospitals was almost unthinkable, as a century of failed reform efforts bore witness. Yet prominent clinicians and managers in these hospitals saw their institutional futures constrained by the bureaucracy of existing funding mechanisms. The NHS's internal market therefore appeared attractive since it offered institutions freedom to draw patients from any location; it was anticipated that the quality of services they offered would entice patients in sufficient numbers to sustain these institutions.

In practice, the central London hospitals were particularly vulnerable to the withdrawal of contracts for patient treatment. Purchasing authorities soon began to switch contracts to cheaper suburban hospitals, threatening to destabilise whole systems of institutions. It rapidly became clear that unchecked market forces could threaten co-operative links between hospitals built up on an informal basis over 100 years. The most quoted example was medical research: elements of this often involved cross-institutional linkages, the costs of which could not easily be factored into contract negotiations for specific types of treatment in individual hospitals. This demonstrated the limits to the neoliberal approach implicit in the reforms. Those elements of hospital activity which could not be captured in a crude cost calculus were vulnerable. It was clear that the result would be *ad hoc* attrition of hospital capacity and this forced the Conservatives into a U-turn, in the form of an inquiry into the future of London's hospital services (Tomlinson, 1992). This reimposed a degree of planning on an unstable situation, guaranteeing certain hospitals a (comparatively) secure future. As Hutton (1995) argues, the NHS reforms demonstrated both the ways markets do not (as their apologists suggest) lead to stability and equilibrium, and the need for careful regulation to secure socially desirable goals.

The reforms also introduced competition into primary care via the innovation of General Practitioner Fundholding (GPFHs) whereby practices of a given size were given control of a budget for elective surgery. Without lapsing into an excessively functionalist interpretation one can argue that this could have been designed to placate the Conservatives' electoral support in the South East. For the practices most likely to be eligible – large, multipractitioner outfits – were to be found largely (though not exclusively) in middle-class areas and it was not surprising, therefore, that initial take up of fundholding was most rapid in the more prosperous parts of the country. A second factor may also be plausible although evidence on this is less definitive. GPFHs were given budgets to purchase elective (non-urgent) surgery, much of which is done in the private sector. A GP with a high proportion of patients who are privately insured could therefore save money by seeking to maximise referral to private hospitals. Given the existing distribution of the privately insured this is another reason behind the rapid advance of fundholding in the more prosperous regions, especially in the outer metropolitan area.

The development of markets in health care has not, therefore, been a painless process; it has taken place on an uneven socio-economic landscape; and it has arguably widened divisions of welfare, contributing to the rapid rationalisation of hospital care in inner city areas while insulating those in middle-class areas through the GPFH system. There were many complaints about an emergent 'two-tier' health care system as patients of fundholding practices received preferential treatment, simply because their practices had sums of money which they could deploy flexibly for the benefit of individual patients. Such were the problems of fundholding that Labour was forced swiftly to replace it with a more collaborative form of organisation in primary care. In hospital care, too, the

emphasis has swung back towards elements of planning, to avoid the destructive consequences of competition.

However, it is in education that we may see the most pervasive consequences of market reforms. For some two decades education policy had aimed at the creation of a system of comprehensive education, in which all children would attend their local school. As with health care, neoliberal critics bemoaned the lack of choice and competition in this system. The 1988 Education Reform Act (ERA) was the result, introducing parental choice and open enrolment for secondary schools. In these circumstances schools perceived as failing, or inadequate, would suffer the financial consequences as enrolment fell; conversely, successful schools would benefit from greater income. Choice is exercised against the background of an educational landscape which, even within the state sector, has become more fragmented and stratified. This follows from several measures introduced by the Conservatives, including: local management of schools (LMS); provision for opting out of local authority control and becoming a grant-maintained (GM) school; and the creation of specialist City Technology Colleges (CTCs). Bradford (1995, 1602) suggests that GM schools form part of the blurring of the private–state divide, part of a deliberate attempt by the Conservatives to create an educational hierarchy. At the top of this hierarchy there are private schools receiving assisted places, then there are the GM schools. These are followed by a two-tier structure within the LEA sector in which there are market-oriented and 'minimalist' schools, the latter denoting schools serving pupils in disadvantaged areas but attracting few from outside. However, geography complicates this analysis (figure 9.1): there was clearly an uneven take-up of GM status, with inner London, the metropolitan counties and parts of the North seeing low adoption. Subdividing according to the balance between public and private education, it could be argued that in some counties (Kent, Essex, Lancashire) the GM schools could be regarded as filling the relative gap in private provision. Elsewhere low take-up was associated with low levels of private provision

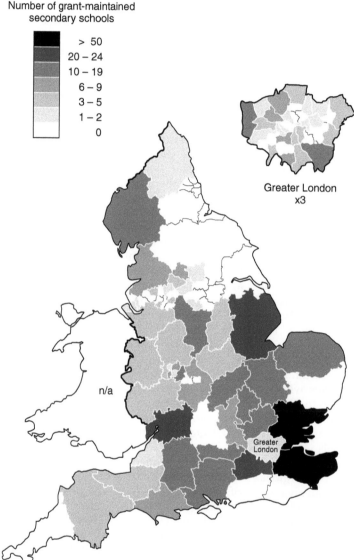

Number of grant-maintained
secondary schools

> 50
20 – 24
10 – 19
6 – 9
3 – 5
1 – 2
0

Greater London
x3

n/a

Greater
London

FIGURE 9.1 Distribution of grant-maintained schools in England, 1993 (Bradford 1995)

(Cleveland, Durham), but there were also counties with high private provision but no GM schools. The creation of this implied hierarchy, with an emphasis on parental choice, led to increased competition for places in popular schools. Taylor (1997) demonstrates this with respect to the geography of appeals by parents against refusal of places in their school of choice. The areas where appeals are most concentrated seem to be in the middle-class London suburbs; as Ball *et al.* (1995) show, middle-class parents appear to be acutely aware of landscapes of (perceived) educational quality and will go to considerable lengths to secure places in desirable schools. In contrast the preferences of working-class parents tend to be much more localist.

All parents exercise choice against an uneven background and choices are not unconstrained. Bradford and Burdett (1989a and b) demonstrate the uneven geography of private education and, even within the state sector, admissions policies (selective admissions tests; interviews with parents or children; geographical priorities) constrain choices. Children who do not attend designated schools or live in the designated catchment will find it difficult to gain a place in desirable schools. Hence housing markets are crucial because, Byrne and Rogers (1996) suggest, part of the 'life strategy of the middle classes involves picking areas of residence and primary schools in order to maximise their children's chances of accessing good schools'. By allocating schools, via their postcodes, to electoral wards, Byrne and Rogers were able to produce a cluster analysis of school types for England and Wales and, for the Northern region, of the 205 wards in which secondary schools were located (an admittedly imperfect approximation to the school catchment). The results (table 9.3) demonstrated a significant degree

of segregation; there is a relationship, albeit an imperfect one, between social deprivation and the performance characteristics of the schools. (There are some anomalies which result from the position of schools near the edges of wards when in practice they are primarily serving adjacent wards in quite different socio-economic circumstances.) Byrne and Rogers (1996) argue that we now have a Darwinian landscape (albeit one constrained at the margins by physical limitations on the expansion of successful schools). The 'fittest' survive by attracting both enough pupils and enough pupils who will perform well in public examinations; the poorest schools suffer both because they have no choice about who they recruit and because falling rolls undermine their financial viability. Focusing on Northumberland and Tyneside, Byrne and Rogers identified three clusters: the first

TABLE 9.3 School clusters by social deprivation cluster, Northern region, 1995

	School achievement cluster** Number of schools			Social characteristics	
	High	Middle	Low	% Households in social housing	% Econ. active males unemployed or on a scheme
Social deprivation cluster*					
High	2	15	33	55	28
Middle	7	47	28	32	16
Low	9	45	16	10	9
Total no. of schools	18	107	77		
School characteristics					
% 16 year olds obtaining 5+ GCSEs at grades A–C	77	39	16		
% private	49	10	3		

Source: Byrne and Rogers (1996), tables 1, 2, 3.
Notes: * Schools were placed in these categories as a result of a cluster analysis of the socio-economic characteristics of the wards in which they were located.
   ** School achievement clusters principally reflect exam results.

contained a city technology college (CTC) and most of the private schools, in all of which at least 75 per cent of pupils obtained at least 5 A–Cs at GCSE. The second cluster contained most of the voluntary-aided and county schools: its mean score was 42 per cent. Finally the third cluster contained 13 comprehensives, one voluntary aided (VA) school and two private schools; its mean score was 19 per cent and most of the comprehensives within it were located in areas of high deprivation. Put another way, 'the poles of division in residential space and schooling do essentially correspond'. Consequently, any assertions about school management or teacher quality which disregard the effects of social context 'are essentially vacuous' (Byrne and Rogers, 1996, 13). This analysis is supported by Gibson and Asthana (1998) who show a clear link between an index of social disadvantage (using several census characteristics directly related to children) and school attainment, for a set of 249 non-selective schools (figure 9.2). Their index sought to overcome the problem of defining school catchments by allocating each pupil to an enumeration district (ED) with their postcode and calculating a school score based on the weighted values of the deprivation indicators for each ED containing a pupil. Such methods had enabled the authors to account for up to 75 per cent of between-school variance in GCSE pass rates.

Both Byrne and Rogers (1996) and Cutler and Waine (1997, 22) are in no doubt that, because educational credentials confer advantage in the labour market, there is a 'strong potential constituency for educational inequality'. The middle classes have a vested interest in this and policies have been so structured as to promote inequality. While to date the UK does not have the educational apartheid of the USA (Kozol, 1991), the processes described here can only lead to greater segregation, not less. However, there is little sign, despite Labour's criticisms when in opposition, of an alternative to selection and competition. Instead policies continue to emphasise issues of school management, rather than the social context in which schools operate, so that 'failing' schools or education authorities will be penalised and, ultimately, prevented from operating.

9.4 FUTURE SCENARIOS FOR THE WELFARE STATE

Several commentators have argued, implicitly or explicitly, that emerging geographies

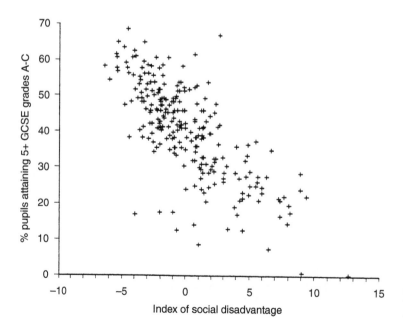

FIGURE 9.2   Performance at GCSE and the socio-economic status of school populations (Gibson and Asthana, 1998)

of production will have correspondingly new geographies of welfare. Bennett (1989) sees geographies of welfare being a function of the success of places in competing for new investment. Others see the replacement of the Keynesian welfare state (KWS) with a two-faced, Janus-like state: an entrepreneurial state for the well-off and a 'soup-kitchen state' for economic casualties. At a higher level of abstraction there are claims about an emergent 'Schumpeterian Workfare State' (SWS) which, it is speculated, is functionally compatible with post-Fordism (Jessop, 1993). Such arguments portray state strategies towards welfare not as *ad hoc* responses to economic crises, nor as solely political manoeuvres, but rather as efforts to secure the conditions for a successful transition from one regime of accumulation to another. If Fordism was exhausted, efforts had to be made to find a new stable relationship between production and consumption, and some commentators suggested that developments in the welfare state symbolised this. Stoker (1989, 141), for example, interpreted Conservative welfare policies as efforts to 'transform the production process, the pattern of consumption and the arrangements for political management' in ways compatible with the enterprise culture and flexible economic structures (see also Hoggett, 1987). There are certainly parallels which are relevant – for example, the expanded role for the private sector, the entrepreneurial character of the state, and the evidence of labour market flexibility in public services (Mohan, 1995b, chs 6–8) – and these have been the subject of vigorous debate (e.g. see Burrows and Loader, 1994, and the collection of papers on this broad theme in *Environment and Planning A*, vol. 27, nos. 10, 11).

Jessop (1993) also identified important discursive shifts. Whereas the KWS was identified with bureaucracy, hierarchies and planning, the emergent emphases in British welfare policy are on flexibility and entrepreneurialism, involving greater privatisation and deregulation, as well as the imposition of commercial criteria in the residual public sector. Jessop (1993) also refers to a 'hollowing out' of the state (chapters 3, 11) but it is not

clear how far this might apply to the NHS, for example. There has been decentralisation but the service is run on a tight centralist rein; one reason for this is that political intervention was necessary if the competitive strategies of the NHS reforms were not to destabilise systems of health care provision. Hence Labour has proposed an end to unbridled competition in health care. Finally, a defining feature of the SWS is said to be the subordination of welfare policy to the dictates of competitiveness, implying a system in which the welfare services available to individuals depended substantially on their labour-market position. Despite substantial growth in private provision and cuts in state benefits, the UK is still some way off this 'workfarist' position. Thus, while there are important similarities between the putative Schumpeterian workfare state and recent developments in the British welfare system, these reflect pragmatic adaptations to the (so-called) 'realities' of globalisation, coupled with selective political strategies designed to benefit strategic groups of supporters (Jessop *et al.*, 1988). The difference between Labour and Conservatives, given these constraints, will be limited, although Labour is pursuing some redistributive policies.

Yet this is unsatisfactory in several respects. There remains scepticism about existing welfare arrangements, tempered by acknowledgement that the welfare state is necessary for two main reasons. First, it is an essential foundation of individual autonomy which 'is valueless if exercised in a community that is denuded of the inherently public goods which create worthwhile options and thereby make good choices possible' (Gray, 1993, 122). The recent impacts of economic insecurity on the mythical 'Middle England' have brought this home to many people; the middle classes need the reassurance that collective services will continue to be provided. Second, few would dispute the importance, in a global economy, of certain minimum standards of public provision below which neoliberal solutions simply do not work. Britain (and other EC states) simply cannot compete with the workfare states of South East Asia on wage levels, so some kind of floor to the labour market is necessary.

Accepting these fairly minimalist arguments is fine as far as it goes, but there is still scepticism about the costs of large-scale, standardised provision, and concern at the impersonality and rigidity of welfare state structures, and their tendency to promote passivity and dependency. There have been several responses to this. Hirst (1994) proposes an attempt to revive the voluntary tradition, which promoted participation and democracy, by devolving public welfare and other services to voluntary self-governing associations. They would obtain public funds to provide services to their members, who would have the right to move between providers of services. This might counter the tendency for large commercial organisations to gain control of welfare services, but Hirst perhaps overestimates the democracy and accountability of the voluntary tradition. Nevertheless, mechanisms to provide greater participation in welfare delivery are highly desirable.

A related set of arguments sees the reconstruction of welfare organisations as part of a strategy to promote civic solidarity and 'social capital'. This concept draws on Putnam's (1993) study of civic traditions in Italy. He argued that the prosperity of certain regions derived from their strong traditions of civic participation and community association. The Commission on Social Justice defined social capital as:

> the institutions and relationships of a thriving civil society – from networks of neighbours to extended families, community groups to religious organisations, local businesses to local public services, youth clubs to parent–teacher associations, playgroups to police on the beat. Where you live, who lives there, and how they lead their lives – cooperatively or selfishly, responsibly or destructively – can be as important as personal resources in determining life chances ... communities do not become strong because they are rich; rather, they become rich because they are strong.
>
> (CSJ, 1994, 307–9)

The Commission argued that instead of welfare solutions being imposed from outside,

initiatives were required which aimed at recreating a sense of community and building a capacity to cope with change. This entails creating what Lipietz (1992, 99–106) terms a 'socially useful third sector', the trajectory of which is not determined solely by market forces. This would undertake tasks which were not being carried out either by the market (because they are unprofitable) or by the state (for lack of funds and/or political will). These would include: basic medical care or care for convalescents; child care; environmental improvement; recreational services; renovating public housing projects. There are arguments about the relationship of this third sector to the formal sector of the economy, but Lipietz argues that it would achieve socially useful goals, promote social integration and help overcome problems of resentment at apparent benefit dependency. At present there are numerous small-scale examples which, if expanded, might form the basis of a social economy, though there is as yet no sign of a firm commitment to supporting them (CSJ, 1994; Amin et al., 1998).

9.5 CONCLUDING COMMENTS

Human geography's main contribution to the analysis of social policy has been the mapping of spatial distribution and consideration of the extent of territorial justice in welfare systems. This is valuable, but ultimately limited. I suggest that there are reasons why geography matters to the analysis of welfare states, namely: the impacts of globalisation on the capacity of national states to deliver welfare; the undermining of the social bases on which the welfare state was constructed, and the consequent fragmentation of support for collectivism; and, as a consequence, the likelihood that future welfare arrangements will incorporate a greater degree of decentralisation and localism, a development which has both positive and negative aspects.

There is little doubt that global competitive pressures reduce the scope for manoeuvre available to nation-states wishing to advance social cohesion and welfare, but there are disputes about the precise impact of globalisation on welfare policy. Townsend argues forcefully that social polarisation and political instability have reached 'threatening proportions because powerful forces within the international market are operating in conditions of diminishing democratic restraint' (1995, 138). The only viable solution from this perspective is an internationalist one (pp. 149–50). There appears, it must be said, little prospect of the implementation of Townsend's ideas and for others the debate is not about rejecting globalisation. Rather, it focuses on whether the current regressive restructuring of welfare states is '*necessary* (summoned by an inexorable logic of globalisation), *conditional* (on the perception that such a logic is at work) or altogether *contingent*' (Hay, 1998, 528; see also Taylor-Gooby, 1997). Hay argues that the parameters of welfare policies are limited not by globalisation *per se* but by perceptions of what globalisation entails. In his view, therefore, there is scope for a positive restatement of the case for the welfare state against those (within New Labour) suggesting that there is no alternative (see also Hutton, 1998).

However, this must be tempered by a recognition that the social base of collectivism has been fragmented by recent socio-economic changes (chapter 8). Various studies show emerging fault lines between those in fairly secure occupational niches, dependent to a decreasing degree on the state, and those either excluded from or on the margins of the labour force. It follows that the scope for redistributive policies is limited by such electoral parameters. The new government has thus been subject to continual pressure to pursue a more egalitarian strategy. It has generally resisted the temptation with the exception, perhaps, of the 1999 Budget, which has widely been seen as an attempt to target resources on children in poverty. Elsewhere, the competitive policies of the Conservatives remain, albeit with minor modifications, and for sound electoral reasons. A central argument here is that, particularly with respect to education, divisions in welfare contribute towards the reproduction and extension of other social divisions. Competition between schools stimulates migration of middle-class families to areas where schools are perceived as better; this in turn is likely to drive house prices upwards in areas containing desirable schools, and possibly downwards in other less-favoured locations. Given a local taxation system based on property values this can further weaken the capacity of poorer areas to provide services. Hence tendencies can be seen towards greater social segregation caused, in part, through processes operating within the welfare state. This is not a promising basis from which to ask, as Baldwin (1990) does, why the privileged should give up some of their wealth to support the disadvantaged: the incentive structures are pointing in the opposite direction. This explains much of Labour's reluctance to reverse processes of selection and competition in the welfare state. The parameters of welfare policy are thus set not only by globalisation but by a segregated socio-economic landscape in which the suburban middle classes hold the key to electoral success. Greater redistribution is largely ruled off the political agenda. The welfare state is thus transformed into a competitive arena as 'normative claims to justice ... [become] selfish claims to desire' (Young, 1990).

We can therefore expect a trend towards localism in which – taking all (public, private and voluntary) services into account – the wealthier parts of the country steadily diverge from the rest. In the latter, a residualised state sector plus a struggling social economy will struggle to contain the costs of widening social inequality. Insidiously, the assertion that problems of service delivery reflect not a lack of resources but managerial failure or poor-quality staff will be used to justify a punitive approach which permits central government to abdicate responsibility for what happens locally. Only in the most deprived areas will we see positive discrimination, for example in Education Action Zones or Health Action Zones, though even these rely on drawing funds from other programmes and attracting resources through various partnerships. Politically, such initiatives cast welfare problems as discrete and local events, confined to small areas of concentrated social exclusion, and thus manageable by limited, but carefully targeted, action. Significant injections of public finance are unlikely. Yet the costs of inequality remain, and as a growing body of evidence makes clear, threaten social cohesion. The welfare state ought to be part of the solution to that, not part of the problem.

Note

1 Preliminary analyses of the British Household Panel Survey suggests a figure around 25 per cent for the South East (excluding Greater London), but this (and the associated national figure of 16 per cent) seem out of line with other estimates, so should be regarded with caution. I am grateful to Liz Twigg for these figures.

10

PLACE AND POLITICAL MOBILISATION

10.1 INTRODUCTION

How does place have an impact on politics? This question goes to the heart of contemporary debates in geography which have sought intellectual credibility by suggesting that space and place make a difference to the operation of social processes. One school of thought, evident in some studies of voting behaviour, is that spatial effects are purely compositional: in other words, a place votes in a particular way because of its social-class composition. An alternative is that there are contextual effects, over and above – and independent of – compositional influences; something about the characteristics of a place exerts an influence on political behaviour therein.

However, links between place and mobilisation are not unproblematic, as two examples make clear. The coal dispute of 1984–85 provides a good example. The miners have historically been among the most militant sections of the working class, so, confronted with a government clearly determined to run down the industry, the variability of the response to the call for strike action was surprising. For a full explanation of this we cannot simply 'read off' industrial militancy from social composition. Instead explanations must draw upon localised analyses of the history, working traditions and political traditions of specific places (Griffiths and Johnston, 1991; Sunley,

1990). Second, the urban 'riots' of the 1980s and early 1990s are sometimes cited as examples of links between place and mobilisation. But there are real difficulties with such accounts. Identifying and measuring a 'riot' are difficult enough. Even if 'riots' could be identified unproblematically there were many deprived places which did not witness riots, while those individuals who contributed to disorder could not easily be stereotyped (Keith, 1987). The proportion of 'rioters' was always small in relation to local populations and so explanations were more likely to lie in the spilling-over of conflicts which had, for some reason, become unmanageable, as in some parts of London where riots reflected and symbolised the gradual breakdown of relations between the police and sections of the local community, rather than economic disadvantage (Keith, 1993).

Byrne (1997, 24–5) speaks of a continuum of collective political action stretching from mainstream political parties at one end, through protectional (e.g. trade unions) and promotional interest groups, to social movements and riots and civil disobedience at the other. At one end of the spectrum action is motivated by a reformist ideology, is conducted through a relatively formal organisation, and employs conventional tactics. This contrasts with social movements which may be (superficially) less formally organised and employ unconventional tactics to promote their radical aims. Here, I start with an analysis of conventional elements of political behaviour (electoral geographies) before considering the geographies of nationalist movements. Given the burgeoning literature on social movements, the various strands of environmentalism are important; the extent to which they can be considered a social movement or a protest movement is discussed. I conclude by considering the different relationships to place of these forms of political mobilisation. Considerations of space prevent discussion of union membership and strike action (Martin et al., 1996), riots (Campbell, 1993; Keith, 1993) and non-participation (e.g. variations in electoral registration and turnout: Dorling et al., 1996).

10.2 ANALYSING ELECTORAL GEOGRAPHY: DOES PLACE MATTER?

Regional divides in voting patterns represent 'one of the most significant developments in postwar British politics', although their precise origins are disputed (McAllister, 1997, 641). By the mid-1980s the two major parties were either dominant in or largely absent from much of the UK, leading to speculation that region rivalled class as a political cleavage.

Historically, the assumption has been that because the two major parties were closely aligned with particular classes, spatial variations in voting patterns reflected the social composition of particular places; moreover, even when there has been a swing from one party to another, it has been assumed to be uniform. This theory of class alignment has been challenged. First, Sarlvik and Crewe (1983) postulated a process of class dealignment, as a result of which voters no longer voted consistently for the party associated with their class. Much discussion of dealignment focused on the decline in support for the Labour Party. Heath et al. (1985) argued, however, that this decline was a function of the changing socio-economic structure; after allowance was made for this, the classes were as electorally distinctive as ever. Subsequently, Heath et al. have modified this position,

conceding that some dealignment had taken place (Johnston et al., 1993, 68).

Given this evidence of dealignment, how does the electorate decide how to vote? One view is that retrospective and prospective voting takes place, involving conscious voter evaluation of the past performance and future promises and prospects of a party. Individuals are influenced by both national economic trends and by perceptions of their own personal economic situation. Voting is therefore related more closely to a party's performance and policies on salient issues. There is evidence, certainly from the 1983 election, that economic evaluations were more important than voters' social characteristics as predictors of voting.

A second view is that spatial polarisation is taking place. Initially dated to the late 1950s, it has been convincingly shown that this trend is not simply a function of changing social composition (e.g. differential migration). Figure 10.1 shows the scale of regional variations: at the level of standard regions it can be seen that in 1987 the mean deviation from the national share of the vote for the Labour Party was 9.3 per cent and 8.6 per cent for the Conservatives. Regional variations in voting peaked in 1987, but remained significant for Labour and increased for the Liberal Democrats.

This tells us that for each party there are high average deviations in the share of its support, but there is clearer evidence of variability and of strong associations between voting patterns and uneven development. Johnston and Pattie (1992a and b) show, using a geographical regionalisation, that over the 1979–92 period, the Conservatives consistently performed better in rural areas and in the south of the country. The North–South polarisation was especially marked during 1983–87, with Conservative support dropping by substantially more than average in the industrial North and Scotland. In the 1987–92 period the Conservatives performed relatively badly in the 'South', with the exception of inner London, while doing better than average in much of the North. The Labour Party's performance was the converse of the Conservatives: it performed better than average in much of the North (although it only

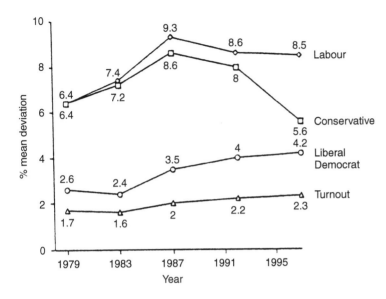

FIGURE 10.1 Regional voting and turnout, 1979–97 (McAllister, 1997) *Note*: Estimates are the mean deviation in the role (or turnout) from the Britain-wide figure, by region.

increased its share of the vote in four regions (Merseyside, Greater Manchester, the rest of the North West, and industrial South Wales)) while performing worse than average in the Conservative heartlands.

A functional regionalisation highlights some potential connections between uneven development and voting behaviour rather more clearly. In table 10.1 the regions are ordered so that the most depressed areas during the early 1980s are at the top. The figures are deviations from the grand mean in a multiple classification analysis of change in support for the Conservatives. Negative deviations indicate that, compared to their national performance, the Conservatives performed relatively poorly, and this is particularly clear for deindustrialised and inner city locations. On the basis of a similar analysis of the Labour and Alliance vote, Johnston and Pattie (1992a, 1503) argue that there was a North–South divide in voting patterns between 1979–87, but it did not survive to 1992, partly because of the Conservative recovery in Scotland and their relatively poor performance in South East England. The rural–urban divide came and went over the same period.

While the broad pattern suggests a strong association between voting behaviour and the extent to which places had (or had not) benefited from Thatcherite economic policies, this

must be qualified in several respects. Examining other contextual influences on voting in 1992, Johnston and Pattie (1992b) argue that much of the reduction in Conservative support took place in constituencies in the South East, due to rapidly rising unemployment prior to the General Election. The reductions were greatest, of course, in places with large Conservative majorities so that they lost comparatively few seats. Voting patterns also revealed that the electorally unpopular poll tax was a liability for the Conservatives in locations where it was high; this tax was symbolic of the Conservatives' disdain for local government and was highly regressive. However, the electoral cushion built up during the 1980s, in the form of large numbers of very safe seats, was not significantly affected nor eroded.

The consistency of the association, during the 1980s at least, between uneven development and voting, led Johnston *et al.* (1993, 67) to suggest that a paradox was emerging: greater voter volatility (dealignment) was co-existing with electoral stability (spatial polarisation). How could this be reconciled? There are problems in doing so because the data required to evaluate voter behaviour (e.g. whether retrospective or prospective voting is taking place) are usually only available from national surveys, which have insufficient respondents for allocation to

TABLE 10.1 Geographical and functional regionalisation of the Conservative vote, 1979–92 (a) Geographical regionalisation: deviations from the grand mean, by region, in the multiple classification analyses of Conservative percentage points change

Region	1979–83	1983–87	1987–92	1979–92
Grand mean	−1.81	−0.66	−0.54	−3.00
Strathclyde	−3.73	−3.70	2.52	−4.72
East Central Scotland	−1.61	−2.94	1.77	−0.39
Rural Scotland	0.32	−3.67	2.09	0.23
Rural North	2.21	−0.89	−0.46	0.71
Industrial Northeast	−0.58	−2.59	2.34	−0.30
Merseyside	−3.56	−5.56	0.77	−8.39
Greater Manchester	−1.81	−0.48	0.41	−1.58
Rest of Northwest	−0.78	−0.92	0.28	−2.14
West Yorkshire	0.79	0.31	1.30	2.71
South Yorkshire	−0.62	−2.07	2.08	−1.45
Rural Wales	1.85	−2.82	−0.11	−0.04
Industrial South Wales	−0.85	−0.28	0.13	−1.30
West Midlands Conurbation	−2.28	0.66	0.82	−1.08
Rest of West Midlands	1.24	0.42	−0.63	0.90
East Midlands	2.00	1.59	−1.67	1.66
East Anglia	1.93	1.88	−0.85	2.44
Devon and Cornwall	−1.56	−2.50	−1.86	−3.97
Wessex	1.30	0.37	−2.84	−0.92
Inner London	−1.85	1.17	0.60	0.68
Outer London	0.15	3.65	−0.87	2.40
Outer Metropolitan	1.41	2.86	0.04	3.82
Outer Southeast	1.08	0.92	−1.79	−0.25

individual constituencies. Accepting this, Johnston et al. (1993, 71) argued that dealignment and evaluative voting could be used together to account for spatial polarisation. If class dealignment is taking place, voters will seek clues as to how to vote; some of these will relate to national-level events (e.g. the alleged influence of the 'Falklands factor' in 1983), while others will relate to evolving patterns of uneven development. Perceptions of prosperity and of government performance will therefore be expected to vary spatially. Much of Johnston and Pattie's recent work has therefore been concerned with linking individual-level data, on the opinions of the electorate, to aggregate data on voting patterns.

First, they produce estimates of geographical variations in dealignment, by combining national survey data on the electoral preferences of different social groups, with census data at constituency level. The process generates maximum likelihood estimates of voting for each party in each constituency by each of six occupational classes. These estimates then allow analysis of spatial variations in voting, independent of the effects of social class, for the 1964–87 period. If class dealignment and greater evaluative voting were occurring, there should be evidence of spatial variation in the voting patterns of socio-economic groups. Figure 10.2 appears to bear out this supposition. It shows the trend over time in the ratio of Conservative: Labour voting in each of 22 regions for the skilled manual occupational class. The range of ratios has clearly increased: in 1964 there was about a threefold disparity

TABLE 10.1 (Continued)

(b) Functional regionalisation: deviations from the grand mean, by type, in the multiple classification analyses of Conservative percentage points change

Type	1979–83	1983–87	1987–92	1979–92
Grand mean	−1.81	−0.66	−0.54	−3.00
Metropolitan inner	−2.28	3.32	−1.91	0.80
Industrial or immigrant	−1.38	0.55	−0.57	0.18
Poorest immigrant	−3.59	4.36	−0.96	0.73
Intermediate industrial	−0.34	0.11	0.23	0.86
Old industrial or mining	−0.75	2.41	−0.56	1.81
Textile	−0.16	0.11	0.73	1.80
Poorest domestic	−1.52	2.70	0.51	2.33
Conurban local authority housing	−1.75	−0.23	0.24	−0.75
Black Country	−2.13	2.90	0.33	2.71
Maritime industrial	−2.50	−0.84	0.26	−1.59
Poor inner-city housing	−2.09	2.59	−1.18	0.15
Clydeside	−4.07	−0.71	1.84	−2.61
Scottish industrial	−1.86	−1.59	1.37	−1.26
Scottish rural	1.04	−3.68	0.77	0.22
High-status metropolitan	−2.09	−0.09	0.54	−2.23
Inner metropolitan	−3.21	1.37	0.22	−0.65
Outer London	0.33	2.16	−0.10	1.22
Very high status	1.14	0.54	0.43	0.61
Conurban white-collar	−1.02	−1.57	−0.18	−3.20
City: service employment	−1.20	−2.81	−1.33	−4.65
Resort or retirement	1.70	−1.10	−1.43	−2.15
Recent growth	0.21	−0.42	1.77	1.30
Stable industrial towns	−0.16	0.94	0.34	0.92
Small manufacturing towns	1.76	−1.55	−0.79	−1.35
'Southern' urban	1.81	0.99	1.00	2.58
Manufacturing or commuter towns	0.73	−0.78	0.14	−0.89
Metropolitan industrial	1.21	2.30	−0.34	2.93
Urban Scotland	−0.72	−4.51	3.93	−0.79
Rapid growth	3.03	1.42	0.23	3.52
Prosperous towns	2.12	−1.65	−1.34	−2.33
Agricultural	2.71	−1.05	−0.81	0.41

Source: Johnston and Pattie (1992a).

between the lowest and highest regions but the difference was sevenfold by 1987. Geographically, the northern regions have experienced 'trendless fluctuation' – there is little evidence of change in their voting patterns. By contrast the regions at the top of the diagram – those in which the skilled manual workers moved towards the Conservatives in ever greater proportions – cover all of South East and South West England, plus East Anglia, with the exception of inner London. In region 17 (the South West) this class gave roughly equal support to the two main parties in 1964; by 1983 the ratio was

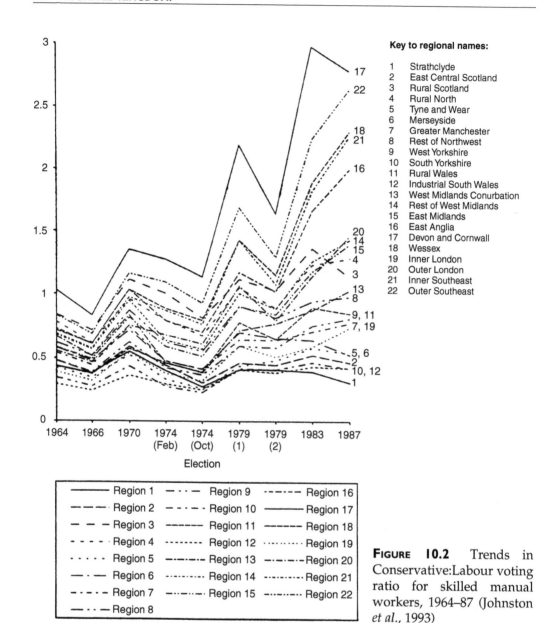

Key to regional names:

1	Strathclyde
2	East Central Scotland
3	Rural Scotland
4	Rural North
5	Tyne and Wear
6	Merseyside
7	Greater Manchester
8	Rest of Northwest
9	West Yorkshire
10	South Yorkshire
11	Rural Wales
12	Industrial South Wales
13	West Midlands Conurbation
14	Rest of West Midlands
15	East Midlands
16	East Anglia
17	Devon and Cornwall
18	Wessex
19	Inner London
20	Outer London
21	Inner Southeast
22	Outer Southeast

FIGURE 10.2 Trends in Conservative:Labour voting ratio for skilled manual workers, 1964–87 (Johnston *et al.*, 1993)

approximately 3:1 in favour of the Conservatives. These estimates offer strong circumstantial support for a regionally specific process of dealignment – at least if, by dealignment, we mean that particular classes vote in different ways in different regions. However, Weakliem and Heath (1995, 644) suggest that theories of dealignment imply that different classes have become more similar in their voting behaviour. If so, we would need to analyse not the changes over time of voting patterns in a particular class, but rather the voting behaviour of different classes. If there has been a regionally specific process of dealignment we would expect convergence between classes to have proceeded further in some regions than in others. In addition, simply focusing on the ratio of Conservative to Labour votes can be misleading; voting for third parties, which varies

greatly from place to place, can have a significant effect on this ratio.

A second important contribution by
Johnston and Pattie has been their attempts
to link survey data (e.g. the British Election
Survey) to data on voting behaviour or intentions. Voters' evaluations of their local
economies were not simply reducible to their
individual backgrounds. Voters of comparable social position were more likely to feel
that the regional economy had improved if
they lived in the south of the country rather
than in the north. Controlling for several
socio-economic characteristics of individuals,
it was clear that perceptions of the state of the
region's economy did significantly affect voting behaviour. Voters in more depressed
regions were more likely to perceive economic deterioration and project it into the
future; in contrast, voters in booming regions
were more likely to be positive than negative
about local economic prospects (Pattie and
Johnston, 1995, 4–5). Such divergent perceptions were even more marked where voters
held the government responsible for regional
decline. For example, those who felt that
their region had experienced relative decline,
and who held the government responsible,
were almost 5.6 times less likely to support
the government than those who thought the
regional economy had improved but who
did not believe that government policies had
had an impact. This implies that 'voters were
thinking in terms not just of their personal
well-being ... but they were also thinking in
terms of their local contexts' (Pattie and
Johnston, 1995, 26–7). The regional cleavage
in voting behaviour appeared to persist even
after controlling for individual characteristics
and regional economic evaluations: voters in
Scotland, Wales and northern England were
significantly less likely to support the Conservatives even after these characteristics
were taken into account. Such analyses do
seem to demonstrate the independent effect
of regional context on voting.

Nevertheless, despite their technical
sophistication, these arguments perhaps
leave unresolved the question of precisely
how place impinges on political behaviour.
Several possible explanations have been postulated. First, there is the argument that

voting patterns reflect the geography of
social class, but the evidence summarised
here does not support that view. Second,
some have postulated a neighbourhood
effect; the social milieu in which people live
leads them to vote in certain ways. There is
some empirical support for this: working-
class people living in middle-class areas are
much less likely to support Labour than
working-class people in working-class areas
(Abercrombie and Warde, 1988). However,
the processes whereby this happens are
opaque: even in socially mixed neighbourhoods, classes do not interact on the scale
necessary to achieve 'conversion by conver-
sation'. Third, there is the notion of local
political cultures. Most frequently used to
explain associations between rurality and
Conservatism, or between mining areas and
support for Labour, this concept is ahistorical
and lacks purchase on contemporary features
of the social structure (Savage, 1987)
although there is some strong evidence for it
in, for example, the persistence of high levels
of strike activity in certain regions (Gilbert,
1996).

We must look elsewhere for explanations
of the influence of context on voting. There
are differences of opinion regarding the most
appropriate spatial scale for such an analysis.
McAllister (1997, 646–50) presents an
analysis of standard regions and shows that,
controlling for class, there are significant
regional variations, but this does not get us
very far: we need to show how economic
change at the level of large regional units
might impinge on decisions of voters.

Savage's (1987) assessment remains one of
the most significant elucidations of the effect
of 'place' on voting patterns. He argues that
individuals can earn money through both
housing and labour markets, and that they
will be acutely aware of their *position* in
both, and of the *dynamics* of both (e.g. trends
in property prices or unemployment levels).
Two localities with an identical occupational
profile may be going in opposite directions – one attracting new investment in
services, the other losing jobs through
deindustrialisation – and these dynamics
may not be captured by ecological variables
for one point in time (note, however, that

some of Johnston and Pattie's work does incorporate change, e.g. in unemployment levels). Voting is determined by people's monitoring of local developments, especially in the labour and housing markets, and their perceptions of the quality of local services and the environment. This might help to explain why individuals in the same occupational class vote in quite different ways in different places. Johnston and Pattie's explorations of the links between perceptions of a community's economic situation and voting behaviour are one attempt to investigate this. However, there are objections to this. Heath *et al.* (1991) argue that, rather than an individual's vote being determined by their degree of optimism about the local economy, their political beliefs determine their degree of optimism. Thus, discovering an association between perceptions of the local/regional economy and voting behaviour is therefore simply a way of redescribing people's political beliefs, not explaining them. Irrespective of one's position on this question, what is suggested by these analyses is a more nuanced and subtle account of voting. Instead of reading off votes from class position, or attributing spatial variations in voting to (assumed) external and invariant characteristics of places, the stress is on voting as an outcome of conscious political choice. Established political alignments have been eroded; individuals may adopt a range of attitudes according to the issue in question. These attitudes are not simply determined by their social or geographical location; nor is the way individuals vote.

Finally, these comments are supported by the results of the 1997 election (Pattie *et al.*, 1997). This might seem surprising given the massive national swing to Labour, but it was the regionally specific pattern of anti-Conservative voting which damaged them most. The Conservatives suffered few losses in Wales and Scotland, where their vote was already low. By contrast they experienced substantial losses in London and the South East, where their vote declined by 14 per cent and 13 per cent, respectively (McAllister, 1997, 642–3, table 2). Even so, and providing additional support for Johnston and Pattie's contentions, the Conservatives were still over-represented in the South East; despite a national collapse they still had a bedrock of support there. Other region-specific influences which had a disproportionate effect on the outcome included the potential for tactical voting, which was regionally concentrated. Conservative seats where there was either a Labour or Liberal Democrat candidate in second place were disproportionately found in the South East (96 per cent of Conservative seats with potential for tactical voting), South West (77 per cent), and East Anglia (87 per cent). There were fewer such seats in other regions or (due to nationalist candidates) in Scotland or Wales. There were particularly high levels of tactical voting not so much in constituencies where Labour was second as in areas where the Liberal Democrats were second. In the most marginal seats, where the Liberal Democrat challenger had been within 5 per cent of the Conservative MP in 1992, the Liberal Democrat vote was 22 per cent higher than would have been expected, while Labour's vote declined. Thus future election campaigns may have to 'deal with highly sophisticated regional electorates who vote instrumentally in order to achieve a very specific outcome' (McAllister, 1997, 655).

10.3 NATIONALIST MOVEMENTS

Many commentators suggest that the nation-state is outmoded, and that, due to pressures from above and below, we will witness a break-up of existing nation-states. Nairn (1977) argued that peripheral nationalisms would be the 'grave-digger' of the British state. He was, of course, writing at a time when nationalist movements had achieved a high degree of visibility, and at a time when several West European states seemed to be struggling to contain separatist demands. What is the current strength of contemporary peripheral nationalism in Britain?

Nationalism has been described as perhaps the key social movement of the twentieth century. The goal of nationalist movements is to achieve a correspondence between nations and states. In this process, we should acknowledge that nations are essentially 'imagined communities' (Anderson, 1983); they are not primordial entities which have always existed, based on ties of ethnicity and language. Nationalist movements typically develop myths about the nation, and this gives nationalism its Janus-faced character (Taylor, 1989, 200); nationalism usually involves both looking backwards to a romanticised past and looking forwards to a better future.

The rise of nationalism has been explained in various ways. Nairn (1997, 66) sees it as a 'necessary response to industrialisation' because nationalist ideologies could create a sense of common purpose in circumstances of 'generally and chronically uneven development'. This is one explanation for the spate of nation-building movements in the nineteenth century, and for the resurgence of nationalisms in the former communist states. It has also been used to explain territorially based protest movements in Western Europe. However, uneven development on its own cannot explain such movements; it is not difficult to think of economically advanced regions in which separatist or autonomist demands have been articulated; conversely, many lagging regions have not developed territorially based movements.

An alternative, proposed by Hechter (1975), argues that nationalism in the Celtic fringe also results from uneven development, as a consequence of internal colonialism. The centralising colonial powers are insensitive to, and indeed actively exploit, cultural differences; ultimately there is a reaction against this process from the peripheral nations. One theory, then, relies upon uneven development, the other on a combination of uneven development and cultural marginality. Both may be too simplistic in terms of understanding recent nationalist movements in the UK. In particular, in order to understand *peripheral* nationalism, we need to understand *British* nationalism. Relative economic decline made appeals to the national interest a matter of urgency (McCrone, 1992, 208). But such appeals were based upon a sense of nationality which was firmly rooted in the English South, and which was defined, furthermore, by Britain's external relationships to its

Empire. This imperial identity had helped to counter competing national identities from the Celtic fringe, but was no longer able to provide such integrative force as it once had. Against this background we may understand the re-emergence of nationalism not simply as the assertion of peripheral discontent but as a reflection of the contradictions of a centralised, unitary state, and of a lack of correspondence between nation and state; the key to understanding therefore lies as much at the centre as on the periphery (McCrone, 1992, 211).

The resurgence of peripheral nationalisms in the UK from the late 1960s contained elements which appeared to bear out both of the foregoing theses. Linguistic decline in Wales was one catalyst: another was a more general response to the perceived threat of cultural domination due to the importation of mass (American) culture. Economically, the decline of formerly prosperous regional economies, and increased penetration by multinational companies, produced growing awareness of the degree of external control of the economy, and a realisation of the limited powers of established regional elites. The politicisation of regional problems, through the establishment of regional planning councils, contributed to awareness of regional uneven development especially when the state proved unable to solve the problems it was addressing (Anderson, 1989, 43–4).

The case of Scottish nationalism illustrates some of the difficulties of establishing connections between place and politics; according to conventional theories, it should never have happened at all. There are no fundamental linguistic or religious divides between Scotland and England, nor is Scotland distinctive in terms of ethnicity or culture. In terms of Hechter's thesis, it is difficult to argue that Scotland has suffered from the imposition of an alien culture because, while Scotland did not retain statehood after the Act of Union, it nevertheless preserved distinctive and autonomous institutions (e.g. its legal and education systems) which have been jealously guarded. The breakthrough by the Scottish Nationalist Party (SNP) in the late 1960s was inspired more by concerns about the benefits of

North Sea oil. This does not accord with theories which attribute nationalism to underdevelopment.

As a consequence, support for the SNP has varied considerably and is somewhat unpredictable. The party achieved a low share of the vote during the 1950s and 1960s, but experienced a surge in support which peaked at 30 per cent in the general election of October 1974. The failure of the devolution lobby to achieve a breakthrough in the 1970s was followed by a rapid decline in SNP support, to 11.3 per cent in the 1983 election (figure 10.3; Brown et al., 1998, 154). The subsequent revival of nationalist fortunes has been based on a reorientation of strategy and tactics. The SNP's rise in the 1960s was attributed to their appeal to upwardly mobile individuals who no longer had strong class loyalties. By the 1980s and 1990s, SNP support was more working class, as a result of two trends: the SNP had shifted to the left, in competition with Labour; and Scottish politics had moved in a generally anti-Conservative direction. Thus the principal electoral battle in Scotland has been between the SNP and Labour for the working-class vote; both parties drew over 50 per cent of their support from manual workers. In attempting to broaden its appeal, Mitchell (1996, 294 and 299) suggests the SNP has emphasised creating a 'decent Scotland'. There was a widespread perception that the distinctive Scottish components of civil society were under attack from a centralising administration opposed to welfarism and collectivism. It is little wonder, therefore, that piloting the socially regressive Poll Tax in Scotland generated such resistance (McCrone, 1991). Mobilising opposition to Thatcherism therefore often took the form of efforts to preserve Scotland's distinctive institutions. This could become what Harvie (1994, 217–18) describes as a civic, as opposed to an ethnic nationalism, which would emphasise 'good government, responsibility, self-discipline, and organisations to make a richer common culture and a sense of independence'. There remain challenges to this vision, from fringe nationalist organisations which adopt a virulently anti-English platform, but these are no more than echoes of nationalisms's atavistic heritage (Mitchell, 1996, 299).

The extent to which the SNP have been successful is evident only partially from its share of the vote, which rose slightly between 1992–97, from 21.5 to 22.1 per cent; this placed them ahead of the Conservatives but was still half Labour's share of the vote. Support for the SNP is not structured very strongly by the socio-economic characteristics of communities; in general the party draws similar support in all types of constituencies. This says much about the cross-class appeal of the party, but under the present electoral system this prevents the SNP from winning many seats (Denver, 1997, 27). Stronger indications of demands for

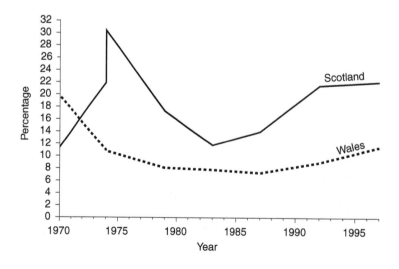

FIGURE 10.3 Nationalist voting in Scotland and Wales, 1970–97

autonomy and support for the SNP's agenda were provided by the referendum of 1997: of those voting, 74 per cent supported a Scottish Parliament and 63 per cent voted to grant it tax-raising powers. Support for both questions on the referendum was strongest in central Scotland, and the Western Isles, and weakest in rural Scotland and the Borders (Pattie *et al.*, 1998, 10–12).

Support for nationalism and autonomy in Wales has always been more equivocal than in Scotland. Linguistic decline and external control of the economy motivated an upsurge in nationalist sentiment in the 1960s but these factors appealed to quite different groups in different parts of Wales. Plaid Cymru's main support at that time came from the Welsh-speaking rural west; the party failed to make headway in industrial South Wales. Since the south dominates Wales in terms of population, this has meant that Plaid's share of the vote has rarely exceeded 10–11 per cent (figure 10.3), and the 1979 referendum saw a decisive rejection of devolution proposals. The 1997 referendum result thus 'signals a significant shift in the political mood of Wales' (McAllister, 1998, 149). In 1979 only 20 per cent of those voting were in favour of devolution; in the 1997 referendum a narrow majority (50.3 per cent) were in favour, albeit on a 50 per cent turnout. Significantly, certain areas in (de-)industrial South Wales voted strongly in favour (e.g. Neath/Port Talbot, Rhondda, Cynon, Taff: McAllister, 1998, 161) in contrast to 1979. For this reason, efforts to interpret the pattern of support in terms of a 'three Wales' model must be treated with caution. This view sees Wales as divided into three areas: British Wales (the Borders, Pembrokeshire, Newport and Cardiff); Welsh Wales (the southern valleys plus Swansea); and the Welsh-speaking heartlands (Giggs and Pattie, 1992). However, numerous vagaries and anomalies in the pattern of support cut across such straightforward geographical splits, not least because minor shifts in votes in some locations would have changed the whole result. Reflecting the relative strengths of devolution demands in Scotland and Wales, the proposed Welsh Assembly will have fewer powers than the

Scottish Parliament. However, the arguments for the Assembly – its emphasis on democracy and partnership – share common ground with equivalent arguments in Scotland, being rooted in critiques of Conservative centralisation and patronage.

The question of Ulster is also inextricably bound up with national identity and nationalist (and Unionist) aspirations. The exhaustion of military attempts at reconciliation, and the Herculean efforts involved in the peace process, illustrate the difficulties of reconciling the aspirations of all parties. Both sides are divided: Nationalists between those seeking a united Ireland and those prepared to reach an accommodation with the UK; Unionists between 'integrationists' (who see Northern Ireland as a physically separate but fully integrated region of the UK) and 'devolutionists' (who see Ulster as a distinctive place deserving a degree of self-determination) (Graham, 1997, table 10.1). Perhaps, paradoxically, the character of Unionism is most interesting. For Unionists neither desire unity with what they would perceive as a sectarian Irish state, nor do they wish to rely indefinitely on a Britain that is ambivalent to – indeed has little strategic interest in – Northern Ireland. This dilemma has forced Unionists to search for symbols of identity around which to mobilise, such as evidence of the cultural distinctiveness of Ulster which would justify the claim of Protestants to that territory. That search, however, has thrown up as many problems as solutions: it has, for example, pointed to the Irishness of Ulster Protestants, and connected them to a narrative of separateness from England (Graham, 1997, 208). Other social cleavages (notably class, but also urban–rural divides) further complicate and cut across the primary sectarian conflict in Ulster.

The intractability of the Northern Ireland problem reflects the legacy of sectarianism in which both sides have become fixated with an exclusive territoriality attained either through an independent all-Ireland republic or a purely British Northern Ireland. In reality these are 'no more than mutually unattainable bargaining positions' (Anderson, 1997, 233), but the result is that nations and states will never coincide, in this case, without unacceptable violent conflicts. Europeanisation and globalisation have weakened, but not eliminated, the nation-state, reducing its appeal as an ideal. The Single European Market has led to calls for greater economic cooperation between North and South. The peace process has also given impetus to cross-border collaboration, which Anderson (1997, 234–5) suggests, with cautious optimism, might prefigure 'new types of institutions which could cope with the contradictions between national territory and transnationalisation'.

With this in mind it is interesting, finally, to speculate on the ways in which the political geography of Scotland, Wales and Northern Ireland might be transformed as a result of devolution. Pattie *et al.* (1998) envisage a 'new style of politics' in assemblies elected by forms of PR. More significantly, the results of the devolution referenda have, at least in Scotland, further boosted separatist demands: by the spring of 1998 surveys showed the SNP level with Labour in voting intentions. Nairn (1998, 41–3) argues that this is indicative of a desire for egalitarian policies and for greater Scottish control over Scottish affairs which, in the absence of further constitutional reform for the whole of the UK, cannot be satisfied within the existing structures of the British state. On this view the break-up of Britain is inevitable, though perhaps for reasons other than those suggested in Nairn's earlier writings.

10.4 ENVIRONMENTALISM
AND GREEN POLITICS

There are conflicting assessments of the scale and significance of environmental issues in British politics. Some argue that evidence for a cultural change in public attitudes is almost beyond dispute, but others point to the tendency for environmental issues to feature on the public agenda sporadically in response to specific events, and to the evident divisions on attitudes to the environment. It is difficult to measure support for environmental issues since so many groups are involved, while action takes many forms ranging from green consumerism, to direct action protests.

There are three major types of environmentalist awareness and activism (Norris, 1997). First, there are many groups concerned with 'old green' issues of countryside preservation, exemplified by the National Trust, the Council for the Preservation of Rural England, and the outdoor recreation and access lobby. Relying on conventional political channels, these groups combine environmental concerns with leisure activities, and usually place faith in technology and government to minimise damage to the rural environment.

Second, there is a 'new green' dimension, reflecting the expansion of more radical organisations such as Friends of the Earth or Greenpeace. These bodies share a concern for environmental protection, and therefore work with established groups, but they take a more radical stance on technological developments and on the need for a more sustainable trajectory for economic development. They are most well known for various high-profile campaigns on specific issues. More recently direct action campaigns often coordinated by fringe environmental groups – Earth First!, or ALARM – have been focused on issues such as road construction (Carter, 1997) and on the transport of live animals and vivisection. Third, Norris identifies an anti-nuclearist strand to environmentalism, exemplified by CND and the protests at the installation of American Cruise missiles in the UK. She also showed that the most significant predictors of support for environmentalism were education and age; by contrast, income, class and gender were insignificant (Norris, 1997, 332).

There is evidence of public concern about environmental issues but the British Social Attitudes surveys reveal that while people might be 'green in word' they are not 'green in deed' (B. Taylor, 1997). Greatest concern is expressed about water pollution (seen as 'extremely' or 'very' dangerous by some three-fifths of survey respondents), air pollution and the greenhouse effect. Exhaust fumes in towns and cities are seen as 'very serious' by 63 per cent of respondents. When asked about threats to the countryside, road building and pollution are seen as of greatest significance, with worries about the former apparently rising.

There appears to be a willingness, in the abstract, to make sacrifices; at least 60 per cent agree that both businesses and individuals need to do more to protect the environment, even at the expense of jobs and lower prices (Taylor, 1997, 119). This support proves fragile when specific measures are considered. For example, despite strong awareness of pressures on rural environments, only a quarter agree that restrictions on the numbers visiting the countryside are needed. This may indicate unwillingness to countenance restrictions on personal freedom. Perhaps not surprisingly, then, attitudes to potential environmental protection measures show that support is greatest for measures which would have direct personal benefits: 79 per cent support public grants to improve home insulation. Least popular are measures which involve reducing energy use or rationing it (e.g. only 13 per cent support rationing petrol). There is, likewise, strong support for improving public transport (68 per cent) but only 28 per cent think it very important to reduce their car use, notwithstanding high levels of concern about vehicle emissions. And even the support for improved public transport wanes when people are asked how it would be financed, with only around a third of the respondents favouring road pricing or cuts in road expenditure to fund improved public transport (Taylor, 1997, 127). Thus, despite the growth in pro-environmentalist attitudes, people were reluctant to match words with deeds.

It may be indicative of the ambiguities of environmental attitudes that this growth in support for environmental issues has not been translated into success at elections, apart from a very small number of parish or local government elections. Typically, in the 1970s and early 1980s, the Green Party only polled a fraction of the votes cast (1.4 per cent in 1987). Thus the Greens' performance in the 1989 European elections, when they received 2.25 million votes (14.9 per cent of the total) was remarkable. This result is not just an historic psephological curiosity; it highlights the reasons why, under the present electoral system, the Greens will make little headway in British politics.

The most obvious explanation for the success of the Greens in 1989 would be a sudden increase in the public's awareness of environmental issues, but the level of membership of nature and countryside protection organisations was well ahead of continental levels from at least the 1970s. This did not translate into radical political activism for several reasons (Lowe and Flynn, 1989): the concern about mass unemployment and deindustrialisation; the fact that the effects of river pollution and acid rain were generally less visible in Britain than elsewhere; and the focus of major environmental groups on a formalised policy process which has effectively incorporated them.

A revival of environmental concern in the late 1980s was linked to the effects of a development boom in South East England. Mrs Thatcher's ostensible conversion to environmental issues in 1988 lent some respectability to environmental concerns, while concerns about the safety of food and water, partly against the background of privatisation, all pushed environmental issues up the political agenda (Rootes, 1991, 291–2). The result of June 1989 was not, therefore, a surprise – though a turnout of 30 per cent meant that only 4.5 per cent of the electorate voted Green. This was seen as a protest vote, which is more likely in a 'second-order' election (such as those for the European Parliament) which does not affect the composition of the government. The Greens took over the role of the Liberals as recipients of protest votes; they were able to exploit 'deep disillusionment' with the Conservatives, and various contingent factors were also at work (see O'Neill, 1997, 288–9).

These factors, however, do not explain the spatial pattern of support for the Greens, which displayed a 'remarkable' consistency: highest in the more affluent, semi-rural parts of southern England, and steadily declining towards the North. Support ranged from an average of 20 per cent in 25 Southern divisions, to 12 per cent in the 21 divisions in the north. The Greens' poorest performance in England – three divisions where they received less than 10 per cent of the vote –

was in the North East. A strong correlation was noted between Green support and the proportion of the middle class. Pattie *et al.* (1991, 286) suggest this indicates support for a 'post-materialist' political agenda. This agenda includes feminism, racism, nuclear weapons and power, and the environment; support for it was claimed to be especially strong among the service class. Regression analysis indicated that constituencies in Scotland, Wales, the North, and Greater London all supported the Greens less strongly than would be predicted from their social composition. This analysis provides support for the view that development pressures and environmental concerns were important in explaining the above-average Green vote in the South East. Further investigation of environmental attitudes confirmed this: regardless of social status, residents of Southern England and the Midlands tended to be more pro-Green on issues of countryside protection, compared to their Scottish, Welsh and Northern counterparts (Pattie *et al.*, 1991, 295–6).

The subsequent decline in the Green vote illustrates the problems it faces in Britain. The Green Party was unable to build on the 1989 result and by 1997 fielded only 84 candidates, receiving a total of 54 000 votes. There are several reasons for this. First, the electoral system makes it virtually impossible for new or minority parties to obtain fair representation and so the British Greens have not emulated the success of their German counterparts. Second, the Liberals have projected themselves as the guardians of the environment and have been instrumental in pursuing environment-friendly policies in local government. However, others contend that Britain is exceptional in that 'new social movements' have not developed to the same degree as in continental Europe (Rootes, 1992, 174). One reason for this is surely, as Bagguley (1995) argues, the weakness of the theories of the post-materialist 'new middle class'. If the service class (and especially the public sector service class) were so committed to a 'post-materialist' agenda, then we would expect much more support for Green parties. As it is, low levels of support indicate that support for the post-materialist agenda

is either not strong, confined to a smaller range of social groups than expected, or takes second place (in general elections at least) to productivist considerations.

Finally, a notable feature of environmental politics in the 1990s has been the growth of direct action, especially over road construction and animal rights, which has involved radical and unconventional (sometimes illegal) tactics, yet has also secured middle-class support. Anti-road protests in eminently 'respectable' communities – Bath, Winchester, Newbury – have seen the unusual spectacle of local residents supporting unconventional actions (tree houses, 'digging-in' of protesters into tunnels under roadworks). Such campaigns have been initiated by radical organisations such as Earth First! which deliberately distances itself from other environmental organisations (Byrne, 1997, 145–50). Protests over animal rights – vivisection, live animal exports, and hunting – have seen a similar tension between conventional campaigning and direct action. Clearly we are moving towards the more radical and unconventional end of Byrne's (1997, table 2.1) spectrum of mobilisation, but whether these new developments represent embryo social movements is debatable. They are better seen as protest movements because of their focus on single issues; by contrast, social movements espouse comprehensive and radical ideologies which reject prevailing political, economic and social norms (Byrne, 1997, 166). On this criterion organisations like Friends of the Earth, which emphasise far-reaching alterations in social organisation, would come closer to qualifying as a social movement.

10.5 CONCLUDING COMMENTS

The foregoing are only illustrations of many possible connections between place and politics. They illustrate some of the ways geographical thought has developed in recent years. Electoral geography, for example, has produced sophisticated quantitative analyses of associations between socio-economic contexts and voting behaviour, but such extensive studies have been criticised for their limited insights into individual voting decisions. One attempt to resolve this has been via intensive, locality-based studies (Savage, 1987), reflecting influential developments in human geography in the 1980s (chapter 1). Studies of nationalisms initially relied on rather deterministic models which linked politics somewhat unproblematically to uneven development. These are now being supplanted by work which pays attention to the construction of political identities and the multiple and overlapping nature of such identities. The consequence is, as environmental politics shows, that individuals may adopt apparently contradictory positions depending on the issues at stake. Numerous authors have commented on the growth of single-issue politics in response to a range of risks to which individuals are exposed. This is in sharp contrast to the aggregate and unidirectional accounts of political mobilisation which have exercised many geographers.

This also raises questions about what a revived political geography might study. Painter (1995, chapter 6) makes a case for the study of new social movements. Instead of emphasising the autonomous influence of contextual factors on political behaviour,

studies of such movements would emphasise the conscious strategies through which resources are deployed to achieve wider goals (Painter, 1995, 178). A notable feature of such activities (whether or not they are termed 'social' movements or 'protest' movements) is that they come together at particular places. They may also involve a fluid politics of mobility as individuals move around to lend support to particular campaigns, in contrast to the somewhat static geographies discussed in this chapter. Examples might include peace protests and anti-road demonstrations (Carter, 1997). By their occupation of threatened environments and skilful use of the media, such groups can dramatise issues and mobilise support which cuts across conventional alignments. Carter (1997, 201) argues that anti-road protests did have an impact on the scaling-down of road construction, but the evanescent character of such protests raises questions about how far they might achieve more widespread support. On the other hand, Grove-White (1997, 117) sees such protests as 'graphic new bearers' of a growing concern about deeper cultural issues at stake for society in the field of 'the environment'.

11

SUBNATIONAL GOVERNMENT IN THE UK

RECONCILING FUNCTION, AREA AND LOCAL DEMOCRACY

11.1 INTRODUCTION

Consideration of local government in texts on Britain's human geography has largely been confined to a discussion of institutional arrangements and administrative reforms, without analysing how and in what ways the political geography of the country conditions the process of change. There are both technical and political reasons why some form of subnational government is necessary, and discussions over the most appropriate spatial form for such institutions demonstrate the tension between these two sets of criteria. Thus administrative maps are never merely spatial structures (Taylor, 1991b), for they raise issues of legitimacy and accountability, issues which are resolved in different ways at different times.

There are sound technical reasons why subnational units are necessary. Local needs simply cannot be assessed from a central government office. There are also political reasons, given that finance for local government services is (at least in part) raised locally; unless one is a committed centralist, it is unreasonable to expect taxation without representation. In determining the scale of subnational units, the problem – as Harold Laski put it – is of reconciling function and area (Sancton, 1976). But there are also quintessentially geographical reasons for changes in spatial structures of subnational government. Structures established in one era may

prove inadequate to cope with emerging patterns of uneven development. For example, during the nineteenth century the pace of urbanisation and industrial change far outstripped the pace of reform of local government structures. A confusing and chaotic assemblage grew up, in which there was variability, inefficiency, duplication and local rivalry, with the concomitant possibility of inequity (Cochrane, 1992, 5). Of course, this was also an era of large-scale municipal enterprise and innovation (Marr, 1996, chapter 2). Nevertheless, there were concerns about the correspondence between this patchwork quilt and emerging patterns of economic activity, notably in the inter-war years. One reading of the post-Second World War history of subnational government would emphasise a technocratic search for rational forms of administration, culminating in the Maud Commission's (1969) report. Local government needed to be organised in units large enough to meet the technical and administrative requirements of the services it administered. Failure to meet these requirements would undermine faith in local government. The Maud Commission therefore recommended single-tier administrations comprising, generally, urban 'cores' and surrounding functional regions. This was rejected by the 1970–74 Conservative government, who substituted a two-tier system, including the (subsequently abolished) metropolitan counties. Although the metropolitan counties were a welcome innovation, the

1974 structure was criticised as an ill-judged compromise instituted with a view to preserving the Conservatives' electoral base. The government was further criticised for interfering in the process of boundary delimitation. Critics felt that the metropolitan counties were not large enough to facilitate their strategic role of administering a conurbation, and that this was an attempt to preserve as much as possible of the historic counties rather than to create appropriate regional units for the late twentieth century.

The criticisms of the 1974 reorganisation underline the point that decisions on the spatial structures of subnational government do not simply reflect the working out of a rationalist logic. For several reasons the past two decades have seen considerable challenges to local government in the UK. There has been political conflict with a government widely seen as centralising and antipathetic to local authorities; economically, questions have been raised about the scope for autonomous local government in an era of globalisation; and the combination of the two has forced a reappraisal of the way local government is theorised.

For these reasons, the chapter begins by examining some contemporary theories of subnational government: these can best be summarised by the phrase 'from local government to local governance'. This is followed by a discussion of recent policies towards local government, and of the character and significance of the non-elected local state. Finally there is a more prospective discussion of the arguments for and against various forms of regional government and devolution.

11.2 FROM LOCAL GOVERNMENT TO LOCAL GOVERNANCE

There is a range of competing theories of subnational government which, for simplicity, can be divided into liberal and Marxist approaches. Both shed light on aspects of recent change in subnational government in Britain.

Liberal theorists – using 'liberal' in its clas-

sical sense, to denote resistance to any restraint on individual liberty – praise local government as a vibrant example of democracy and as a check on centralised state power. In the form of 'public choice' theory, liberal accounts were highly influential on the policies of the Conservatives. For public choice theorists, democracy is best promoted where citizens are essentially customers with market power and are able to migrate to locations which offer packages of goods and services which best meet their needs This is an argument for decentralisation, for large numbers of competing local authorities, and for financing mechanisms which render transparent the costs to the electorate of voting for different political parties. Such considerations weighed heavily with the Conservatives in devising alternative forms of local government financing (see Hepple, 1989). However, public choice theorists have been criticised: decentralisation can lead to fragmentation and therefore inefficiency, while citizens are simply not as mobile as the theory suggests.

Opponents of public choice theory contend that the role of subnational government is not simply to compete in a marketplace to deliver services to the electorate. Instead, theorists emphasise the role of the local state in defusing social conflict (through its welfarist responsibilities), managing uneven development, and insulating the central state from pressures from antipathetic local electorates. A strength of Marxist analyses, in particular, is their emphasis on the relationships between local government and the changing geographies of production. Harvey (1989b) identifies a shift from managerialism to entrepreneurialism, and a parallel transition from government to governance. The roots of this, Harvey argues, have 'substantial macro-economic roots and implications'. A combination of globalisation and the declining capacities of the nation-state have meant that decisions on investment increasingly take the form of a negotiation between international finance capital and local powers doing the best they can to maximise the attractiveness of 'their' locality for capitalist investment (see Cooke, 1989).

These changes undermine the traditional

managerial role of local authorities, and require them to engage in *entrepreneurial* activities designed to attract mobile capital. Harvey also emphasises questions of governance, rather than institutions of government *per se*, because the real power to reorganise urban life lies outside local government and within a broader coalition of forces, in which democratic urban government and administration have only a facilitative and coordinating role to play. He sees emerging a new entrepreneurialism, based on public–private partnerships, integrated with the use of local government powers to try to attract external funds. The activities of institutions of local governance are more likely to be speculative and *ad hoc* in their execution and design, rather than being rationally planned and coordinated. However, their distinctive characteristic is that risk absorption is undertaken by the local state rather than by the private sector. Third, there is a focus on a political economy of *place* rather than of *territory*: this kind of speculative activity is not intended to have a territory-wide effect (though, ideologically, it is usually claimed that its benefits will 'trickle down'). However, such place-based projects typically become such a focus of attention that they divert concern and resources from broader problems that beset the region and territory (see also the comments on urban regeneration policies in chapter 12). What is interesting about recent developments in urban governance in the UK, for instance, is the extent to which a range of unelected, *ad hoc* coalitions claiming to represent particular places now appear to be taking the initiative in promotional activity (section 11.4).

These arguments can be connected to broader debates about an alleged transition to 'post-Fordism', which, some have argued, is producing a 'hollowing-out' of the nation-state, the powers of which are displaced either downwards (to local states), upwards (to supranational organisations) or horizontally (Jessop, 1993; Goodwin and Painter, 1996). This is said to be driven by changes in the organisation of production which require more flexible, decentralised modes of governance. There are some clear parallels between the concept of a regime of 'flexible

accumulation' and recent changes in the public sector in Britain (Stoker, 1989; Burrows and Loader, 1994) but some are sceptical about whether these are such as to support claims that a genuinely 'post-Fordist' local state has emerged (Cochrane, 1992). Others make more definitive statements, apparently discerning a post-Fordist local state in the recent reforms of local government (Johnston and Pattie, 1996). These various theoretical strands are woven into the discussion which now follows.

11.3 CENTRAL–LOCAL RELATIONS SINCE 1979: CENTRALISATION, ABOLITION AND LOCAL GOVERNMENT REFORM

Mrs Thatcher's colleagues 'were frequently shocked by her vehemence' on the subject of local government (Jenkins, 1995, 41). The 1980s were therefore marked by constant conflict between centre and locality. There were two principal strands to Conservative strategies: an attempt to gain strict control of spending, which led not just to greater central control of finance but also, ultimately, to abolition of a tier of local government; and attempts to bypass what were seen as ideologically unpalatable local administrations. This was not without opposition from within the Conservative Party. Thus Edward Heath argued, in the Commons, that he had come into the House on Winston Churchill's pledge to 'set the people free ... not to set them free so that they could do what we tell them' (quoted in Duncan and Goodwin, 1989).

It is tempting to see post-1979 policies as a transparently personal and political attack on left-wing, urban local government in particular, but that is to underestimate the extent to which these policies were motivated by wider strategic considerations. Peck argues that the government intended to reconstruct the institutional infrastructure through which market forces would operate. A good illustration of the general tactics was given by John Patten (sometime Minister for Urban Affairs):

the public sector dominated municipal solutions of the past have simply not worked. ... *because they were based on a misunderstanding of what makes cities successful.* We should draw a line under them. *Cities grew and flourished because of private enterprise and civic pride;* it is private enterprise, backed by helpful, direct and concerted government action, that will renew them.

(quoted in Peck, 1995; emphases added)

Put another way, the market had to be liberated from the restrictive grip of state planning and bureaucracy. This was not just a matter of freeing up the market, but of allowing it to be policed by new institutions in different ways. Other motivations included a concern to ensure that the electoral decisions of communities were reflected directly in the burden of taxation borne by individuals (which motivated a series of changes in the financial basis of local government).

Centralisation

The crucial issue here is control over the financial bases of the local state, and in particular the implications of various forms of local government financing. Prior to the Conservatives' reforms, there had been considerable dissatisfaction with the rating system of property taxation. Rates were regarded as inequitable, since they were based on the imputed value of property regardless of means to pay, and since those who did not own property paid rates only indirectly, through their rents. Thus, the government's argument ran, there was a need to find a more accountable way of financing local government services, such that the electorate had a clear indication of the consequences of voting for a particular party. This argument was, in outline, consistent with the views of public choice theorists.

Duncan and Goodwin (1988 and 1989) argue that in order to maintain the pretence of democracy the Conservatives had to leave local government with its own tax-raising powers, but this led to controversy over the extent of these powers. Initially there were limits on central government grants to local authorities, followed by a system of rate-

capping – preventing local authorities from raising rates above a specified level. The government also introduced a highly centralising measure known as the standard spending assessment (SSA) which carried with it an implicit theory of local government. The argument was that local government's functions are limited to some quite basic ones – education, sewerage, social services, transport, housing – and that therefore in order to induce conformity and financial probity, central assessment was possible of how much each local authority needed to spend in order to provide a given package of services at a standard cost. This meant that the government effectively decided what the role of local authorities was (rather than the electorate doing so through their voting). It also made visible the consequences of any decisions to spend above the level of the SSA. But given that 'excessive' spending was confined to a small number of local authorities, it is debatable whether all the efforts to centralise local government finance were strictly essential.

The community charge (more widely known as the poll tax) was the government's first alternative to the rates. It was a flat-rate charge for each adult, albeit with some rebates for low-income individuals. The implicit theory was that of public choice: the rating system produced perverse incentives, because low-rated (or non-rated) majorities of voters could exploit high-rated minorities of voters and thereby redistribute income. In an ideal public choice world, voters would have perfect mobility and would migrate to jurisdictions which have their preferred combination of taxes and services. There are numerous, and quite obvious, difficulties with all this, given the problems of moving house and inflexibilities in the labour market.

The aim of the poll tax legislation, then, was to make voters feel the full marginal costs of services, so that votes would more accurately reflect real preferences, and accountability would be improved. If a local authority wished to promote any further local income redistribution – e.g. by higher spending on social services – all local expenditure would have to be raised through the poll tax and no additional central government grants would be made available. Consequently the 'gearing effect' of such decisions was substantial. The poll tax was vigorously opposed – most notably in Scotland, where it was piloted, since in addition to its general unpopularity it was perceived as an imposition by a Westminster government with little electoral legitimacy in Scotland (McCrone, 1991). It was demonstrably regressive, and the government were forced to find another alternative, as a result of mounting political opposition and highly visible public demonstrations.

The solution adopted was the council tax, to be paid not per adult but per property, with properties being banded according to their values. Again the use of SSAs to determine levels of grant to each local authority meant that any additional services over and above this standard package were to be paid for through local taxation. The government quite correctly calculated that local electorates would perceive that the high-spending authorities were Labour-controlled councils, mainly in large cities, and would vote accordingly, and (via what some saw as systematic efforts to skew the distribution of government grants in favour of flagship councils, such as Wandsworth and Westminster) the government sought actively to convince electors of the folly of voting Labour.

Implicitly this legislation sought to disavow the principle that local government is in some sense about income redistribution or about collective provision, and replace it with a view of individuals as self-interested consumers in a marketplace for council services. There was no wider vision of local government as an intermediary institution standing between communities and the global marketplace. It is in the financial basis of local government, and not in more transparently 'political' decisions such as abolition of tiers of administration, that we find the greatest restrictions on local autonomy and the attempts radically to transform the character of local government, to the point where Butler *et al.* (1994, 182) suggested that local government survived in 'so emaciated and withered a form that some doubt whether it deserves the name'.

Abolition

The abolition of the metropolitan counties was the most obvious of a series of centralising measures undertaken during the 1980s. These counties had only been established in 1974, and were intended to provide higher-level strategic government in major conurbations (particularly with reference to transport and planning issues). Given the social composition of these conurbations, they soon became regarded as bastions of left-wing administration. Moreover, the more progressive metropolitan authorities pursued what for some were regarded as visionary policies (e.g. high levels of subsidies to public transport) and for others radical and wasteful programmes, which brought them into conflict with the Conservatives. Not surprisingly the Conservatives took exception to the policies of the metropolitan counties – especially those of the populist Greater London Council.

Among the arguments for abolition were that these counties were superfluous, bureaucratic and costly. Somewhat inconsistently, the government claimed that a two-tier system of local government (county and district councils) was appropriate in non-metropolitan areas, but argued that the metropolitan counties only had full responsibility for a limited range of functions, the majority of which could be provided by boroughs and districts. In addition, the government felt that the rate-raising powers of the metropolitan authorities enabled them to raise large sums of money and helped them promote policies in conflict with national policies which were the responsibility of central government. Moreover, some suggest a geopolitical motivation for this attack: the metropolitan counties represented outposts of determined resistance to Conservative hegemony especially outside the Conservative heartlands, and therefore if the 'South' was ever to swallow the 'North' (to quote Gamble, 1989), those bodies would have to be eliminated. The government forcefully and repeatedly claimed that there were conflicts between the activities of the metropolitan counties and national priorities, lending support to suspicions that their motivations were transparently political. As a result of all this, the metropolitan counties were abolished in 1986 and replaced with a complex network of joint boards. These were widely criticised: voluntary cooperation was rarely effective and a bewilderingly complex set of joint boards was established. Particularly in London the result was inertia and an absence of strategic supervision and management of development.

Hence, within a few years of abolition, there were soon efforts to put the governance of the capital on the political agenda. These included (under the Major government) a Cabinet committee on London, a Minister for London Transport, various interdepartmental coordination mechanisms, several private sector initiatives claiming to speak for the capital, and the associations of local authorities (ultimately merged into one). However, the shadow of the GLC hung heavily over any proposals for a revived local authority, and Labour's policy pronouncements avoided such guilt by association.

The eventual proposals for an elected Mayor and a slimline 'strategic' Greater London Authority can be seen, somewhat negatively, as an attempt to avoid reinventing the GLC. More positively they can be seen as an attempt to reinvent government. The proposed Greater London Authority (GLA) will have responsibility for a range of functions (transport, economic development, environment, planning, emergency services, culture and public health). It will absorb responsibilities from a range of agencies; in the case of transport, for instance, it will take over some functions carried out by London Regional Transport, the Metropolitan Police, the Highways Agency, the Government Office for London as well as responsibilities for several other services. Thus an integrated transport strategy may become a reality. The GLA also has a statutory duty to promote sustainable development.

The obvious criticism of a revived tier of government would be its size and bureaucracy but Labour insist that this will be a streamlined authority with only 25 members and up to 250 care staff; direct service delivery will be left to the relevant authorities operating within a framework determined by the GLA. The creation of this authority – one which steers, not rows, in Osborne and Gaebler's (1992) terms – and the innovation of an elected Mayor have allowed Labour to present this as an exciting experiment in modernising urban government rather than a throwback to hierarchical bureaucracies. It remains to be seen whether this comparatively small organisation can shoulder the weight of expectations placed upon it.

Ending tiers: reforming local government

The Conservatives, not content with centralising the financial basis of local government, went still further in implementing a wholesale reorganisation of local government in the mid-1990s. The origins of this are related to unresolved criticisms of the previous (1974) reorganisation, unfinished business arising from the government's own policies towards local government during the 1980s, and contingent political influences, notably the need to effect some positive change in local government after the political disaster of the poll tax (see above).

Traditionalists argued that the 1974 reorganisation had created artificial administrative units which bore no relation to community identity; this criticism was directed mainly at the 'new' counties of Avon, Cleveland and Humberside, but it was also targeted at some district councils. An opposing argument was that the 1974 reorganisation had not gone far enough. The universal adoption of a two-tier system of local government had been a missed opportunity; unitary authorities should have been established throughout the UK (which could also have the beneficial effect, from the Conservatives' point of view, of saving public expenditure). A third criticism, voiced particularly by large urban authorities in non-metropolitan areas (Leicester, Nottingham) was that, by being treated as district councils, their status was actually lower than the unitary former metropolitan authorities or London boroughs. The possibility of a reversion to something like the pre-1974 County Borough system attracted much sympathy.

To these unresolved and contradictory criticisms should be added a set of contingent influences. The metropolitan counties were the only innovation in 1974; their abolition thus signalled that the entire 1974 settlement was open to revision. Moreover, the arguments used to justify abolition could also be used elsewhere (e.g. non-metropolitan counties).

The immediate political background to the review is also pertinent. The political disaster of the poll tax meant that an alternative means of financing local government had to be found. This poisoned chalice was handed to Michael Heseltine, who accepted it on the understanding that he was allowed to lead a review of local government. Wright (1993, 83) thus suggests that it is 'entirely characteristic' that the review of local government was initiated solely as a 'byproduct of the need to do a political fix on the Poll Tax'. Tensions remained: Major was believed to have favoured an agenda based on community identity and historic associations, whereas Heseltine favoured unitary authorities, justified in terms not dissimilar to the 1974 reorganisation (improved coordination, greater understanding, accountability). Some commentators detected political motivations behind Heseltine's actions, arguing that he favoured unitary authorities in order to enhance Conservative control over rural districts, while eliminating unpopular creations of 1974 which also happened to be Labour-controlled (such as Humberside).

The Conservatives requested the Local Government Boundary Commission (LGBC) to review local government boundaries, starting in 1993. Such a review would normally take a broad view of the philosophical issues and questions of principle affecting its remit, but the philosophy underlying the review was far from transparent. For example, no attempt was made to consider both the case for regional government and the relationship of local authorities to a wider European frame of reference. Plainly, were regional government to be established, the case for a two-tier system of local authorities would disappear by default, but this was not considered by the review (even though in some respects it would have added weight to the government's case).

Moreover, the government made its preference for a unitary system clear from the beginning. As Chisholm (1995) points out, the assumption underpinning the review was that the outcome would be unitary authorities. Thus, 'structural change' in local government boundaries was defined as the 'replacement ... of the two principal tiers of government with a single tier'; yet in London and the former metropolitan counties, where a second tier did not exist, the LGBC could not recommend the creation of a second tier. Chisholm gives other examples of the ways in which the drafting of the Act precluded anything other than the recommendation of unitary authorities, and argues that the Act failed to consider the range of permutations that might result from the review. Moreover, Ministers and MPs made no secret of their preference for unitary authorities. By mid-1993 it was clear that the Commission was inclining towards the advantages of two-tier authorities in more cases than Ministers expected and wanted. Revised guidance to the Commissioners (issued by Heseltine's successor, John Gummer) therefore expressed the expectation that a two-tier solution would be the exception, and appeared to indicate a preference for more, and smaller, unitary authorities. Chisholm interprets this as 'an attempt to steer the Commission very firmly in one direction' (1995, 566). In a legal challenge it was pointed out that this appeared to prejudge the outcome of the LGBC's deliberations, and the High Court ruled that the Environment Secretary had exceeded his powers in issuing this guidance to the LGBC. Despite being forced to back down by this decision, the government subsequently issued directions to the LGBC to reconsider their proposals for a number of local authority areas, in what was believed to be a 'transparent move' to ensure that the Commission would come up with the unitary solution desired. (Chisholm, 1995, 566).

In practice, relatively little was recommended by way of change. In several counties public pressure persuaded the

LGBC to retain the status quo and the LGBC recommended a change in the existing structure in only 15 counties; in 17 others a two-tier structure was to be retained.

The recommendations were then considered by John Gummer, who was put under considerable political pressure by Conservative backbenchers and former Ministers. Gummer's principal intervention concerned referring some recommendations back for further consideration, mostly concerning the possibility of creating further unitary authorities in urban areas. The result was the proposed creation of further (mainly urban) unitary authorities as well as the rejection of recommended unitary authorities in some of the more rural components of some counties. The final result has been a 'chaotic assemblage' of local governments. In only three counties (Avon, Cleveland and Humberside) have unitary authorities replaced county councils; thirteen counties have retained a two-tier structure in its entirety. Elsewhere, 21 counties were divided, usually by removing parts to create unitary authorities (e.g. Portsmouth and Southampton have been removed from Hampshire) with the rest retaining a two-tier system. Given the range of outcomes adopted, it is quite impossible to discern a consistent philosophy or principle. The contrast with Scotland and Wales – which have a uniform structure of unitary authorities – could not be clearer. Of course, the Scottish reforms were coloured by suspicions of gerrymandering, because of the suggested creation of small councils, likely to have guaranteed Conservative majorities, with populations considerably smaller than their Labour, urban neighbours (Dawson, 1993). It was also argued that there were technical weaknesses, such as the failure to recognise functional linkages between urban areas and their hinterlands (a verdict which could equally apply to many of the reforms in England). Despite these criticisms, at least the reforms in Scotland and Wales represented the application of a consistent principle. Whether the English reforms represent another uneasy and unstable compromise remains to be seen.

11.4 LOCAL GOVERNANCE: THE UNELECTED STATE, QUANGOS AND PUBLIC–PRIVATE PARTNERSHIPS

There is considerable evidence to support Harvey's (1989b) arguments that there has been a transition in urban governance from managerialism to entrepreneurialism. This has come about in a number of ways.

First, many public services have been commercialised. Public authorities must now seek competitive tenders for a growing range of services, thus removing discretion from them (e.g. Ascher, 1987). Second, in order to expand the scope of market-led provision of services, several services were handed over to unelected quangos (e.g. Training and Enterprise Councils); alternatively, the power of central government to appoint – or select – its nominees to public bodies was enhanced. Thus, public appointments became politicised, and insulated from local democratic influence. This facilitated the pursuit of an agenda of commercialisation, but it also fuelled the accusation of abuse of power in what some termed a 'quango state' (Jenkins, 1995).

Third, whole areas of territory were handed over to undemocratic bodies, the Urban Development Corporations (UDCs), which were initially established in London's Docklands and on Merseyside, and subsequently extended to a number of locations. They were given sweeping powers to permit development in their jurisdictions. These represented a particular critique of the failings of the local democratic state and also an ideological attack on the predominantly left-wing local governments controlling it. The argument was essentially that the bureaucratic, welfarist policies of such authorities represented impediments to the free play of market forces. They consequently had to be bypassed. Moreover, local needs were subordinated to a national political agenda; the LDDC, in particular, took on a symbolic importance to the recovery of the nation as a whole. Overriding local communities was a price worth paying for a project which would help secure national economic growth. The rhetoric of regeneration policies – frequent references to urban 'task forces' or 'Action Teams' – suggested the recapturing of territory which had fallen into alien hands (see Keith and Rogers, 1991). Far from being a technical response to an administrative difficulty, UDCs were part of a wider geopolitical agenda, of diminishing the powers of left-leaning local authorities and expanding the scope for the unfettered operation of market forces. (Assessments of their success as *spatial* policies are considered in chapter 12.)

The foregoing initiatives have been driven by central government. The general effect of Conservative policies was to diminish the scope for autonomous action by local public sector agencies and this partial policy vacuum has been occupied, to a degree, by forms of public–private partnership. This involves the promulgation of a rather different agenda to that of local government.

The characteristics of such partnerships have been described by several commentators (Harding, 1991; Shaw, 1993; Peck and Tickell, 1995b). The general aim (even if not always clearly articulated) is the promotion of a favourable pro-business climate, with a view to attracting mobile international investment. These organisations have played a prominent role in several locations: the 'Great North' campaign in North East England (Robinson and Shaw, 1994); Manchester's promotional activities in bidding for the Olympic Games (Cochrane *et al.*, 1996); and the various agencies claiming to speak for London. They have several features in common: a stress on promoting their 'place' as in some way a distinctive location, to some degree presenting a myth about the region (Sadler, 1993); representatives of key private sector companies who have influence with the government, which is necessary in order to lever in the high levels of public money necessary to finance development; a vigorous and pro-active approach to decision-making, in contrast to the indecisiveness of the local state; and comparatively little local government representation or access. Consequently there is concern that these committees are being used to 'crystallise a local business elite that can claim to be both representative and coordinating all key policy decisions in [an area]. This can then be used

to deflect the arguments for an elected ... government' (Colenutt and Ellis, 1993, 20).

However, the efficacy of these efforts is open to dispute. What is *not* evident from such exercises in regional promotion (what Robinson and Shaw (1994) refer to as a philosophy of 'let's sink our political differences in the interest of promoting the region') is a coherent philosophy or a development strategy. Much energy is devoted to quasi-philanthropic exercises in active citizenship (Kearns, 1992). While such activities may be well intentioned, they lack focus and coherence: local business may 'articulate a critique of the state', but it does not embody 'an alternative rational mode of governance' (Peck and Tickell, 1995b, 75). The really difficult problems of urban governance are left to other public agencies while public–private partnerships continue with their flagship projects.

11.5 REGIONALISM AND DEVOLUTION

Bogdanor (1996, 296) observes that the explanation of why the history of devolution in Britain has been 'largely one of failure' lies deep in the UK's constitutional principles, and crucially, in the unitary nature of the British state. The principle of the supremacy of Parliament is hostile to any possibility of power-sharing or subsidiarity. The image of a centralised, domineering European state is evoked powerfully by those opposed to European integration. None the less, for many years it has proved possible to accommodate a range of relationships (e.g. by administrative devolution to Scotland and Wales) which at least acknowledge the legitimacy of distinct national identities without undermining the unitary British state.

This situation has been subject to pressures from a range of sources. First, there have been direct political pressures for devolution. In both Scotland and Wales, nationalist demands reflect reactions to the centralisation of Conservative rule and the often insensitive way in which these countries have been ruled. More positively, such demands for autonomy reflect the importance attached to place and identity as a reaction to the homogenisation and cultural colonisation induced by global media (see chapter 10). The experience of nationalist movements in other European states, which has resulted in greater autonomy being granted to certain regions, has added weight to these demands. The experience of operating within the EC, with its emphasis on subsidiarity, has also been instructive, not least because a condition for assistance from the EC's Structural Funds is a demonstration that money is being spent within a coherent regional framework.

The other arguments for regional governance coalesce around its role in promoting economic competitiveness. The most deterministic statement of this position is Leonardi's suggestion that the 'secret of regional performance lay largely in the institutional infrastructure of an autonomous regional government' (quoted in Hebbert, 1993, 714). The argument draws on Amin and Thrift's (1994) suggestion that a 'thick institutional tissue' is a necessary condition for regional success in an era of globalisation. More conventionally the suggestion is that the centralised nature of the British state has led to a neglect of – and a failure to maximise – regional economic potential. However, Evans and Harding (1997) argue that while gaps might exist in institutional capacity for promoting development, there is no agreement on whether regionally elected institutions would close those gaps. Moreover, it is too simplistic to ascribe regional economic success to the presence of democratic institutions at a regional level (p. 28); subnational institutions elsewhere in Europe have varying degrees of autonomy and have not always been created in response to economic imperatives. While Murphy and Caborn (1996) suggest that there is an economic imperative for regional government, others are more sceptical. Examples of the success of the development agencies representing Scotland and Wales do not make the case, as these are motivated by a strong sense of nationhood, which is absent from most of the English regions.

The combination of deregulation and cen-

tralisation pursued by post-1979 Conservative governments has also revived interest in devolution and regionalism (John and Whitehead, 1997, 12). In the absence of strategic planning at the regional level, the costs of deregulation became all too apparent to residents of South East England during the Thatcherite boom years of the 1980s. Despite their abandonment of regional physical planning (with the abolition of Economic Planning Boards in 1981) the Conservatives were therefore forced to reintroduce some form of controls (regional planning guidance: Thornley, 1993) on development. Related to this, the *de facto* deregulation of, and competitive approach to, spatial policy (in the form of competitive schemes such as City Challenge and Regional Challenge) created a situation in which competitive bids for regeneration funds could be submitted from within the same region without taking account of strategic priorities (see chapter 12). A further consequence of the hollowing-out of the state has been that various agencies vie for local and regional leadership, against a background of a multiplicity of local economic initiatives. Various reports pointed to the absence of effective coordination which ensued (e.g. Trade and Industry Select Committee, 1995).

Many of these arguments have found favour with the Labour Party and the devolution legislation has granted different degrees of devolution to Scotland and Wales, principally because of the different levels of support for devolution and associated powers (chapter 10). The priority granted to devolution raises various questions for the English regions. The important issue raised is how devolution – particularly Scottish devolution – can be taken forward without further exacerbating existing constitutional imbalances. For example, Scotland is over-represented in Westminster, which makes it more rather than less possible for Scottish MPs to affect the balance of power at Westminster, while both Scotland (and Wales) enjoy a generous share of public expenditure (see chapter 7). It is difficult to imagine the English regions accepting greater devolution to these countries without a reduction in this disparity.

In fact, devising appropriate solutions for the English regions is the major challenge posed by the devolution legislation. Scotland, Wales and London have gained but the position of the English regions remains unclear. The principal innovation in regional governance in England has been the creation of the Government Offices for the Regions (GORs) (Mawson and Spencer, 1997). These brought together civil servants in the Training, Enterprise and Employment Division (TEED) of the Departments of Education and Employment, plus the Departments of Environment, Transport, and Industry; these were made accountable to one regional director. The GORs are aimed at ensuring that there is coordination between all public programmes, and maintaining links with those departments without a specific regional presence. The key focus is obviously programmes relating to the environment, infrastructure, regeneration, and economic development. Other departments and programmes do not have such a clear-cut regional focus and in any case allocate expenditure on the basis of national formulae relating to population structure. The GORs have been welcomed as a departure from the compartmentalised structure of the civil service, and for stimulating the creation of partnerships and networks within the regions. Against this, they have been criticised as being immune from local democratic pressures; they do not operate within the framework of a formal policy forum, in which the key regional institutions arrive at an agreed regional view (Mawson and Spencer, 1997, 80).

This tension, between a democratic agenda and one focused substantially on economic development, has not been resolved as yet. Labour has established Regional Development Agencies (RDAs) in England. These were seen as local bodies for coordination of inward investment, raising skills and improving competitiveness. They were presented as regional economic powerhouses, which would develop and promote strategies for regional development. They are partnership organisations with members from business, local government, unions and the voluntary sector.

For some supporters of regionalism this represents a half-way house because, while in opposition, Labour had proposed regional chambers which would determine the strategic framework within which RDAs would operate. Even the idea of regional chambers fell some way short of directly elected regional assemblies. Tomaney (1999, 78–9) points to two main unresolved problems. First, given that Labour had criticised the democratic deficit and quangocracy which had developed under the Conservatives, business-led RDAs hardly seemed an adequate response, particularly as they are accountable to ministers in London. The focus of RDAs is on raising 'regional competitiveness', an area in which (given criticisms of competitive and overlapping initiatives in recent years: chapter 12) there is certainly a role for regional agencies. However, economic prospects remain dependent on decisions taken by centralised institutions, largely in London. Such institutions are more likely to be responsive to demands from political voices, such as those from Scotland, Wales and London, which will be far louder than those available to the English regions. Lacking powers or democratic legitimacy, the RDAs 'may struggle to maintain political credibility' (p. 79).

Second, Tomaney argues that regional government extends beyond a narrow economic development agenda to incorporate elements of social, cultural and civic development. Regional government could stimulate strong regional political cultures and participation, but the government's proposals demonstrate their unwillingness to concede too many powers. In fact, the government's argument that directly elected regional assemblies could develop where there is a clear popular mandate may lead to a worst-case scenario. First, Scotland, Wales and London, with their directly elected bodies having both legitimacy and resources, will clearly have an edge over other regions. Second, there may be demands for directly elected assemblies elsewhere (the North East is a case in point), usually reflecting a strong sense of regional identity. These demands may in due course be satisfied but those regions will still lag behind those named above. Finally, what of regions which perhaps lack a strongly cohesive identity: will they be left behind or engulfed in a (southern) English backlash, one possible result of which might be the creation of an English Parliament? (This would not solve the problems of the excessive concentrations of institutional and political power in the South East.)

The result of all this has been an asymmetric regionalism: areas best placed to benefit (especially London and Scotland) have the strongest institutions; the English regions lack them. This *ad hoc*, pragmatic approach is not really underpinned by a coherent, unifying vision. This is a weakness but it may also be an opportunity for advocates of regional government. Constitutional change will generate its own momentum, as in Scotland where the SNP's support has increased substantially. Once the devolution genie is out of the bottle, the likelihood is of growing challenges to metropolitan domination (Tomaney, 1999, 82). Whether Labour seeks to resist or accommodate such challenges remains to be seen.

11.6 CONCLUSIONS

Many recent developments in subnational government have been instituted for largely negative reasons (witness the Conservative governments' 'nationalisation' of Britain: Jenkins, 1995) or have arisen in an *ad hoc* manner, either to fill gaps created by the hollowing out of the state, or (even more prosaically) to conform to requirements for EU funding. Motivated by pragmatism or prejudice, this does not suggest a grand design or strategy. Nor does it appear to represent a coherent vision of what subnational governance might look like in a putative 'post-Fordist' era, though there are emerging trends – the 'enabling' local authority, seek-

ing to ensure that services are delivered rather than doing so itself – which point in this direction. Against this, the degree of centralisation in many post-1979 policies might be inimical to post-Fordism's emphasis on flexibility and localism. There have been attempts to privatise services, to increase the scope of market forces, and to involve business in urban governance, but it is not clear that these amount to a clearly articulated vision (cf. Peck and Tickell, 1995b). Much more support may be found from the British evidence for Harvey's (1989b) notion of entrepreneurialism. Few localities now lack some form of public–private partnership, while the political *volte-face* executed by cities such as Manchester is ample evidence of the transformed political climate.

There is still considerable turbulence in subnational government, as the 1996 reforms and the devolution legislation work through. However, there is also widespread agreement that greater dispersion and decentralisation of powers are necessary, both from a democratic standpoint and arguably for economic success. The recent erosion of subnational government has removed or severely weakened important intermediate institutions which formerly stood between communities and the forces of global capitalism. Both at the national and regional level, commentators have attributed economic success to the strength of democratic institutions and civic participation (Putnam, 1993; Hutton, 1995). If these arguments have substance, more will be required than *ad hoc* changes and incremental reform. McCormick and Alexander (1996, 99) suggest that without wider constitutional changes – a project to modernise and democratise the structures of the British state, in other words – there is a risk of undermining the credibility of devolution and perhaps of local democratic institutions.

MANAGING UNEVEN DEVELOPMENT
PHILOSOPHIES, POLICIES AND IMPACTS

12.1 INTRODUCTION

Initially conceived as temporary and limited expedients, introduced in response to the severity of the inter-war depression, regional policies have become an enduring feature of the UK's political geography. Though not large in terms of public expenditure, their ideological and political significance is disproportionate since they are a key index both of state commitment to less prosperous places, and of the extent to which – and the ways in which – governments have been willing to interfere with the natural operation of market forces (Lewis, 1984). Regional policy is a terrain, therefore, on which competing ideologies vie for supremacy. However, policy changes are not determined purely by ideological shifts, and external events – notably, the diverse impacts of globalisation – have influenced the scope and character of policy. One also has to consider a range of criticisms of the philosophical basis and practical application of spatial policy.

Thus, whereas some form of spatial policy appeared to form part of Keynesian demand-management strategies for many years, the persistence of 'regional' problems, and the failures of Keynesianism itself, eventually undermined faith in the capacity of such demand management to 'solve' economic and social problems. Combined with some far-reaching critiques of regional policy's instruments, this created the space within

which the Conservatives were able substantially to retract the geographical extent, scope and *modus operandi* of spatial policy measures and expenditures. In turn, however, realisation of the limitations and costs of such neoliberalism has combined with academic and political awareness of new forms of territorial development to promote a revival of the case for regional policy, albeit in a quite different guise.

In exploring these issues the chapter begins with a brief historical introduction, which considers arguments for and against spatial policies and then summarises the aims and methods through which conventional regional policy was implemented (see also Gudgin, 1995). There follows discussion of the demise of conventional regional policy, its replacement by a politics of enterprise, and the growth of local economic development initiatives (LEDIs). Recent initiatives in both urban and regional policy are examined and the chapter concludes, following Lovering (1999), by relating current policy developments to contemporary theoretical debates.

12.2 A BRIEF HISTORY OF SPATIAL POLICY

Arguments for and against policy

For neoliberals, regional problems are the creation of market forces, and can only be corrected by those forces. Government's legitimate role should be confined to creating the conditions for profitable production and providing incentives to stimulate entrepreneurial activity (Hayek, 1944). Market forces will eliminate regional inequalities because the net effect of growth in prosperous areas is ultimately to inflate the costs of production, leading to decentralisation and a tendency, in the long run, towards equilibrium. State intervention in this process causes inefficiency, since firms are induced to establish plants in locations which they would not otherwise have chosen. Any interference with market forces must, therefore, produce suboptimal results (Minford and Storey, 1991; Barnett, 1995). Finally, it is proposed that regional problems have occurred because workers have been inflexible, being unwilling either to move house or to accept wage reductions; the solution is deregulation of the labour and housing markets and cutting state benefits, the better to promote flexibility.

Against neoliberalism, it is argued that market forces do not and cannot promote balanced spatial development. Myrdal (1957) suggested that the process of cumulative causation reinforces regional inequality. A region growing more rapidly than another can, through its initial advantage, attract more growth. While decentralisation of economic activity might take place (e.g. as the costs of production in the core region become excessive) Myrdal was convinced that this would not outweigh tendencies to concentration and so government intervention was necessary.

Moreover, markets do not tend to equilibrium, because neither labour nor capital is perfectly mobile. Geography makes a substantial difference to the operation of labour markets (see chapter 6). Inflexibilities in the housing market (notably with respect to the availability of rented accommodation), added to the costs and risks of moving, inhibit labour mobility. Similar arguments apply to firms: they tend to prefer the certainty of their present location, and show a low propensity to move except when under severe pressure to reorganise production. The continued emphasis in economic geography on 'sunk costs' and agglomeration

economies bears witness to this. Finally, patterns of regional development in the UK are influenced by institutional arrangements which impart a bias towards South East England, in the form of public expenditure and taxation policies, and via the location of key institutions (chapters 2 and 7).

There are, as a consequence, several economic arguments for regional policy. First, excessive growth can lead to *inflationary pressures*: in tight labour markets, workers can bid up wage settlements, especially when their skills are in short supply; land prices can increase dramatically, especially where land supply is restricted; and above-average wage settlements feed through into higher demand for consumer goods, often sucking in imports and worsening the balance of payments. Second, there are the *costs of economic concentration*: private sector decisions are taken solely on economic grounds, and do not allow for the hidden costs of those decisions, such as environmental impacts (pollution and congestion). Furthermore, excessive growth can deplete the human capital of regions more swiftly than it can be replenished, so that skill shortages can become a constraint on regional growth. All these costs of growth were evident in SE England during the 1980s and were implicated in the re-emergence of inflation (Martin, 1989b: see also Crang and Martin, 1991). More balanced development might have avoided these problems.

The converse of these arguments is that uneven development leads to underutilisation of both human and social capital. Reserves of labour represent a valuable and underused resource, and it is plainly in governments' interests to ensure that physical infrastructure (roads, schools, hospitals) is fully utilised. Securing some redistribution of population might help achieve this goal, and this objective gains more force from recent concerns about sustainable development.

Such arguments, however compelling, have to command assent for regional policy to be implemented and so political considerations influence what is actually done. Britain has rarely seen vigorous territorially based protest movements, though nationalist movements had some effects on the strength

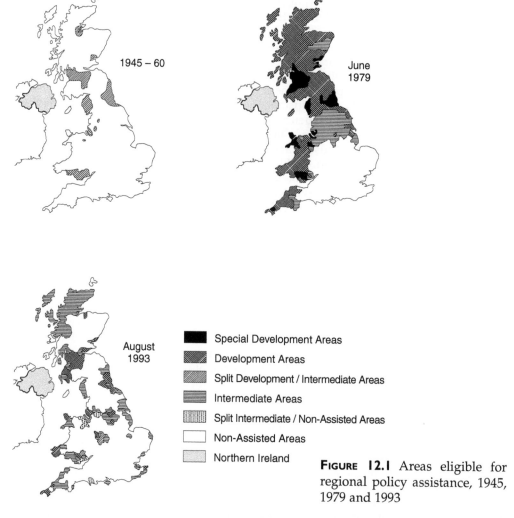

FIGURE 12.1 Areas eligible for regional policy assistance, 1945, 1979 and 1993

with which regional policy has been pursued; however, a strengthening of regional policy in the mid-1960s has generally been viewed as a way of placating Labour supporters in areas which were to suffer from pit closures. Social considerations have been more influential; both Conservative and Labour governments have at various times justified regional policy on social grounds. However, the major political parties have differed on their vision of the rationale for policy. At one end of the spectrum, Labour sought, in the 1960s, actively to use state powers to promote the restructuring and modernisation of technologically backward industry. Conversely, regional policy was

reduced during the 1980s to the status of a social welfare measure, pursued with little enthusiasm.

Aims, areas, and instruments of policy

Historically, the origins of regional policy lie in the inter-war collapse of the basic industries in the peripheral coalfields, which produced enormous job losses and mass unemployment. However, the problem was diagnosed as a limited and temporary one, which would be corrected by the equilibrating forces of the market (Middleton, 1996). Policy measures were circumscribed by a rigid economic orthodoxy, which proscribed financial assistance to private companies. Hence efforts were limited to promoting labour transference and infrastructural improvements.

A more sophisticated diagnosis was provided by the Barlow Report (1940) which recognised the symbiotic nature of the co-existence of depression in the peripheral regions with congestion in London and the South East. The Barlow Report argued that policies ought to promote a more balanced distribution which would avoid the costs of congestion and the wastage of unemployment. This report is generally regarded as the foundation of post-war policy.

The aims of regional policy were arguably never clearly stated. No government has ever explicitly set out to 'solve' the regional problem. Any government claiming to do so within the framework of a capitalist economy would have been overstating its case, to put it mildly, given the constraints on and limits to state intervention. As a consequence, policy aims have been couched in broad, vague terms such as bringing supply and demand for labour into balance, or seeking to help the assisted areas towards self-sustaining growth. It is not really surprising that regional policy has been criticised for failing to achieve anything.

What is notable about the areas eligible for assistance is the consistency with which certain locations – peripheral coalfields – have remained eligible (figure 12.1), reflecting the persistent differentials in unemployment between those areas and the rest of the UK. Also notable is the waxing and waning of the areas eligible for assistance and the complexity of the system. At one stage a three–tier system was in operation, with finely shaded gradations of levels of assistance, covering nearly half the UK's population: was the implication really that half the country was a 'regional problem'?

As to instruments of policy, we should distinguish between efforts to take work to the workers, and to take workers to the work. Efforts to encourage labour mobility have usually been constrained by inflexibilities in the housing market and/or by variations in housing costs. Consequently most policy has

been about the redistribution of economic activity between and within regions. Controls on industrial location were introduced, in the form of industrial development certificates (IDCs) from 1945. These were not normally granted in regions which were regarded as already excessively developed. These were the primary instrument of policy until the 1960s, although the policy was pursued with declining vigour as the volume of mobile investment shrank, and IDCs were eventually abandoned altogether in 1979.

Policies were modified substantially during the 1960s when capital grants and labour subsidies were introduced. Depending on the kind of area one was located in, grants of a substantial proportion of the cost of capital investment could be obtained. Likewise, under Regional Employment Premium (REP), wage subsidies were available which lessened the costs of production in the assisted areas. Instruments have also included a wide range of *ad hoc* bodies concerned with either promoting industrial efficiency and modernisation, or with particular geographical areas. Notable in this regard are the Scottish and Welsh Development Agencies, which have specific regional remits, in sharp contrast to other peripheral regions (prior to the 1997 election).

An indication of the priority attached to regional policy over the past four decades is provided by figure 12.2, which shows expenditure at constant (1990) prices. Spending peaked in 1975–6 (£1.8 billion), as a result of substantial expenditure on labour subsidies. Thereafter expenditure was steadily reduced, apart from an increase in expenditure in the 1980s, which reflects the substantial sums allocated to urban redevelopment schemes, principally in the London Docklands.

Assessments of regional policy have been complicated by the lack of clarity of the aims of policy, and by the need to answer the counterfactual question of what would have happened had regional policy not been in operation. It is generally agreed, however, that policy was at its most effective in two periods: the immediate post-war years, and the 1960s/early 1970s, prior to the rise in oil prices. What these periods have in common is the presence of much mobile investment,

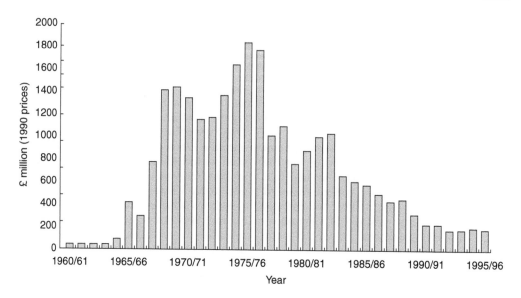

FIGURE 12.2 Expenditure on UK regional industrial assistance, 1960–96 (Taylor and Wren, 1997)

capable of being diverted away from the South East. Additionally, the early post-war years saw all regions benefit from reconstruction, which would have happened regardless of regional policy. At other times (the 1970s, for example) policy has had limited success, and it was virtually abandoned during the 1950s by an avowedly anti-interventionist Conservative government. Though evaluations credit regional policy with creating 325–375 000 jobs between 1960 and 1976 (Moore *et al.*, 1986), it is less clear how this happened, and even during this period of success, regional policy came under criticism which led eventually to its demise.

12.3 THE DEMISE OF CONVENTIONAL REGIONAL POLICY

I argue that regional policy's demise arises from three influences: a growing tide of criticism; changing external circumstances; and changing political priorities.

Criticisms of spatial policy

In general terms regional policy was criticised for resting on an inadequate understanding of firms' locational strategies and

one of the more memorable comments was that policy had been 'empiricism run mad, a game of hit and miss played with more enthusiasm than success' (House of Commons, 1972). More specific criticisms focus on the extent to which regional policy expenditures provided value for money and secured long-term sustainable development. Large-scale automatic capital subsidies were a particular target. Projects often created relatively few jobs at considerable cost. They were often used by firms to subsidise projects which, while increasing output and efficiency, had the net effect of reducing employment (e.g. Robinson and Storey, 1981). Conversely, labour subsidies were criticised for encouraging firms to employ a larger workforce than they might otherwise have done, thereby promoting inefficiency.

The regional development implications of regional policies have also been questioned. The suggestion has been that policy has promoted a branch-plant syndrome and facilitated a particular spatial division of labour in which companies were able to relocate routine, low-skill jobs to peripheral regions. This reflected regional policy's concern with 'distribution not development' (Martin and Townroe, 1992, 20) or, in Morgan and Sayer's (1988) phrase, the effect was to promote development *in* regions rather than development *of* regions. Because of the failures of many 'regional policy factories' to develop strong linkages, they proved highly vulnerable in periods of recession (chapter 4; Fothergill and Guy, 1993).

Regional policy has been further criticised for a lack of concern with the *type* of jobs created. For much of its life regional policy was restricted to manufacturing, arguably an important limitation at a time when the service sector was the main provider of new jobs. Many jobs created with regional policy assistance have been taken by women (indeed, jobs are consciously gendered in this way by firms: Massey, 1997). Regional policy has therefore contributed to a convergence of female activity rates, but it has done less to reduce unemployment among men, particularly middle-aged men with few skills. There were, finally, criticisms of the spatial *scale* of policy, since at one stage more than half the land area of mainland Britain was eligible for aid of some kind. The clear implication seemed to be that Britain was a 'regional problem', and that what was urgently needed was a project of national economic renewal. Awareness of the deep-seated problems of inner cities also prompted a reappraisal of policy.

Many of these criticisms were, of course, taken on board in refinements of regional policy and some contend that, far from having failed, spatial policy has never had the resources necessary to make a significant impact; furthermore, they simply advocate refinements and better targeting of the range of instruments available to policy-makers (Taylor and Wren, 1997). These arguments raise the question of whether, given the transformations in the economic and political climate in the past 25 years, there is a future for regional policy and, if so, in what form.

Changing external circumstances

It would be tempting to suggest that the real challenge to regional policy came with the election of the Conservatives in 1979. These governments were, of course, unsympathetic to regional policy, but external economic changes also had a substantial impact, forcing governments to realise that their ability to control the location of economic activity was very limited (Martin, 1989c).

First, from the mid-1970s the recession and the rise in oil prices exposed the problems of poor competitiveness in the British economy. This meant that the attention of central government was focused on slow growth and high inflation, accompanied by rising unemployment and a declining balance of payments. Reducing public expenditure became an overwhelming priority in the battle against inflation. As a condition for an IMF loan to bail out the economy in 1976, the British government were forced to implement substantial cuts in public expenditure. Consequently, the largest single reduction in regional policy expenditure was made by a Labour government, via the abolition of the Regional Employment Premium in 1976.

Second, faced with increasing energy costs, industry sought to rationalise production, often through investment and technical change (Massey and Meegan, 1982), which had the effect of displacing workers. This was often supported by regional policy assistance in the form of capital grants. Third, in a recessionary climate, the national need for *any* inward investment led governments, both local and national, to seek to loosen restrictions on the movement of employment. This was related to a growing awareness that multinational capital was extremely mobile and capable of relocating at comparatively short notice.

Changing political priorities

Against this sombre background there would have been a reappraisal of spatial policy, regardless of which party won the 1979 election. Given the Conservatives' views on the nature of the British disease, spatial policy was regarded as subsidising inefficiency and distorting decision-making. The regional problem was explained, in the Conservatives' election manifesto, in terms of institutional rigidities: workers and unions had demanded excess wages (i.e. greater than those justified by increases in productivity) which deterred investment and created unemployment. The policy prescription was a firmly neoliberal one: the 'regional problem' would only be solved by rekindling a spirit of enterprise locally as well as nationally; workers should respond flexibly, by pricing themselves into jobs, or by being prepared to move to locations where work was available. Regional policy's role was reduced to an essentially social one; expenditure was scaled down dramatically, and some commentators suggest that the only reason for the continued existence of *any* regional policy was to secure access to the EC's Structural Funds. Revisions to the map of assisted areas in 1983 and 1993 saw a substantial reduction in the proportion of the country eligible for assistance. However, these changes were criticised for relying too heavily on localised increases in unemployment without distinguishing between locations experiencing structural difficulties and those where, despite comparatively high levels of unemployment, economic prospects were essentially favourable. In particular, critics felt that the 1993 revisions of policy were opportunistic reactions to a recession which had pushed up unemployment rates in several coastal towns in the South (Martin, 1993).

Revisions to regional policy in 1983 and 1993 introduced greater selectivity, partly in response to previous criticisms. Thus, automatic capital grants (Regional Development Grants) were abolished in 1988, leaving Regional Selective Assistance (RSA) as the main financial instrument. This is available to manufacturing and some service sector activities investing in the Assisted Areas subject to various conditions, though it is not available to firms relocating within the UK. Other instruments include expenditure on land and buildings, intended primarily to induce new FDI, and some (limited) selective assistance to small firms to assist in the innovation process.

There is some evidence that these changes were followed by an acceptance, by the Major government, that regional policy could play more than a residual social role and could help enhance the competitiveness of the Assisted Areas. Taylor and Wren (1997, 838–8) argue that this was primarily for rea-

sons of national economic policy as well as pragmatism. FDI is very important here as it was believed that the availability of RSA had been an important factor in attracting inward investment which would otherwise have gone elsewhere within Europe. More pragmatically, the EU's Structural Funds became increasingly important as a means of improving the economic performance of the UK's disadvantaged areas and accessing these funds requires member states to designate Assisted Areas. The 1993 revisions defined the Assisted Areas at the maximum permissible (35 per cent) proportion of the working population in order to maximise receipts from these sources. It may also be that the difficulties experienced by the core regions of the UK economy in the early 1990s prompted recognition of the importance of peripheral regions in contributing to national competitiveness. Whatever the reason behind it, there is certainly evidence that the Conservatives' hardline, neoliberal attitude to regional policy had softened by the early 1990s. Nevertheless, by the late 1990s expenditure on RSA was around £130 million a year, or less than one-tenth of levels of expenditure in the 1970s.

To put this figure in perspective, UK government expenditure on regional policy has for several years been approximately half that from various European initiatives (*Regional Trends*, various dates). These funds included the European Regional Development Fund, the European Social Fund and other Structural Funds, plus specific initiatives such as those designed to tackle specific problems in areas suffering from retrenchment in, for example, coal-mining or defence-related industries. The three EU 'regional' Objectives (that is, priority Objectives for the Structural Funds) for which UK regions are eligible are: Objective 1 (lagging regions) – Merseyside, N. Ireland and the Scottish Highlands and Islands; Objective 2 (those in industrial decline) – covering most of the former coalfields, the West Midlands, and isolated outliers such as Plymouth or Thanet); and Objective 5b (rural areas). Together the map of these areas is not dissimilar to that of areas eligible for regional policy during the mid-1960s. There have been criticisms of the overlap between EU policies and those of national governments, while the issue of 'additionality' has attracted much attention (the UK government tended to use Structural Fund receipts as partial reimbursement for net contributions to the EU budget, rather than for their specified purpose) (Townsend, 1997, 226–9). It is clear that, under almost any conceivable scenario for EU integration and enlargement, the UK will lose out substantially to east European states. *All* regions in *all* of the likely new entrants to the EU (essentially, the former Communist countries) will be eligible for Objective 1 status and the UK's regions will therefore have to cope with a significantly reduced share of the Structural Funds (Armstrong, 1998, 202).

12.4 CREATING SPACES FOR ENTERPRISE

With the exception of European initiatives, then, regional policy was thus substantially reduced in scale and scope after 1979. It is clear that the Conservatives intended to replace it with a national politics of enterprise. Predicated on their desire to roll back the frontiers of the state and expand the scope for market forces, numerous efforts were made by the government to stimulate enterprise formation. These revolved around deregulation, privatisation, incentives for small business formation and growth, and general ideological exhortation. Such policies were intended to extend entrepreneurship and broaden the base of ownership of businesses (for example, through increasing the proportion of the population owning shares). Slogans such as 'popular capitalism' (Redwood, 1988) epitomised the changed orientation of policy (Jessop *et al.*, 1988). Enterprise was promoted in several ways – for instance, allowances to individuals setting up new businesses, raising tax thresholds such as those above which VAT was paid, relaxation of planning controls, and the introduction of the Uniform Business Rate. It is important to note, however, that this politics of enterprise operated on a *national* basis

and only very limited efforts were made to skew the assistance available towards places which were arguably in most need of it. Consequently, the impact of enterprise politics was distinctly uneven (chapter 5). Furthermore it was disingenuous to proclaim the regenerative and equilibrating powers of market forces when the countervailing tendencies structuring the pattern of regional development were so strong. These included the geographies of key institutions, imparting greater social power to the South East (chapter 2), and the geographies of the public and private financial systems (chapter 7).

Urban policy also underwent a dramatic change in character. When introduced in the late 1960s, urban policies had been welfarist in character, allocating funds through policies of positive discrimination to support a range of social and educational programmes in disadvantaged areas. Labour's 1977 Inner Cities White Paper recognised the structural causes of urban decline and proposed more interventionist policies, but that programme was launched at a time of severe restrictions on public expenditure. Labour's approach to inner city problems, with the emphasis on the key role of the local state and the bureaucratic organisation of programmes, was roundly criticised by the Conservatives, who nailed their colours firmly to the mast of a model of regeneration led by the private sector (for a detailed account see Atkinson and Moon, 1994).

The Conservatives' faith in market forces was most visibly symbolised by their attempts to create new spaces for enterprise. At times, urban policy seemed to assume the trappings of a quasi-messianic crusade to recapture areas where, it was alleged, enterprise had been stifled by the dead hand of socialism (Deakin and Edwards, 1993). Mrs Thatcher's statement after the 1987 election – 'we have a big job to do in those inner cities' – highlights this, symbolising both the depth of the problems therein, and the failure of economic growth and Thatcherite doctrine to have much impact on such locations (Robson, 1988; Ambrose, 1994). The political priority attached to Urban Development Corporations (UDCs) is indisputable: by 1990 one-third of expenditure by central government on spatial policy was being spent on seven UDCs.

The crucial elements of the Conservatives' diagnosis of urban decline mirrored those of their national economic strategy. The flight of private capital followed from high levels of business taxation, from restrictive planning policies, and the encumbrance of local authority bureaucracy. If the private sector was to be attracted back into the inner city, these problems would have to be tackled. The key innovations were the Enterprise Zones (EZs) and UDCs.

The underlying philosophy of EZs was to create the dynamics of a deregulated Hong-Kong style of economy in the UK. In the original concept, all forms of regulation – planning controls, health and safety, employment legislation – were to be weakened or abolished. This would stimulate enterprise, provide fertile ground in which small firms could take root, and (through spillover effects and linkages) promote development not just in the EZs themselves but also in surrounding areas. In the event, a watered-down version of the EZ concept was introduced, which had two main elements: the *de facto* abolition of development control and a presumption in favour of development; and the granting of rate exemptions (in some cases for up to 10 years).

UDCs could be established if it was held to be 'expedient in the national interest'. Implicit in this was the subordination of local political priorities and social needs to national economic considerations. UDCs became responsible for planning and transport issues within their jurisdiction, but not for housing or education, signalling that their priority lay with economic rather than social questions. No clear definition was established of what constituted regeneration, but the emphasis was heavily on property-led waterfront developments, mimicking American cities such as Baltimore or Boston. Critics have suggested that this exemplified a fairly predictable Conservative prioritisation of financial and property capital over industrial capital (Byrne, 1992; 1994). Moreover, this form of regeneration depended on conditions (particularly vacant land with minimal constraints on development) which do not

obtain in places most in need of new investment. The most spectacular example of the UDC concept in practice was the London Docklands Development Corporation (LDDC), assessments of which highlight the attractions of and limits to this form of regeneration policy. Controversy has focused on the efficiency, in cost-per-job terms, of the UDCs, the question of whether genuinely new jobs are created (as opposed to short-distance relocations), and the extent to which local communities benefit from new sources of employment.

Assessing the cost of new jobs created provokes disputes about what counts as financial assistance from the public sector. UDCs have typically focused on the extent to which public money has 'levered' private investment and there was much talk of impressive 'gearing ratios', i.e. the ratio of private to public investment. However, these ratios look rather less impressive if one adds in the costs of, for example, infrastructure provision (does the new Jubilee Line extension count as a subsidy to Docklands, for instance?) and taxes foregone in the form of rates exemptions. Moreover, many ostensibly 'new' jobs are in fact short transfers rather than long-distance relocations. The most famous examples of this are the newspapers which have transferred their printing operations to the Docklands from Fleet Street in Central London. These are not new jobs and may be associated with net reductions in employment overall, due to the introduction of new technology. Thus the impacts of the UDCs on employment should be given more cautious and critical celebration.

The more general criticisms of UDCs relate to their conception of regeneration. The model of regeneration was market- and property-led; this made the questionable assumption that benefits would 'trickle down' to the rest of the community. This was symbolised by redistribution of funds away from the (needs-based) Urban Programme to the (market-led) UDCs. In fact there is limited evidence that 'trickle-down' effects have worked, raising questions about the efficacy of this model (Turok, 1992; Imrie and Thomas, 1993). More generally, specification of the objects of policy has been vague: there was no real attempt to define the focus of activity in the general field of urban policy. While the market was favoured, it was plainly being underwritten by the state on a massive scale, and the mechanisms whereby markets were supposed to benefit *all* social groups were not clearly spelled out. While some locations have experienced spectacular flagship developments, trickle-down effects have been limited. Perceptions of the limitations of property-led schemes have been aggravated by the lack of effective community strategies on the part of UDCs (Imrie and Thomas, 1993, 18–19), and by the social polarisation associated with the forms of development that have taken place. The disparity, in the Docklands, between affluent residents of new housing developments and existing long-time inhabitants is particularly stark (Crilley *et al.*, 1992).

Writing on the abolition of the UDCs in March 1998, Robson (1998) argued that they had dramatically transformed the urban landscape in numerous locations. They had also added momentum to the process of repopulating the cities. Their emphasis on services, leisure and recreation, and residential development was consistent with 'the most plausible reading of the future roles that big cities will play' (p. 101). Against these achievements, urban problems in a broader sense have not really been tackled and if there has been a middle-class repopulation it has not been associated with social mixing (Crilley, 1990; Brownill, 1992). Some of the lessons from the UDC experience have been incorporated into subsequent policy initiatives, which are considered below.

12.5 ECONOMIC LOCALISM

There has been a rapid expansion in a whole range of local economic development initiatives (LEDIs). They could be defined to include the experiments in enterprise described above, but are usually thought of mainly in relation to the activities of local authorities and, more recently, TECs. In addition, there are the social conscience activities of organisations such as BSC (Industries) and British Coal Enterprises, the aim of which

was to recreate employment in areas devastated by coal and steel closures. In terms of scale LEDIs operate from a regional or city-wide level (such as the promotional activities of organisations like the Northern Development Company, seeking to attract inward investment) down to very small spatial units. Very few localities now lack such initiatives in some form.

There are several reasons for the expansion in these localist endeavours. One argument is simply that they are a necessary response to global changes in the organisation of production. Increased mobility of capital enables investment to flow to wherever local cost advantages can be created, thus increasing the returns for successful local initiative. The emphasis in some recent writings in economic geography on the competitive success of cities and small areas has given legitimacy to local strategies by implying that the 'replacement of a supposedly ubiquitous Fordist economic order by its postfordist sequel becomes a grinding historical inevitability' (Lovering, 1995b, 113). Consequently localities have little choice but to deploy a policy repertoire appropriate to this putative transition. Lovering argues that this is not an inevitable process. A succession of changes in political parameters – principally the withdrawal by central government from the task of economic development – has created a policy vacuum, but this policy vacuum is filled by local institutions with even less power than nation-states to steer developmental trajectories.

Analytically, distinctions have been made between 'moderate' and 'radical' strategies although the distinction might equally be between 'passive' and 'active' strategies. In the 'moderate/passive' camp are what were, until the early 1980s, the principal tactics used: essentially a 'publicity and premises' strategy. Local authorities would vigorously promote the attractions of their locality from a business point of view, and would also socialise some of the costs (principally, buildings and infrastructure) of providing appropriate property for SMEs. The focus therefore was on underwriting the costs of private production in a manner not dissimilar to conventional regional policy. Such approaches are often contrasted with the more 'radical/active' policies pursued by Left local authorities such as the GLC (Mackintosh and Wainwright, 1987). These are said to have advocated restructuring for labour rather than for capital, and argued for using the purchasing power of major local authorities as an instrument of progressive economic policy. Such strategies pursued responsible and equitable employment practices as well as greater workplace democracy and socially useful production. Direct intervention in the economy took the form of enterprise boards; these sought to target and support strategic sectors of local economies and, through planning agreements with firms, strengthen the capacities of firms to survive. As well as direct financial assistance the enterprise boards also gave technical assistance (to which small firms traditionally lacked access).

However, the distinction between these two approaches is less clear-cut than it at first appears. Despite their 'radical' aspirations, interventionist approaches foundered: their funding was limited (designedly so as a result of central government legislation); there was an inherent tension between economic and social priorities, usually resolved to the advantage of the former; strategies were often defensive and reactive; and ultimately economic development strategies fell back on infrastructural provision, socialising some of the costs of private development.

Consequently, Eisenschitz and Gough (1993, 87) argue that the pendulum has swung back between these two opposites, to what they view as a 'consensus' or 'centrist' approach. While this sees markets and enterprise as the basis of a dynamic economy, it also recognises the need for steering market forces in order to deliver growth and avoid social exclusion. There is an emphasis on enhancing the cohesion of localities through promoting cooperative relations between social groups and through stimulating productive linkages. Their argument is that this centrist consensus contains the following elements: small firm support; broadening participation in enterprise through supporting disadvantaged groups; promoting physical renovation and creating premises; technol-

ogy policy (e.g. technology training; university linkages; science parks (Massey et al., 1992)); and active labour-market and training policies. It could be argued that a policy consensus including so many elements is vacuous and that this reveals a lack of clarity about what can be expected of LEDIs.

The impacts of even the most vigorously pursued initiatives may be limited. The former steel town of Consett, Co. Durham, has experienced a range of schemes since the steelworks closed in 1980. However, Hudson and Sadler (1992) suggested that jobs created *since* the closure had not even kept pace with post-closure job losses, much less made inroads into the pool of unemployment left when the steelworks shut. Consett was arguably an extreme case of an undiversified economy; steel and coal interests had, in the past, apparently discouraged efforts to bring alternative employment to the town, in the belief that this would attract men away from these staple industries. But a similar story could be told of efforts to replace jobs in other localities: a strategy based on SMEs would have to see the development of very substantial numbers of enterprises even to compensate for the loss of one pit. Moreover, concerns about job quality and the costs of new employment can as easily be voiced about LEDIs as about more conventional spatial policies. Competition for subsidised and low-wage jobs is almost inevitably a zero-sum process in which jobs created in one location are at the expense of job losses elsewhere.

There are, then, very real difficulties in assessing the impacts of LEDIs. Moreover, many strategies are competitive, not complementary, so there is a real need for greater coordination. In such a competitive climate it has undoubtedly been possible for investors to play local agencies off against one another in seeking the most attractive package of incentives. Furthermore, most economic development initiatives consist, in practice, of exercises in place-marketing: constructing attractive images with which to 'sell' a locality to investors. It is often argued that growing local involvement in economic development indicates an entrepreneurial stance, but Lovering (1995b, 115) suggests that recent developments in many British cities would be better 'described as "commodification", attempting to package and sell what is already there'.

Given these sceptical remarks it might be surprising that LEDIs are viewed in a positive light, but Gough and Eisenschitz (1996) locate them in the wider context of debates about economic modernisation. They suggest that LEDIs are seeking to address several criticisms of the archaic and anachronistic character of British capitalism. Thus LEDIs can promote economic modernisation in several ways. They can make finance available on more favourable terms than those available from the City of London, and they can help counter the anti-industrial biases of the City. They can stimulate the provision of training and facilitate access to new technology, both of which are areas in which the market fails to invest. They may also promote more cooperative social relations and workplace harmony. This optimistic vision of local modernisation must be tempered in several key respects. First, the mobility of capital militates against local strategies, which have found difficulties in persuading large firms to establish closer, more supportive relations with suppliers and customers. Second, local democratic institutions simply lack powers to curtail the freedoms of individual firms, or to arbitrate in inter-firm disputes, while the absence of strong regional banks prevents the kind of leverage on firms that financial institutions in other states can exert. Third, policies are very fragmented (training, finance, technology, marketing, etc.) and overcoming this would require a recentralisation of authority, possibly via local government, which Labour seems unwilling to contemplate although the proposed RDAs may help. Fourth, there is desperate competition between places for available investment. Finally, modernisation projects nationally have always been about prioritising industrial development and long-term investment, but the reality for many LEDIs has been an acceptance of whatever jobs are on offer and a tendency to hit 'soft' targets (like numbers of jobs) without regard to quality. These comments indicate the constraints on local, as well as national,

initiatives in circumstances where there is heightened mobility of capital and a reluctance to raise taxation levels for fear of losing mobile investment. These are the parameters within which a revived spatial policy must operate.

12.6 REORIENTING URBAN POLICY: MARKET-LED PROGRAMMES AND THEIR LIMITATIONS

As well as creating new spaces for enterprise one can also discern efforts to transform the *modus operandi* of existing public policies as well as to redistribute funds within programmes. As a consequence of the massive expenditure on the UDCs, for example, there was a clear redistribution from regional to urban aid. However, of perhaps greater long-term significance is the transformation of the mechanisms whereby funds were allocated, from a needs-based system to one driven by competition. This has clear parallels with the simulation of market forces elsewhere in the public sector (see, for example, chapter 9).

The main innovation here was the switch of funding from the Urban Programme to City Challenge. Conservative critics argued that the Urban Programme was bureaucratic, lacked incentives for innovation, and was state-centred. Consequently, under the City Challenge scheme (from 1991), a process of competitive bidding for funds was instituted. Furthermore, because City Challenge was based on a concept of partnership between local authorities and other community organisations (principally businesses and TECs, but also educational, health and voluntary agencies), it was less state-focused than the Urban Programme, and it was hoped that this approach would produce innovative solutions geared to local needs A similar competitive approach to bidding was also adopted for the Single Regeneration Budget (SRB; see page 192).

Despite the language of partnership, City Challenge proved to be heavily private-sector led and indeed property-led; much depended on the availability of substantial parcels of land within reasonable proximity to major commercial or office developments. While the rhetoric of partnership was welcome (community organisations had had comparatively little input to urban policy initiatives previously), in practice economic and labour market issues have taken priority over housing and social issues. Finally, there was no objective relationship between levels of need and funding allocations; for instance, in the first round, identical sums of money were awarded to all successful bids; of the local authorities regarded as the most deprived in England, only five received funds in the first two rounds of the scheme. The process of competition meant that the most spectacular bids, or those making greatest claims for leverage of private funds, tended to win, and these were not necessarily the most needy authorities, as analyses of relationships between levels of deprivation and success in City Challenge bids made clear.

The Single Regeneration Budget (SRB), announced in 1994, was a response to forceful critiques of the failures of urban policy. These included: the absence of a coherent policy; the overemphasis on the private sector; the marginalisation of local government; and the absence of effective channels for participation (Robson *et al.*, 1994). The SRB brought together 20 programmes from five central government departments (the Department of the Environment, the Department of Trade and Industry, the Department for Education, the Home Office and the Employment Department) under the overall control of the Department of the Environment to secure greater coordination among programmes that had implications for urban areas. Bidding guidance for SRB funds emphasised the need to use resources in a strategic manner, moving away from the previous situation in which a multiplicity of unrelated schemes operated in a locality. Initially there were no new funds for the SRB as all money was committed to ongoing projects; subsequently, however, additional funds became available and were allocated (as was the case with City Challenge) on a competitive basis. Unlike City Challenge, SRB allowed all areas in England to submit

bids; this was defensible to the extent that deprivation is not confined to the priority areas at which City Challenge was targeted. However, the effect was to move funds away from the largest and most deprived cities, identified as the locations which were doing least well in terms of attracting regeneration funds or indeed in terms of 'regeneration' (Atkinson, 1998a, 10). Furthermore there was a clear relationship between the chance of a bid being successful and its regional location, a relationship determined by the funding allocated to each region. Thus, bids of similar quality did not have the same likelihood of succeeding.

Thus SRB represented an advance on previous initiatives but was still open to criticism. Furthermore, although SRB bids could in principle be led by any agency (statutory, private or voluntary), in practice most bids have been led by local authorities and there has been limited time in bid preparation for involving community organisations. Hence the risk is that much-vaunted participation is purely symbolic. Finally, SRB's reliance on competitive bidding, and a continued emphasis on leverage ratios, mean that bids may focus on areas with development potential rather than on disadvantage, thus replicating the disadvantages of property-led regeneration. Nevertheless, as SRB was broadly welcomed by Labour in opposition, it has been retained, though with modifications to the criteria by which funds are allocated. Thus, guidance issued to applicants has emphasised the importance of complementing the government's manifesto commitments such as the welfare to work programme. Funds will also be concentrated on comprehensive regeneration schemes in areas of greatest disadvantage, with 80 per cent of new resources being concentrated in the most deprived areas. This is a welcome move away from a process of purely competitive bidding. Bidding guidance also requires evidence of the mechanisms for securing community involvement in schemes, and demonstrations of the links with other new area-based initiatives and strategies (health or education 'action zones' for instance).

Finally, mention should be made of one of Labour's new policy initiatives, the New

Deal for Communities, for which £800 million has been allocated. This builds on the work of the SEU and will offer intensive help to the most disadvantaged areas. Among other things it will attempt to avoid the fragmentation of previous policies, whereby 'a joined-up problem has never been addressed in a joined-up way' (SEU, 1998, 9). A key diagnosis of the SEU report is of a lack of coordination at neighbourhood level. The intention is to create integrated approaches, supervised by cross-departmental working groups in Whitehall, to overcome these problems. There are undoubtedly ways of integrating the work of different agencies – expenditure on housing construction and renewal could, if properly targeted, contribute substantially to the process of regeneration. The phased release of receipts from sales of council houses will be helpful here. However, as Atkinson (1998b) observes, the fragmentation of urban policy remains only too evident, and rhetoric about community partnership can be vacuous in practice.

A full assessment of the impacts of urban policies has been provided by Robson *et al.* (1994). However, a crucial point is that despite substantial resource inputs, much more money has been taken *out* of disadvantaged places through restrictions on local government expenditure. Moreover, despite the free-market rhetoric, innovations such as EZs or UDCs actually represent highly subsidised enterprise.

Shaw and Robinson (1998) make several further relevant points. They echo previous criticisms of the failures of regeneration policies (the limitations of property-led developments; the absence of trickle-down effects; the limited impacts of a scattergun approach to policy). They stress, as do many critics, the need for holistic, integrated approaches; this has been reflected in changes to administrative arrangements for programme delivery, evidenced in a broadening-out of the scope of City Challenge and SRB programmes. These are now in many cases addressing issues such as crime, education, health and social exclusion, in a manner which responds to needs identified by the communities concerned and which seeks actively to engage communities in devising solutions to social problems. They argue that broad-based coalitions are essential because the exit of funding agencies (UDC; City Challenge) will have a destabilising effect on programmes solely geared to maximising existing resource 'streams'. Such partnerships must also have clear aims and realistic objectives, rather than the vacuous and overblown rhetoric associated with many urban renewal projects (see Edwards, 1997). Urban policy has been driven by 'periodic knee-jerk reactions to failure' (Wilks-Heeg, 1996, 1264) and as a consequence policies are tailored to fit the requirements of grant-making agencies rather than reflecting local needs. In particular, economic and physical indicators (leverage ratios, office floorspace) have been given prominence and precedence over social indicators. This reflects the difficulty of 'measuring' social aspects of regeneration and the severe challenges associated with regenerating people rather than simply reconstructing the physical fabric (Shaw and Robinson, 1998, 60).

12.7 REVIVING THE CASE FOR REGIONAL DEVELOPMENT AGENCIES

Two sets of pressures, originating in quite different contexts, produced demands for some form of regional agency to coordinate development. First, the overdevelopment of the South East produced demands for more coordinated planning and, in particular, for a strengthening of strategic planning. The constraints on development in the region during the 1980s are well known (Breheny and Congdon, 1989). Some went so far as to claim that the rapid growth of the South East's economy and the associated 'regulatory deficit' were harbingers of a post-Fordist mode of social regulation and indeed of the collapse of Thatcherism (Peck and Tickell, 1995b). There is little doubt that such a regulatory deficit did exist in the region, and that this signalled some of the limitations of Thatcherism as a mode of social regulation. Whether the problems of unbalanced growth can, even in part, be held responsible for the downfall of Thatcherism seems much more debatable. Nevertheless, these problems of unbalanced growth were recognised by the Conservative Party's representatives both in Parliament and local government, and their complaints and those of others prompted renewed thinking about the merits of regional policy, evident particularly in debates about the case for strategic planning, especially for London and the South East, to fill the policy vacuum which existed.

The second reason why some came to regard regional government as an 'economic imperative' relates to the lack of coordination of economic development activities in the regions (Murphy and Caborn, 1996, 187–9). Competitive bidding for funds such as City Challenge inhibited cooperation between neighbouring local authorities. Effort was focused on eye-catching, novelty projects rather than long-term strategic planning. There have also been problems arising from the complex map of areas eligible for assistance in various forms (European Structural Funds, Assisted Areas, areas eligible for urban policy funds, Rural Development Areas). As a result there has been a 'bewildering profusion' of economic development bodies, leading to duplication and waste in use of the available funds (Trade and Industry Select Committee, 1995). The creation, in 1994, of the Government Offices for the Regions (GORs) was one attempt to respond to such criticisms; Labour's Regional Development Agencies (RDAs) take this process further.

The RDAs were created by central government and will be accountable to Parliament. To allay fears that they will be business-dominated quangos, they will include representation of statutory bodies and community organisations, although those with a business background will predominate, appearing to contradict earlier statements about community representation, and causing concern about democratic accountability (see chapter 11). The purpose of the RDAs is to 'promote sustainable economic development and social and physical regeneration, and to coordinate the work of local partners in areas such as training, investment, regeneration and business support' (DETR, 1997).

Their novelty, according to the government, lies in the 'integration of economic, social, environmental and democratic elements', the implication being that the GORs were only a half-way house. The RDAs will be responsible for coordinating business support services, training and other labour market programmes (including ensuring that further education provision reflects the needs of the labour market), regeneration programmes (including SRB and European Structural Funds) and tourism promotion. Their activities will therefore comprise both monitoring and advising other organisations, and direct delivery of services in partnership with other agencies. They will take over various responsibilities from other bodies – for example, SRB administration (from English Partnerships). Reflecting concerns about a separation of urban regeneration policy from economic development, they will seek to integrate the two. A further novel feature of their remit is an emphasis on sustainable development. There is a stress on the quality of the natural and built environment as a potential resource, both in attracting inward investors and in promoting tourism. There is also an emphasis on the economic benefits of sustainability and the potential for waste minimisation and the application of new technologies. References are also made to the need for RDA strategies to form an input into reviews of regional planning guidance, which affects strategies for the provision of housing, transport and other infrastructure.

In this context there is a potential conflict between the (laudable) concerns of the RDAs to promote sustainability, and the (equally laudable) intentions of the government to promote competitiveness by encouraging the formation of industrial 'clusters'. The 1998 White Paper, *Our Competitive Future* (DTI, 1998), focused on the idea of a knowledge-driven economy and the implications this has for determinants of growth and the organisation of production. This document argues strongly for the creation of 'clusters' of related activities, claiming that clustering creates a 'critical mass of growth, collaboration, competition and opportunities for investment' (para. 3.25). The arguments for clusters owe much to debates about new industrial spaces which stress the importance of repeated contact and collaboration between firms as a means of stimulating flows of tacit knowledge and innovation (chapter 5). Several other forms of collaborative relationships between firms are also mentioned in the White Paper, all of which require spatial proximity. An example given is biotechnology; embryo clusters have been identified around Oxford, Cambridge and Dundee. The first two locations are among the most severely pressurised localities in the UK and it is difficult to see how they could accommodate many more rounds of investment. Yet the Competitiveness White Paper insists (para. 3.29) that reconciling the demands of existing or likely future clusters with 'wider environmental objectives' is feasible, even though the same paragraph speaks of 'reducing planning delays' and 'reviewing how the planning system can best help promote the needs of businesses in growth industries'. Somehow this circle seems unlikely to be squared, not least because elsewhere it is suggested that government cannot create clusters; it can only 'create the conditions which encourage their formation and growth' (para. 3.26). The White Paper does not spell out in detail how governments might do so. However, recent developments in economic geography appear to have been influential on some spatial policy initiatives.

12.8 THE FUTURE OF REGIONAL POLICY: THEORY LED BY POLICY, OR POLICY LED BY THEORY?

One response to criticisms of regional policy has been to argue for administrative reforms which simply refine the existing *modus operandi* of policy. Thus Taylor and Wren (1997, 844) call for, *inter alia*: greater targeting on new and existing small firms with growth potential; expanding the sums available to very small firms; relaxing cost-per-job limits by permitting the inclusion of 'indirect' jobs created upstream or downstream of assisted projects; and, of course, creating Regional Development Agencies. Others advocate

changes more in line with contemporary thinking on regional development and the competitive success of regions. Morgan (1997) argues that the old paradigms of Left and Right are exhausted, the former by an excessive faith in the powers of the state, the latter by the demonstrable failings of neoliberal experiments in the 1980s. He argues that a new paradigm is emerging, referred to variously as the 'network' or 'associational' paradigm, which provides an alternative to markets or hierarchies as a way of mobilising resources for innovation and economic development. In terms of realising the potential for the creation and diffusion of economically exploitable knowledge, markets and hierarchies have limitations; in hierarchies, knowledge is power while in markets there is minimal trust. By contrast networking indicates a predisposition towards 'reciprocal, preferential and mutually-supportive interactions' (Cooke, 1995, 13). In Morgan's view this process of networking might be achieved most effectively at a regional level, provided that regional policy is conceived as 'a dimension of innovation policy, rather than just a social welfare measure'; the reason for this is that regional policy has been excessively preoccupied with symptoms (like high unemployment) rather than causes (such as low innovation potential). He argues this with reference to contemporary theories of the innovation process, which emphasise interaction, networking, and the ways innovation is shaped by a 'variety of institutional routines and social conventions' (p. 493). There are links here with Putnam's (1993) concept of social capital, which refers to 'features of social organisation, such as networks, norms and trust, that facilitate coordination and cooperation for mutual benefit', the point being that innovations are more likely to emerge in localities with strongly developed social capital.

Informed by recent developments in theorising about the comparative success of regions, new thinking on regional policy centres on three points: the need to create endogenous growth capacity via the promotion of innovation in technologically advanced sectors; the importance of networks which promote relationships of trust and reciprocity, rather than relationships mediated purely by price mechanisms; and the development of new forms of governance below the level of the nation-state, in recognition of the contemporary significance of regional economies.

Storper's (1995) recent work indicates some important connections between location theory and regional policy. He argues that the resurgence of regional economies results not simply from the economies gained through transaction cost mechanisms *à la* Scott, but also from territorialisation – economic activity is dependent on resources which are territorially specific. These resources take the form of 'untraded interdependencies' – the sharing of tacit knowledge, conventions, research and design ideas, etc. – which arise in circumstances where agglomerations of industry exist with dense inter-firm linkages of both formal and informal kinds (for a British example, see Pinch *et al.*, 1997; and chapter 5). Successful industrial clusters therefore comprise 'aggregates of firms and non-firm institutions that supply external economies of scale through a capacity for optimisation of learning practices' (Cooke, 1995, 11). Such clusters are, according to Cooke, characterised by cooperative rather than purely competitive inter-firm relationships. The former involve reciprocity and custom and are sustained by the need to preserve a firm's reputation and integrated by trust. Competitive relationships, on the other hand, are regulated purely through a rational cost-benefit comparison of alternatives by firms; rather than collaboration there is minimal exchange or disclosure of information by one firm to another. The challenge is therefore to develop an interactive, even transparent, high-trust business culture. In addition to the economies of scale associated with agglomeration, the real benefit of such networks is the promotion of a 'learning capacity', and Morgan demonstrates this at a number of spatial scales, from the nation-state to the firm. The advantage of promoting a learning capacity is also that it can develop flexible responses to economic change, to avoid becoming locked into particular resources or branches of production. As

Cooke (1995, 239) puts it, in an economic climate characterised by higher levels of risk and uncertainty it is only regions which possess institutional reflexivity, or the capacity to anticipate and plan for change (rather than merely reacting to it *post hoc*) which will succeed (see also Grabher and Stark, 1997). Places which simply react after the event (post-closure reindustrialisation strategies) or which unreflectively borrow elements of ideas from elsewhere and seek to transplant them, will have limited chance of success. Critics of this view regard it as a bland statement of the obvious (Lovering, 1999).

Morgan proceeds to outline one form that regional policy might take (Morgan, 1997; see also Cooke and Morgan, 1993; Cooke, 1995). Essentially it is about developing a *regional innovation process*, bringing together regional stakeholders to define a common agenda and strategy sensitive to regional needs. This entails promoting at least three kinds of competence: technological competence, i.e. mastering particular technologies relevant to firms' needs; entrepreneurial competence, i.e. the capability to integrate relevant technologies with the wider corporate strategy of the firm; and learning abilities, which means structuring organisational and management routines within firms, enabling them to absorb information on changing markets, technologies and organisational structures. Morgan describes the preliminary efforts of the EC-sponsored Regional Technology Plan exercise, launched in June 1994. In his view the exercise is about stimulating a collective learning process, acknowledging that the impetus for regional renewal must come from within the regions and that this 'turns on the region's networking capacity'. The exercise is engaging with what Morgan regards as the 'right targets', namely the institutional and institutionalised inertia which characterises so many peripheral regions, in contrast to the 'glorified subsidy regimes designed to attract mobile capital' which previously typified regional policy. However, it must be questionable whether networking within the region, between already existing firms, can do much to create new employment.

For this process to work, attention needs

to be paid to the institutional frameworks through which networking will take place. Following Hirst (1994) Cooke calls for 'regional associationalism', a bottom-up version of democracy in which the functions of democratic governance are devolved to the greatest extent possible, including the granting of powers to voluntary self-governing associations. This is a rather more radical conception of decentralisation than that currently on offer for the British regions. Other critics suggest that regardless of the extent of decentralisation of powers, the weakening of the nation-state will produce structures of governance even less capable of acting as effective counterweights to global capital (Hudson, 1997b). An alternative criticism is that strategies such as those proposed by Cooke and Morgan cannot solve the key problems facing the peripheral regions of Britain, namely mass unemployment and social exclusion (Lovering, 1996). Dealing with such problems may well require much more active labour market policies than currently exist, as well as determined efforts to redistribute work. Finally, if Morgan and Cooke's arguments are correct, decades of relative and absolute decline cannot be reversed overnight, and the rich networks of social capital characteristic of Europe's regional success stories arguably took centuries to form. Moreover, Britain lacks the powerful regional partnerships that have distinguished the growth regions of West Germany, and it is pointless to pretend they can easily be constructed.

A much more fundamental critique can be put forward, which is that the so-called 'new regionalism' in economic geography rests on a partial understanding of contemporary patterns of regional development. Lovering's (1999) discussion of applications of the new regionalism to Wales could equally be applied to other areas of the UK. First, there is a partial focus on high technology and innovation, especially as applied to manufacturing, at the expense of other sectors of the economy. Amidst all the publicity about FDI in Wales, the largest contributor to recent employment growth in Wales has been the public sector. Second, the distributional consequences of change are glossed over:

technological change and innovation characteristically destroy more jobs than they create. However, Lovering makes more fundamental points than these. The resurgence of regional economies has been proclaimed by many writers often from quite different starting-points. The policy prescription is that the regional scale can facilitate the interactions necessary for flexibility and responsiveness in the face of globalising markets. However, there is a lack of clarity about the mechanisms through which 'the abstract "region" posited in these theoretical adventures relates to the actual regions in which real people live and work'. For Lovering, vague statements about 'institutional reflexivity' and 'learning regions' are inadequate substitutes for careful conceptualisation and abstraction. The result is the production of 'vaguely specified analytical boxes' which are filled in an ill-fitting way with whatever empirical examples come to hand.

There is also an 'obsessive focus on the supposed logic of informational economies' and enormous weight has been placed (as in the Competitiveness White Paper) on innovation. Lovering argues that the emphasis on the need for permanent innovation is taken as read, and as something to which firms and regions must adapt. By contrast, he contends, innovation has to be related to two decades of neoliberalism and fiscal austerity, which have precipitated particular forms of competition – of which innovation is one. Moreover, innovation and new technology have been used largely to cut jobs rather than to meet unmet needs, create new product markets, and support employment growth.

A related problem concerns the focus of 'new regionalists' on 'competitiveness'. There are immense problems in transposing the concept from the level of the firm to the level of a territorially defined collectivity (a nation, or a region); the economic position of such a collectivity cannot be reduced to something measurable by a single indicator. Lovering argues that the focus on 'regional competitiveness' can be used to justify a bias towards larger firms, international business, and 'hi-tech', regardless of how many jobs are actually created in these sectors. This bias is not accidental and, of course, reflects an implicit political choice.

This brings us to Lovering's final point, which is that the 'new regionalism' is influential not because of its intellectual merits but because of its utility to the burgeoning economic development industry and to related 'powerful industrial, state and social constituencies'. The ideas in the 'new regionalism' are there 'because they seem to resonate ... [when considering] the scope for policy initiatives at the regional level ... the policy tail is wagging the analytical dog'. Lovering's view is that this fit between policy and theory is because theorists have accepted – or ignored – many of the parameters of the debate, rather than challenging them.

Thus, one of the reasons regions and localities are becoming more significant is precisely because of the rolling back of the nation-state. This was not an historical inevitability, but resulted from quite deliberate policy choices (e.g. in Britain and the USA) substantially to weaken and fragment national redistributive policies. Against this background regions have had little choice but to become more pro-active but this is not the same as being good at it, or being the most appropriate tier of government for such activities (see also chapter 11). Thus, Lovering argues, a regional policy based on competitiveness and innovation and informed by the new economic geography, will be flawed and partial. It may produce some limited successes but the idea that regions such as South Wales can emulate European regional success stories is hopelessly optimistic. Such innovation-based policies will invariably lead to greater (and probably unsustainable) interregional polarisation in the absence of strong macroeconomic policies to promote an expansion of employment and a commitment to progressive social policies.

13

SUSTAINABLE GEOGRAPHIES?

13.1 INTRODUCTION

Many texts on the UK's human geography have, in the past, emphasised the connections between physical (especially mineral) resources, patterns of economic activity, and population distribution. It might be argued that a predominantly service-based economy has been (at least partially) liberated from the constraints of its physical resource base. But existing patterns of development (and especially of consumption) threaten to reopen the issue of physical constraints, while changes in agricultural practices and the inexorable rise in energy consumption raise entirely new issues of risk. Hence a discussion of sustainability is central to the concerns of this book.

The Bruntland Commission on Environment and Development is widely credited with placing sustainability at the centre of political debate. This Commission argued that, in order to allow all humanity to attain Western consumption levels, the resources of ten Earths would be required. Moreover, when even current levels of energy consumption are credited with responsibility for global warming, and when renewable resources such as forests and fisheries are believed to be on the verge of irreversible decline, it must be acknowledged that 'humankind is getting within touching distance of the determining factors for global carrying capacity' (McLaren *et al.*, 1998, 12).

In the past the result might have been localised famine and economic collapse; 'today the whole planet is local' (ibid.).

The problem, of course, is that environmental problems are the result of decisions by millions of individuals to consume resources. Driven by imperatives to increase economic growth, the promotion of personal consumption has become 'probably the single most important objective of modern politics' (Jacobs, 1997, 47). The obsession of politicians with 'consumer confidence' and the elusive 'feelgood factor' makes this clear. Governments are not going to advocate a reduction in living standards. Inability to check consumption ultimately is therefore implicated in causing the greenhouse effect through the high levels of energy consumption associated with Western lifestyles and the depletion of mineral and timber resources, while demands for cheap food lead to agricultural intensification and, in turn, pollution and pesticide hazards, to say nothing of exposure to new risks such as BSE.

Environmental economists emphasise that, however rational existing patterns of consumption may be for an individual consumer or business, they have externalities which impinge on other people, societies, or the entire world. Environmental resources are generally 'undervalued or ignored, [or] regarded as free goods' (Blowers, 1997, 37). Consider waste disposal. Rational entrepreneurs find it cheaper to act as 'free riders' and discharge wastes rather than purify them, as to do otherwise would erode their competitive advantage. Each individual will carry on in this way until the global environment becomes so damaged that their individual profits are affected, by which time it may be too late to take effective action. This underlines three important points: the conflict between short-term economic gain for individual units in the system, and the longer-term interests in collective survival; the need for intervention of a kind capable of reconciling business and public interests, including inter-state cooperation; and the tendency for impacts to become manifest only once irreversible damage has been done (Blowers, 1997, 39). Whitelegg (1997) provides a classic example: a clothing company, based in Switzerland, sends cloth to Portugal for assembly into finished garments, because the savings in labour costs are three times those of the costs of transport. The externalities of such excessive energy consumption are not reflected in the existing regulatory environment.

Although externalities are widely acknowledged, assessing harm to the environment is the source of dispute between those adopting a 'precautionary' principle and those taking a more robust stance, requiring firm proof before taking action. The precautionary principle acknowledges that because the effects of actions on the environment are not determinable within a reasonable margin of error, regulatory measures need to be cautious and, in a sense, 'go beyond science' (or what is scientifically proven) to allow for unforeseen consequences. However, the UK government has not subscribed to this principle (Munton, 1997, 150; Winter, 1996, 268). Environmental problems have generally been assessed 'within frameworks of technical evaluation and formalised "risk assessment"'. Such frameworks were perceived to be necessary to justify the 'costs' of environmental improvement to powerful government departments. The consequence was a restricted definition of problems as 'essentially physical, and thus … tractable in principle to scientific, managerial and economistic methods of control' (Grove-White, 1997, 111–12). Official responses to

particular issues or emerging risks therefore insisted on the need for sound scientific evidence in advance of action. Such responses were increasingly at odds with a deepening public unease about risk, particularly the cumulative and unpredictable aggregate impacts of human activity on the environment. Protests over road construction symbolise these issues very well: justifications for new roads are usually framed in terms of an economistic cost-benefit analysis which simply cannot capture the deeper cultural issues at stake. This produces considerable frustration at the human and cultural insensitivity of transport policy.

If there is disagreement about the impacts of human activity on the environment, there is even more controversy over the meaning of sustainable development. The Bruntland definition is widely repeated: 'development that meets the needs of the present without compromising the ability of future generations to meet their own needs'. However, such a broad concept has a remarkable capacity to gloss over differences of opinion (Evans, 1997, 7), creating the illusion that growth and environmental protection are mutually reconcilable. Interpretations which stress the maintenance of 'environmental capital' have become increasingly influential (Cowell and Owens, 1997). This implies bequeathing a stock of capital with an ability to produce well-being at least equivalent to that enjoyed by the present generation. Some functions of the environment are vital and irreplaceable, and so human activity must be managed so as to conserve 'critical environmental capital'. This shows some affinity with earlier discourses of limits to growth. McLaren *et al.* (1998, 6) have attempted to show what these ideas might mean in practice by using the concept of 'environmental space': the amount of various environmental resources that we can use at one time without breaching environmental limits. Sustainable rates of resource use are not only determined by the carrying capacity of ecosystems, the recuperative ability of the natural environment, and the availability of non-renewables, but also by assumptions about substitutability and patterns of consumption. This approach incorporates assumptions about the share each country takes of environmental space; implicit in this is the judgement that, for sustainability to be achieved, the West's share of environmental space will need to be reduced. As well as these economic and political dimensions of sustainability, questions of culture and values come into play, especially in the area of conservation, where there are debates about preserving particular environments for their aesthetic or historic value.

This chapter can do no more than illustrate different facets of debates on sustainability. Informed by a concern about whether the current development trajectory is sustainable, the focus is on urbanisation and the planning system, transport trends and policies, and the environmental impacts of agriculture. These raise different but related issues: the costs of deregulated urban sprawl; the problems of challenging powerful vested interests; and the difficulties of assessing risk and apportioning responsibility. In respect of each policy area attention is given to current and future trends, and to the attempts by government to regulate in the interests of sustainability.

13.2 URBANISATION AND PLANNING

The containment of urban growth has been a long-standing problem and the need for some sort of balanced regional growth informed early town planning and regional policy legislation (chapter 12). The issue also surfaced in the context of debates about local government reform and innovations in regional governance (chapter 11). Urbanisation poses several challenges for sustainable development, notably land take, availability of construction materials, disposal of waste, and transport requirements. Given the priority attached to owner-occupation and the long-term drift of population away from cities, the question arises of how to meet demands for new housing. Neoliberals argue that the problem of land availability arises because the planning system restricts the supply of available land, and bids up the price above the free-market equilibrium, contributing to problems of affordability and inhibiting labour mobility. Opponents point to the environmental and social costs of development, which are not borne by developers. There is clearly a tension between these views, and the associated debates reveal much about the ways sustainability has been interpreted in Britain.

The broad trends can be summarised as follows. First, while the UK population is expected to stabilise, the number of households is expected to increase by approximately a quarter: the most widely quoted estimate is of an additional 4.4 million households, the majority of which will consist of one person. Approximately 1.7 million additional households will live in SE England, already the most pressured region in the country. Second, there will be considerable pressure on rural areas, where the number of households is expected to increase by 25 per cent between 1991–2011. This implies a need to construct up to 2 million homes in rural areas. Third, projected growth patterns vary substantially (figure 13.1): whereas in much of northern England under nine new housing units would be required per 1000 population, the figures for a belt of counties from Dorset to Cambridgeshire are typically double that. In total approximately 170000 ha would be converted to urban use during 1991–2016, with the largest increases in absolute terms being in Devon (7700 ha), Essex (7000 ha), Hampshire (6600 ha), Kent (6000 ha) and Cambridgeshire. The resulting development pressures depend on whether or not constraints exist – such as areas subject to national policy designations (Green Belts, Areas of Outstanding Natural Beauty (AONBs) and so forth) – so it is more realistic to express projected urban growth as proportions of undesignated and currently undeveloped land. Nationally this figure is only 2.3 per cent, but in Greater London projected growth would absorb all undeveloped land. The area of undeveloped/undesignated land in Berkshire, Surrey and Hertfordshire would have to fall by 11.2 per cent, 14.8 per cent and 6.3 per cent, respectively. Other areas subject to major development pressures will include Avon (10.8 per cent) and the former

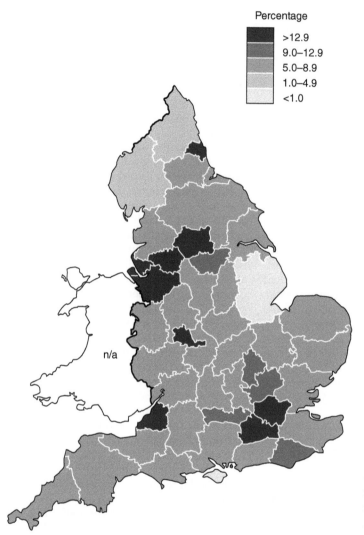

Percentage

- >12.9
- 9.0–12.9
- 5.0–8.9
- 1.0–4.9
- <1.0

n/a

FIGURE 13.1 Urban growth as a percentage of land in rural uses which is undesignated, 1991–2016 (Bibby and Shepherd, 1997)

metropolitan counties (24.9 per cent in Greater Manchester to 6.3 per cent in S. Yorkshire) (Bibby and Shepherd, 1997, 120–1). Finally, as some of these county-level figures indicate, there will be a continuation of the long-term trend of urban–rural migration. Grouping areas in Britain by their degree of 'urbanness', Champion (1997, 78) shows that the six most 'urban' types of area all experienced net out-migration in 1990–91; the other seven experienced net gains (his statistical source only gave information for the 12 months prior to the 1991 Census). The greatest losses (1.24 per cent) were in Inner London; the highest gains (0.73 per cent) were in remote rural areas. Metropolitan England (London plus the six principal conurbations) experienced a net out-migration of 1.25 million people for 1981–94. These figures provoke debates not just about rural development pressures but also about the future of cities. Pessimists see absolute out-migration from large cities (at 5–6000 per week) as a long-term threat to urban viability. For optimists, the *net* migration figures suggest that the balance between urban and rural areas could be tipped back in favour of cities if they were made attractive enough places to live in

Demand for new housing has conventionally been dealt with by a 'predict and provide' approach. Local authorities have been charged with ensuring that land and construction materials are available to meet anticipated demands. Some of the controversies relating to housing development have been well rehearsed, such as: protest against new house building in certain counties from established residents, who are often accused of NIMBYism ('not in my backyard'); disputes between central and local government, with central government sometimes overriding local authorities to impose higher targets (Short *et al.*, 1986); and proposals for entirely new greenfield developments on the grounds that existing settlements cannot absorb anticipated growth. The absence of a strategic framework in which to consider proposals for new housing developments during the boom years of the 1980s was the subject of considerable complaint from backbench MPs (for example, Mohan, 1995a, 89). This led to a modest revival of interest in regional planning. It also led to revised planning guidance on out-of-town shopping developments, with more stringent criteria being applied to proposals, including assessments of their impacts on existing retail facilities and transport patterns. Of course, as several hundred such proposals were in the process of development when this guidance was announced, the horse had already bolted (Hall, 1997, 1322–3) and the government subsequently approved an enormous out-of-town regional shopping centre (Bluewater Park, Dartford) principally on the grounds of employment creation. The legacy of such policies is the sprawl of low-density housing developments throughout much of the South East, ill-served by public transport and therefore highly car-dependent. Such pressures pose important challenges, which cannot be dealt with solely by the planning system: planning decisions only affect 1–2 per cent of the built environment per annum (they only deal with *changes* of use), so they are swimming against the tide. Moreover, without clear national guidance on transport pollution and biodiversity, planning decisions are inevitably taken in a vacuum.

The sheer availability of physical resources is one element of this, with water abstraction as a good exemplar. Although at present the UK is self-sufficient in water, predictions suggest that even medium growth in demand would create water deficits by 2021 over much of southern England (Slater *et al.*, 1994) assuming current levels of per capita consumption. However, absolute shortages for human use are only one facet of the problem. Excessive abstraction is reducing levels in aquifers and leaving river flows too low to sustain wildlife populations in many locations. This is a potential threat to biodiversity. Climatic change may exacerbate the problem if, as anticipated, temperatures increase while rainfall is reduced in the most populous regions (McLaren *et al.*, 1998, 186–9). Dealing with this challenge may require technical and spatial fixes such as long-distance transfers from areas of water surplus, although such solutions are questionable on ecological grounds.

Debates on extraction of minerals for construction purposes exemplify the problems posed for the physical environment by development pressures. They raise questions of how to minimise the direct impacts of quarrying, and also, more generally, the definition of environmental capacities. There are connections to wider elements of uneven development since one 'spatial fix' to the problem is to exploit resources in regions far from where materials are going to be used. The presumption in the British planning system is in favour of development. This places the burden of proof on objectors to show why particular proposals should *not* proceed. Moreover, the planning system is supposed to 'balance' different material planning considerations. However, this simply means deciding on the appropriateness or otherwise of particular land uses in particular locations; it does not involve judgements as to whether developments are sustainable or not. This is obviously problematic if we take a 'strong' view of sustainability, in which 'critical environmental capital' must be handed down to future generations; at the very least we require analysis of what we value in the environment and why. Sustainability clearly relates more easily to physical resources (land availability) than to the amenity and aesthetic elements of the environment. Land use planning has arguably had more to say about the latter – preserving natural and built environments – than the former, though this may be changing (Cowell and Owens, 1997, 19). If planning is to move towards an environment-led system, then several issues will emerge. There will be problems in constructing defensible arguments about environmental capital and capacity constraints. In turn these will expose, at an earlier stage in the planning process, conflicts that the concept of sustainable development was meant to reconcile, for example, by limiting future growth and pre-empting consideration of individual development proposals on merit. These are intensely political issues which challenge ideologies of social need; they are also technical issues, concerned with managing adverse environmental impacts. Finally, there is the issue of whether, by constraining development in one locality, 'unsustainability' is simply 'exported' to another place.

House and road construction require substantial volumes of materials such as aggregates. Control of the extraction of aggregates is achieved through minerals planning guidance, which translates national estimates of anticipated demand into regional guidelines; regional aggregates working parties (RAWPs) then agree allocations for individual counties. These are often contested; usually counties argue that aggregates extraction will have adverse impacts on other environmental assets. Cowell and Owens (1997) show how Berkshire CC's proposals to limit the extraction of aggregates were based on an implicit notion of environmental capacity constraints. Berkshire had argued that the county would have to reduce aggregates extraction by 3 per cent p. a. from the projected (1996) figure of 3.5 million tonnes, and suggested that this figure (apportioned to Berkshire by the RAWP) was not sacrosanct. However, this was rejected, on the grounds that if all local authorities adopted a similar stance, accommodating the needs of the construction industry would be impossible. The 'need' for aggregates is not questioned; developers are not obliged to prove the need for proposed developments. This demonstrates the importance of an ideology of

property rights in the planning process: 'Demands that manifest themselves as consumer preferences continue to be seen as legitimate needs, while demands for environmental quality ... frequently become negotiable' (Owens, 1997, 299–300). Consumer sovereignty or the supply of 'essential' commodities are given and cannot be challenged, and it is the role of the planning system to manage resulting environmental pressures.

Berkshire's use of the concept of environmental capacity was also rejected. The council had sought to define areas in which mineral working would be acceptable; this had involved subjectivity, leading to a more restrictive definition of acceptability than national guidance would suggest. Such subjective assessments, according to the Inspector's report on the inquiry into the plan, could not provide an appropriate basis for determining production levels, which therefore continued to be based on apportionments of regional targets. Berkshire also attempted to promote sustainability through proposals to reduce demand and to make more use of recyclable materials. However, the minerals lobby rejected the council's proposals as over-optimistic and were reluctant to question any challenge to a market-based allocation of resources. The role of the local authority was thus reduced to 'regulating spatial and technical fixes for accommodating their aggregates apportionment' (Cowell and Owens, 1998, 802).

It is, however, arguable that recent decisions to reduce the regional target for extraction in South East England indicate an awareness of the pressures being experienced, and a desire for a 'spatial fix', exporting the problem to other areas. This is the origin of proposals for superquarries in remote locations, such as Rhoinebhal, on the Island of Harris in the Hebrides. The implication is that concentrating extraction in high-output sites would produce scale economies in transport and affect fewer people – and those in communities which would trade off environmental impacts for new jobs. The implicit assumption is that environmental impacts are proportional to the numbers of people affected; this 'channels extraction to lightly-populated areas and downplays the importance of wild land-scapes' (Cowell and Owens, 1998, 802). While the developers stressed economic benefits, opponents drew on cultural representations of the Hebrides as zones of purity; tarnishing Harris's image as a symbolically clean space would make it difficult to resist further attempts to dump unwanted land uses there. The implicit prioritisation of English over Scottish interests was also a source of objection.

The Harris and Berkshire cases clearly demonstrate that sustainability is interpreted in quite different ways in different locations. Hence there is no unambiguous definition of sustainable development: rather, the planning process is a forum in which it is contested and defined. As environmental values transcend specific locations, and as decisions to prevent activity in one area generally imply permitting it elsewhere, the process of determining what is sustainable, and where, 'introduces important new dimensions to debates about territorial jurisdictions, subsidiarity, and social fairness' (Cowell and Owens, 1998, 805). There are clear limits to accommodating additional demands through spatial fixes, implying that government policy must confront and restrain demand or exert rigorous control of planning policies to ensure that regional and local targets are accommodated, thus reducing local autonomy. Both options have political drawbacks.

The second debate examined here concerns whether the anticipated growth in household numbers can be accommodated through 'greenfield' or 'brownfield' development. There is little doubt of the housebuilding industry's preference for the former. Recent guidance has emphasised the need to construct at least 50 per cent of new housing on previously used sites; the DoE suggested aspirational targets of 60 per cent (RCEP, 1997, 91), while the UK Round Table on Sustainable Development suggested 75 per cent. It is debatable whether such targets are feasible – for example, much remaining undeveloped urban land is contaminated – but more fundamental questions are raised, concerning the relationship of settlement density to sustainability.

Breheny (1995) argues that protagonists of brownfield development subscribe uncritically to the view that compact cities will reduce energy consumption. High density urban development is also supported on other grounds (reducing impact on wildlife or undeveloped land) but he believes that projected energy savings underpin most arguments for high density development. There is certainly a relationship between urban size, density and energy consumption, with low density and suburban environments having the highest rates of consumption but Breheny questions whether *increased* energy consumption is a result of urban decentralisation *per se*. He estimates that, compared to 30 years ago, weekly transport energy consumption is now about 2.5 per cent higher than would have been the case in the absence of decentralisation. This is a 'trivial' contribution, implying that efforts to prevent further decentralisation would have minimal impacts. Other policy measures (improvements in fuel efficiency; taxation) would be more effective. He argues that excessive expectations have been raised about the energy-saving effects of urban containment.

Owens (1995) points out that adopting the rhetoric of the 'compact city' (e.g. in hortatory planning guidance) is politically motivated; it avoids the opprobrium attached to any attempt to confront patterns of production or consumption which lead to traffic growth. She also questions Breheny's tendency to present urban decentralisation as an autonomous process of consumer choice. Instead, population movements are responses to an opportunity structure created by state policies interacting with powerful corporate interests to promote a high-consumption, high-mobility lifestyle in suburban or semi-rural environments. Focussing on transport energy consumption is therefore isolating one factor among many. The real issue (e.g. Breheny, 1995, 99) is whether decentralisation tendencies can be contained. This would require measures going well beyond physical constraints on new development. It would require efforts to make cities more liveable in; most fundamentally it would imply changes in cultural myths and aspirations concerning the desirability of home ownership and a rural lifestyle. Robson (1994) has argued for the crucial role of viable cities – whether in terms of innovation or social cohesion – but reversing the drift away from them requires a strong vision to counteract an anti-urbanism deeply embedded in British identity (chapter 2). This, perhaps as much as putative energy savings, is a key argument for creating cities which are sustainable not just in energy consumption but in socio-economic terms.

13.3 TRANSPORT AND SUSTAINABILITY

It is no accident that a key chapter of Hutton's pre-election book, *The state to come* (1997), opens with a discussion of the problems posed by a deregulated public transport system. Transport policies currently prioritise self-interest above collective interests, favour irrational and unnecessary consumption of resources, and weaken capacities for indigenous, community-based forms of development. Social gains cannot be accommodated within a market system with no ability to express what is collectively rational. Transport is a unique exemplar of the challenge of sustainability because individual choices generate externalities on a grand scale but major penalties await those individuals (or, less plausibly, nation-states) who contemplate reducing the amount of transport resources they consume.

Of course, there are major problems in assessing what constitutes a sustainable transport system and in determining at what point a critical situation has been reached. However, the trends are unmistakable. The average distance travelled per person in Britain each day is now 18 miles, an increase of 75 per cent over the past 25 years. People make more journeys; the average length of these journeys has risen; and, significantly, the great majority of journeys are by car: there has been a ten-fold increase in car traffic since 1952 (RCEP, 1994, 9). These trends are not just a reflection of patterns of employment, because the most rapid growth has been in journeys for shopping, 'other personal business', and leisure, which account for two-thirds of all journeys. Although the proportion of work and education-related journeys is the same as 30 years ago, the average length of such journeys has increased by 83 per cent, reflecting a greater spatial separation between home and workplace, and possibly more flexible working patterns. However, the benefits of car ownership and use are unevenly spread. Almost one-third of households do not own a car while one-quarter own two or more.

Freight transport has also seen rapid expansion: the weight of goods carried has fluctuated but has not risen much beyond its 1968 level of 2 billion tonnes, but the total tonne-km show an increase of 141 per cent on 1952 levels. This was largely due to an increase of 128 per cent (from 37 km to 84 km) in the average length of freight trips by road. Hence McLaren *et al*. (1998, 113) emphasise the growing *transport intensity* of the UK economy: while the volume of freight and numbers of people being moved have not risen much per unit of GDP, 'the movement of vehicles has increased by 25 per cent per unit of GDP'.

The pro-transport lobby asserts a causal connection between these trends and economic growth, and advocates increased spending in the national interest especially on road construction. However, this causal link is debatable. Movements of people have grown more rapidly than GDP since the early 1950s, and when international travel by UK citizens is added in, the growth of movement over that 40–year period has been about 50 per cent above that of GDP (RCEP, 1994, figure 2-IV). The principal influence here has been the growth of car ownership, in turn a function of growth in personal incomes. Public policy has clearly prioritised road traffic over other modes in the belief that economic growth will follow but the evidence is ambiguous: 'a network of uncongested motorways in … Merseyside has not been sufficient to overcome the influence of other factors which inhibit economic growth' (RCEP, 1994, 15).

A number of interrelated influences are responsible for trends towards greater ownership and use of cars and other vehicles. Settlement patterns have become more dispersed (section 13.2), generating transport demands that are unsuited to fixed-track or semi-fixed-route public transport. Innovations in production methods, notably 'just-in-time' flexible production, rely on transport as a pre-requisite without which production chains come apart. Because of customisation of end-products, batch sizes are smaller. Inevitably the rail system cannot cope with this and so freight switches to the roads, usually in the form of large numbers of small vehicles, often delivering daily. The extent of adoption of such methods is disputed but there is little doubt of their impact (Whitelegg, 1997, 66–7). Other processes have generated additional transport requirements. Technological change costs more in old, ill-suited buildings; rather than invest *in situ*, new construction is preferred. Most cities now have large amounts of empty (Victorian) office space in their cores coexisting with new property on the urban fringes. Drawing labour from all around the conurbation this inevitably generates a new pattern of travel and one which cannot be met by public transport except at the margins. Flexible working patterns do not help; small numbers travelling to or from workplaces at unpredictable times are not profitable from the point of view of private transport companies. Moreover, workers may respond to labour-market flexibility (see chapter 6) by travelling longer distances to work or via adopting new commuting patterns (e.g. weekly commuting). Some of this demand for additional travel is met by public transport (one effect is an increase in passenger-km travelled by rail) but the general effect is to increase pressure on the roads. Flexible working patterns may help to account for the increase by a third in the average number of journeys made on business (RCEP, 1994, 16).

Retailing is also implicated. More than half of all new retail floorspace since 1982 has been in out-of-town sites; some 40 per cent of retail sales is now in out-of-town locations, compared to 5 per cent in 1980. The creation of massive regional shopping centres (Metro Centre, Gateshead; Meadowhall, Sheffield; Cribbs Causeway, Bristol; Thurrock Lake-side, Essex) is the latest manifestation of this trend to ex-urban retailing. Such developments have a symbiotic relationship with road construction. They rely on the trunk road and motorway network, but once established they ratchet up road use, as consumers make still longer journeys which become more feasible due to faster roads. Ultimately, of course, the ensuing congestion leads to demands for further construction. These developments have been underwritten by relaxations in planning controls which have favoured greenfield or urban-fringe development, and by massive road expenditure which has underwritten the distribution systems on which contemporary supermarkets rely. Such centres represent more profitable investments, with the certainty of quick returns, than brownfield sites, thus appealing to the short-termist financial system (Hallsworth, 1996). These systems draw food in to centralised depots and then send it out again, so that crops from Scotland may travel to the English Midlands and back again before being sold a few miles from where they were grown. The result is that the same amount of food is travelling 50 per cent further around the UK than 15 years ago (RCEP, 1997, 76). In order to facilitate the associated movement of lorries, new road construction is deemed necessary. This in turn creates further potential locations for retailing developments on the edges of urban areas. The net result is that without access to a car, obtaining life's basic necessities at a reasonable price may be well-nigh impossible because of the closure of local shops (Piachaud and Webb, 1996). Consumers are thus forced to make longer journeys to out-of-town locations, almost certainly by car in the absence of public transport links.

Retailing is only one facet of broader trends towards consumerism resulting from increasing affluence and assiduously encouraged by the emphasis, in the neoliberal consensus, on freedom and choice. Mobility is a central part of this so that individuals are prepared to travel further to obtain a particular service. Thus average distances for shopping and leisure increased by 24 per cent and 23 per cent between 1986–91, although the number of journeys for these purposes only rose by 10 per cent and 16 per cent, respectively (RCEP, 1994, 17).

All these developments impart a powerful upward dynamic in road use, and the last published forecasts, in 1989, anticipated that between 1988 and 2025 road traffic would increase by between 83 per cent and 142 per cent; subsequent revisions incorporated the realisation that 1988 traffic levels had been higher than estimated, implying levels of traffic in 2025 which would be a further 13 per cent higher than previously anticipated. Growth on such a scale will have serious consequences for sustainability: it will use up non-renewable reserves, contribute substantially to emission of greenhouse gases, have serious health implications (notably through increased respiratory problems arising from particulate pollution), damage natural habitats and the landscape, and promote land-use patterns which will have the same effects (RCEP, 1994, chapters 3 and 4; Taylor and Taylor, 1996, 238–44). This will require major changes in policy and behaviour, since expert opinion seems to accept that a continuation of the present trends is not sustainable. One result has been an end to the 'predict and provide' approach to road planning, which assumed that the only way to deal with congestion was a continuous programme of road construction and improvement. The Conservatives claimed to have performed a U-turn on this in reviewing the programme of road construction in 1993, but most schemes dropped were fairly small and the growing expense of motorway widening meant that the scale and nature of the trunk road programme changed little. While public spending on roads fell by 27 per cent between 1995–98, this has been offset by a programme of privately financed roads (RCEP, 1997, 83). In this as in many areas of environmental policy, desirable objectives are subordinate to Labour's concern not to offend those voters who flocked to it in 1997.

Congestion continues to increase, however, raising question about how to restrain car use. Here the problem is a twofold one: devising fiscal incentives to reduce car use while ensuring that enough investment is made in public transport to create an attractive and viable alternative to car use. The Conservatives and Labour both accepted a commitment to increase fuel duty annually in real terms, but because of variable oil prices the increase was less than was anticipated (1.5 per cent p. a. in real terms compared to 4 per cent) (RCEP, 1997, para. 2.45). This is nowhere near the doubling of fuel prices by 2006 thought necessary to effect the reduction of some 25 per cent in traffic suggested by the 1994 RCEP report.

This shifts attention to economic instruments aimed at making those who use cars bear a higher proportion of the associated externalities, such as road pricing, tolls, or tax regimes which would limit incentives to excess use of company cars. There are proposals for privately financed roads on which users would be charged (e.g. the Birmingham Northern Relief Road). However, the concerns about generalised motorway tolls relate to displacement of HGVs onto less appropriate routes. Urban road pricing appears some way off for technical reasons and this proposal, as with others, would bear particularly heavily on those least able to afford it. Direct taxation, rather than indirect taxes on road use, would be a more equitable way to achieve a transfer of resources to public transport. The inherent political risks may explain why the exercise of such powers as Labour has granted local authorities has been left to local discretion.

Such sticks will be ineffective in the absence of more attractive 'carrots' such as improved public transport, and a shift of freight from road to rail. However, deregulation of bus services has led to a more fragmented system in which fares increased sharply while many services were cut and networks became less stable. Where there has been competition it has increased congestion on popular routes. The net effect was a steady decline in bus patronage. Privatisation of rail services has so far brought few obvious benefits – integration between companies is problematic, while more complex fare structures are a disincentive to travellers. There are also concerns that levels of investment in rail will be low due to the uncertainty facing holders of service franchises. In both cases improving standards will require additional expenditure; Labour has provided some of this (some additional grant aid for rural public transport) but significant extensions would require higher taxation in some form. Significant switching from road transport *can* be achieved, as the Manchester Metrolink rapid transit system demonstrates: it carries some 13 million passengers a year, of whom 20 per cent previously travelled by car (McLaren *et al.*, 1998, 112). However, this solution is not feasible everywhere, due to physical constraints; few out-of-town shopping centres are likely to obtain such transport links (Meadowhall (Sheffield) is an exception); other locations are served by infrequent rail services (e.g. the Metro Centre, Gateshead). Privatisation has also weakened previously integrated transport systems, such as the Tyneside Metro, which is now exposed to competition from bus operators running into the centre of Newcastle (previously bus journeys were focused around feeding suburban Metro stations). Consequently the RCEP rejects the premise that competition will secure an improved public transport system, and calls for much more intervention on the part of local authorities. Greater integration between services and modes of transport is some way off and significant improvements to services will depend on higher levels of investment.

Reducing the impact of freight, and achieving a shift away from roads, require deliberate decisions to reduce transport intensity, for example, by more local sourcing of food. However, at present this is purely voluntary and this situation will not change without greater use of ecological taxation. Improved interchange between road and rail is also necessary; here, commercial pressures have led to the disposal of many disused railway facilities which could have served this purpose. Expansion of rail freight also depends on competition with profitable passenger services for slots on congested routes. However, reducing freight intensity is working against the grain of powerful socioeconomic trends.

The RCEP argue that three key forms of intervention are necessary to help secure an integrated transport system: promoting improved technology; identifying and remedying market failures in order to provide the right incentives; and establishing effective institutions for regulation and planning. These would take the state's role even further away from that of providing basic infrastructure (its principal function through much of the Keynesian era) towards a regulatory and facilitative role, in line with many developments noted elsewhere in this volume. Whether such developments would be feasible and effective is another matter.

Improved technology does not always have desired effects (e.g. greater fuel efficiency may be traded for larger vehicles) but might facilitate the management of traffic volumes. Effective institutions implies moves away from competition and firmer regulation, in the public interest, of private bus and rail companies. The proposed Strategic Rail Authority, and firmer coordination at regional level, are essential. But these measures would not be effective without economic incentives which reflect the true costs of transport choices. This requires higher rates of increase in fuel duty, abolition of the favourable tax treatment of company cars, and greater powers for local authorities to help create an integrated transport system (including powers to raise revenues from charging for roads or taxing parking facilities). The concerns of the mythical 'Sierra man', invoked by Blair in the election campaign, seem likely to guarantee a cautious approach; the absence (so far) from the legislative programme of any major transport bill is ample evidence of this.

13.4 A GREEN AND SUSTAINABLE LAND? AGRICULTURE, POLLUTION AND BIODIVERSITY

Considered purely in terms of self-sufficiency there is little doubt about the UK's ability to provide adequate levels of food for its population, although this does presume a shift in consumption away from meat and dairy products. Concerns are, however, expressed regarding the environmental impacts of current agricultural practices, and about the risks to which consumers are being exposed. A related issue is whether, given current trends in agriculture, the rural environment can be preserved as a repository of symbolism and meaning; what the urban majority want (and are prepared to pay for) is not necessarily compatible with existing agricultural practices.

Several commentators emphasise the *productivist* regime of post-war agricultural policy (Marsden *et al.*, 1993, 58–61). Post-war food shortages necessitated a rapid expansion of domestic food supply; this was achieved through a new coalition between agricultural and industrial interests. Key elements of the regime included: security of land rights and land use, not least by the exemption of farming from planning controls; financial security through guaranteed prices which, given an unregulated land market, facilitated the concentration of production; political security, because of the privileged access of the farming community to government; and ideological arguments for agricultural support including an identification of farming with the national interest and stewardship of the rural environment, and promoting an image of farming as modern and technologically advanced. Agricultural policy was so shaped as to promote accumulation and growth in the intensity of production.

This settlement was eventually undermined by several global and European influences. Agricultural output rose at a rate which outstripped the capacity of domestic markets to absorb the increase. This produced a crisis of legitimacy for agriculture, since it required an alternative social justification for the levels of public support for it. Resolving this was problematic for the Conservatives, caught as they were between their antipathy to subsidies and the traditional Conservative attachment to land and the countryside. Negotiations with the EC therefore entailed political difficulties: the government strongly endorsed efforts to cut public subsidies but did not wish to be seen to undermine the interests of its own farming community. These dilemmas are not confined to one party.

The consequences of the productivist era of agricultural policy are highly visible in landscape terms. Guaranteed prices plus capital grants for efficiency-raising improvements brought about a transformation: output per hectare increased rapidly while unit costs of production shrank. Grassland was converted to arable land on a large scale. The capital-intensity of farms, expressed in the use of machinery or agro-chemical inputs, has increased substantially (for a review of trends see Bowler, 1991; Allanson and Whitby, 1996, 4–8). The traditional pattern of mixed farming, so essential to the preservation of the rural landscape, has been largely superseded: there is increasing specialisation both within individual farm businesses and between regions, with a concentration of cash cropping on the best farmland in the lowlands and of permanent pasture elsewhere. This specialisation has been accompanied by landscape standardisation as hedgerows and trees have disappeared, and as semi-natural habitats have been converted to intensive agricultural use.

The productivist era has in a narrow sense been sustainable in that the UK is now (at least in principle, making assumptions about exports) self-sufficient in a number of products and this was not formerly the case. In other respects sustainability is questionable. Two issues are discussed here: the growth of agricultural pollution, and the effect of agricultural change on biodiversity.

The intensification of agricultural production has been achieved through increased use of fertilisers and agrochemicals resulting in damage to ecosystems and the contamination of ground water and watercourses. Livestock production has also been intensified: carrying capacities have increased due to the development of higher yield grasslands and of silage. This in turn is associated with new pollution hazards due to discharges of slurry.

Debates about nitrate contamination and pesticide pollution illustrate well the use (or non-use) of the precautionary principle as a basis for policy. It was clear from the mid-1970s that increased fertiliser applications over the preceding 40 years (an estimated 800 per cent increase, to some 1.6 million tonnes by 1985) had produced significant increases in nitrate levels. Nitrate levels in groundwater therefore exceeded EC standards in many areas, especially where intensive arable agriculture was practised. Likewise, given the regulatory regime prevalent in agriculture, pesticide application was seen by farmers as essential if yields were to be increased. Farmers held sincere convictions that weeds were a major threat to high yields, but were much less certain that chemical pesticides posed environmental risks. In contrast to the 'easily identifiable economic threat' posed by weeds, agrochemical pollution risks seemed 'long-term, distant and unproved' (Ward, 1996, 53).

These doubts are echoed in debates about the regulation of pollution. Scientific uncertainties leave framing of the issue open to contestation. One view, based on the precautionary principle, is that pesticides have no place in water; there is uncertainty over the risks attached to individual pesticides, and the 'synergistic effects of consuming water containing a mix of several pesticides … have hardly been investigated'. An opposing view is that EC standards are 'unscientific' because they do not relate to a toxicological analysis of the risks attached to individual pesticides. This view attracted government support, leading to pressures to relax the EC Drinking Water Directive. From this perspective, pollution is a minor issue rather than a fundamental problem with the organisation of agricultural production, and the response has been to clean up drinking water at water works. Water companies can pass these costs on to their consumers, in a perverse overturning of the 'polluter pays' principle. It would be far cheaper simply to pay farmers to use non-polluting alternatives than incur the costs of water treatment.

Pollution from dairy farming has also been increasing. This results from processes analogous to those in arable farming, namely specialisation and intensification. Dairy farms produce increased quantities of animal waste which takes the form of semi-liquid slurry requiring storage and separate disposal (via tankers) as the amount produced by enlarged herds exceeds that which can safely be spread on fields. Previously such waste would have been mixed with hay from cattle sheds to produce manure and recycled onto arable fields (thus reducing the need for chemical fertiliser inputs). Other develop-

ments, such as concreting of yards or building floors, facilitate waste disposal but also increase impermeable surfaces, which (after rain) in turn increase the volume of contaminated liquid to be managed. Lowe *et al.* (1997, 28–9) argue that systems designed to store and dispose of waste are often inadequate. These problems are compounded by the growing production and use of silage, which is necessary to cope with larger herds, but which also generates effluent that is extremely polluting. While substantial direct discharges into watercourses are rare, the long-term impact of small but persistent discharges is much more significant. By the late 1980s, farming accounted for a third of all water pollution incidents, of which the great majority were attributable to dairy farming. Lowe *et al.* (1997) argue that farm pollution was caught up in 'agriculture's moral economy'. Growing awareness of pollution, combined with scepticism about the consequences of agricultural productivism, led to criticisms of the environmental impacts of agriculture. The result was to shift perceptions of farm pollution, from being 'a technical side-effect of efficient food production to [being] an environmental crime' (Lowe *et al.*, 1997, 207). And with this shift came changed attitudes to farmers: far from being 'respected national heroes', they came to be seen as villains. An 'urban moralism' finds farming wanting; that the agricultural community feels beleaguered is evident from the huge support attracted by the Countryside March of 1997. At issue is the representation of rural space, which is no longer pure, unpolluted and a repository of true British values (Lowe *et al.*, 1995, 63–5; see chapter 2).

If agriculture cannot be relied upon to eschew pollution, what of conservation? There is plainly a conflict between productivism and conservation, as evidenced by the loss of habitats to agriculture. Drawing on national land-use surveys Adams (1996, 34) notes substantial losses of boundaries (especially hedges) and loss of species in several habitats. These surveys related to the 1980s: estimating the impacts of previous periods involves reliance on *ad hoc*, local case studies, which typically reveal dramatic losses of semi-natural habitats. Conservation

policy has typically relied on spatial designations to protect valued landscapes, such as
the Sites of Special Scientific Interest (SSSI),
introduced in 1949. Reliance was placed on
voluntary cooperation – site owners were
advised of the merits of SSSIs and on what
activities would cause damage. If owners
wished to carry out these activities they were
offered compensation for profits foregone. A
sceptical view is that farmers are thereby
compensated for desisting from activities
that they only carried out because subsidies
were available (Harvey 1997, 14). Even so,
the value of SSSIs is questionable. About 13
per cent of SSSIs are damaged each year; the
worst incidents relate to inappropriate development, but much short-term damage is due
to agriculture (e.g. overgrazing). A more significant issue is that the policy does not control the predator, but isolates the prey
(Harvey, 1997, 54): in relation to total land
area SSSI designations are minuscule (8 per
cent of the land area), and they are too far
apart (especially given losses of hedgerows
or woodland) to facilitate species migration
(e.g. in the event of climatic change: Adams,
1996, 37). In any case SSSI designations can
be overridden when development is proposed if, for example, planning inspectors
determine that job creation takes precedence.
There are also concerns that this form of zon-
ing is of little value if, outside designated
areas, agricultural practices which have negative impacts on biodiversity continue.

Finally there are concerns that management payments for farmers in SSSIs should
be linked to specific conservation activities,
rather than compensating farmers for actions
which they might not have done anyway
(Adams, 1996, 41–2). This principle has been
taken up in English Nature's Wildlife
Enhancement Scheme, but there remain concerns that SSSIs offer inadequate protection.
SSSIs deemed to be under particular threat
can be protected by a Nature Conservation
Order, but during the 1980s the government
argued that such orders could only be used
to protect a small number of elite sites, which
were deemed to be of national importance.
Only selected sites therefore had orders made.

Adams (1996, 121) suggests that an alternative approach to conservation should 'look

beyond the boundaries of protected areas to
the land around them'; isolating SSSIs or
nature reserves undermines linkages *between*
such places and implicitly legitimates continuation of existing agricultural practices. He
therefore (1996, 131–6) argues that the crisis
of agricultural overproduction in Europe,
coupled with pressures for reform of the
CAP, provides an opportunity to reconstruct
the countryside: there is both a surplus of
food and of land, and a very large budget for
agricultural support.

The first schemes to offer payments to
support conservation on farmland were
those available in Environmentally Sensitive
Areas (ESAs) established in the mid-1980s.
These now cover substantial areas of lowland
and upland throughout the UK; they are
areas whose environmental quality was
threatened by agriculture but whose character could be maintained if appropriate farming practices were adopted. Farmers receive
flat-rate payments if they adhere to environmentally appropriate practices. However, the
voluntary nature of the scheme makes it
uncertain what *new* conservation benefits
accrue: it probably attracts farmers whose
practices already meet the guidelines, and
the broad-brush management criteria allow
for landscape and husbandry changes that
may well affect biodiversity without contra-
vening ESA guidelines. To promote a more
positive approach, the Countryside Stewardship programme (in England) offers land-
holders flat-rate, but discretionary, payments
for *changes* in management that will *promote*
conservation, including efforts to recreate
habitats or establish wildlife corridors and
provide 'buffers' between habitats and agricultural activities.

The obvious criticism of such initiatives
(and of one of the other main responses to
agricultural surplus, 'set aside' schemes) is
that farmers are free to intensify production
on land not covered by ESA agreements
(Winter, 1996). The agricultural industry
needs to be moved away from intensification
and there have been agri-environment measures which seek to achieve this goal. The
1992 CAP reforms appeared to promise some
moves towards 'extensification' of agriculture, but in practice their impact was limited.

A policy of 'rotational' set-aside meant that land was not diverted permanently to conservation purposes. Payments for set-aside were more generous than those available under ESAs. Limited environmental conditions were attached to payments. The net effect was that the intensity of agriculture was only marginally reduced (Winter and Gaskell, 1998).

This experience does not augur well for the latest EU attempt at reforming agricultural policy, the Agenda 2000 proposals launched in 1997 (further revisions to the CAP were under discussion at the time of writing). Despite calls from EU Agriculture Commissioner Fischler for a rural policy underpinned by the notion of sustainable rural development, the measures proposed in Agenda 2000 are weakly developed. There is a commitment to more severe price cuts, which might prompt farmers to reduce input intensity, but the effect will depend very much on trends in world prices. There is an attempt to fix, and thereby limit, the area of arable land, but this encourages a more intensive use of that land involving permanent cropping and chemical inputs. This discourages mixed and integrated farming. In dairy farming, milk quotas will be retained although prices will be gradually reduced. Given rising yields this ought to offer potential to reverse the intensification of dairy farming, but there have been few signs of extensification since 1992. This is also the case for beef cattle, where regulations on stocking densities have had little effect (Winter and Gaskell, 1998, 226–31). Agenda 2000 does talk of a drive to promote low-input farming systems in marginal areas, which might incorporate environmental protection and management centrally within the CAP. But the pace of change will be slow: swifter action might well destabilise agriculture completely in many marginal locations and encourage further concentration. Neither of these would be favoured by environmentalists, and the process of transforming agriculture in a more sustainable direction will be a very slow one. Yet this is very necessary, for as Lowe (1996) indicates, 'our society looks to the countryside as a reservoir of environmental and cultural resources and values'.

Debates about sustainability highlight the importance of the supply and replenishment of renewable resources (plants, animals, soil, air, water) and continuing resources (wind, solar energy, water power). The emphasis is increasingly on the maintenance of environmental capacity rather than maximising resource flows; arguments against agricultural intensification and pollution make this point very clearly. Hence, in policy terms the 'overriding public interest in rural areas is in safeguarding, managing and enhancing their function as an environmental reservoir' (Lowe, 1996, 201). From this perspective agricultural policy could become 'rural environmental management', in which continued agricultural support payments would have to be tied much more closely to public benefits in the form of desired environmental outcomes. This raises broader issues relating to pursuit of the goal of sustainable development.

13.5 TOWARDS SUSTAINABLE GEOGRAPHIES?

In some ways it is surprising that *The state we're in* (Hutton, 1995) makes few, if any, references to sustainability, as the endemic short-termism of the British economy is a principal underlying cause of environmental problems. Individuals, companies and farmers alike suffer if they do not adopt strategies which are rational from their isolated point of view, but the collective effect is resource depletion, environmental damage, and unbalanced patterns of development. These problems are exacerbated by a corresponding political failure to invest (in infrastructure or in education and training) which further weakens competitiveness.

Thus the commitment of the British government to sustainable development has widely been regarded as lukewarm. Public statements on environmental issues, by both the Major and Blair governments, have been notable for insisting that continued economic growth can and should be maintained, and is compatible with sustainability. Labour's consultation paper on the subject goes on to add references to broader questions such as quality of life, employment and social exclusion but the emphasis is still heavily pro-growth. Sustainability is capable of engendering wide support without exposing government to public scrutiny for failure, because the definition of sustainability has been stretched so far as to be 'practically meaningless'. As a consequence environmental policy 'collapses into a series of *ad hoc* measures driven by political considerations' (Helm, 1998, 17; see also Munton, 1997).

It follows that measures to promote more sustainable forms of development have been very limited, reflecting governments' fear of the electoral consequences. Many polluting activities arise in the context of 'basic social primary goods' (heating, lighting, transport, water) posing a trade-off between social and environmental policy. Energy prices are a good example. The political consequences of increased taxation on fuel are immediate, while the impact on pollution is longer term. Adjustments in social security could address some of the distributional consequences, but constraints on public spending prevent this being an option. If the definition of sustainable development is widened to include social considerations, then politicians can present political compromises as being consistent with environmental policy (Helm, 1998, 12) but this is basically an excuse for inaction or a justification for less visible forms of price regulation which are less easily linked, in the public mind, to government action. Devising environmental economic instruments is formidably complicated, therefore (see Helm, 1998, 10–14 for a discussion of the problems).

A commitment to sustainable development would require, at a minimum, three developments. First, it would require challenging the assumption that the primary duty of governments is to deliver conventional economic growth – conventional, that is, as defined in terms of ever-higher levels of personal consumption. This might involve redefining wealth creation in terms of improved quality of life. Second, it requires target setting: this has been resisted by central government, but is done for the inflation rate – so why not for the environment (McLaren *et al.*, 1998, 301)? Without targets it

is difficult to persuade all government departments to pull in the same direction. Third, although there are difficulties in designing economic instruments, greater regulation is necessary. Market signals may produce changes in behaviour but only if market rules are structured appropriately, while the setting of standards is also essential to discourage free-riding. Energy regulation provides a good example: in several American states, utilities have provided energy-efficient lighting and low-cost loans for other energy conservation measures, thus obviating the need for new generating capacity. Such measures could be combined with ecological tax reform which would shift the burden of taxation onto resources, waste and pollution, and away from other production costs or inputs, such as labour. Currently, elements of the tax system promote the overuse of environmental resources: support to company cars, or agricultural subsidies which encourage use of agrochemicals. There are grounds for believing that ecological tax reform could achieve employment and environmental benefits with relatively little impact on competitiveness (McLaren *et al.*, 1998, 269–75).

These comments collectively suggest a greater role for the state, both nationally and locally. One effect of the hollowing-out of British state structures, and of privatisation, has been to weaken the government's leverage on environmental issues. The re-emergence of a regional tier of government is testimony to the belated recognition of this, even by the Conservatives. Crucially, the local actions, on which sustainability depends, will not happen in the absence of a relaxation of constraints on local government. Despite the enthusiastic reaction of local authorities to the Rio Earth Summit's 'Agenda 21' (which emphasised the importance of local initiative and participation), central government has given relatively weak backing to 'Local Agenda 21' initiatives (Young, 1998, 182–3; Agyeman and Tuxworth, 1996). Effective regional planning mechanisms are also a necessary condition of sustainable development. There is also a need for the fashionable 'joined-up' government beloved of New Labour, if the actions of one Department are no longer to conflict with those of another. The productivist concerns of the Ministry of Agriculture, Fisheries, and Food (MAFF) are the most obvious example. There is *a* Department *of* the Environment; we still lack *the* Department *for* the environment (Carter and Lowe, 1998, quoted in Bradbeer, 1999).

Whether such developments are politically feasible is another question. The 'easy environmental plums' may have already been picked (closure of inefficient, polluting industrial plant): further progress will require unpopular new forms of regulation or taxation. More generally, sustainability offers a challenge to enduring features of the British polity and society. A deeply ingrained anti-urbanism, for instance, would need to be challenged by a strong vision of the merits of urban life. The priority attached to landed property and agricultural interests would be called into question as well as the role of farmers in countryside stewardship. The overcentralised British state clearly cannot deal with local environmental problems. We might consider other dimensions of sustainability: for example, are current levels of social segregation either politically tolerable or economically affordable? This is in part a contributory factor to the drift of population away from cities, and needs to be reversed. The overdevelopment of the South East likewise: skill shortages, infrastructure pressures and environmental costs all act as a drag on the economy, and more sustainable patterns of regional development are required.

Generating the necessary public support for an appropriate political programme will be difficult. One reason is the widely shared distrust of government's capacity to deal with environmental issues – the BSE issue being a case in point. More importantly, despite the growth in environmentalism of various kinds, it has very little impact on the everyday lives of the majority, and one reason is an alienation from and scepticism about the 'discourse and activities surrounding *official* promotion of environmental values' (Grove-White, 1997, 118). However, in the varied patterns of associational life in Britain, not just in environmental movements but in new cultural movements around phenomena such as leisure, alternative medicines and philosophies, sport, crafts and so on, lie resources of resilience and inventiveness. Many of these have implicit or explicit connections with environmental or quality-of-life issues; they could be mobilised around a suitably attractive vision of a sustainable future. Adams (1996, 170–4) argues that this will have to be a compelling vision: 'unless people feel the value of nature, it will be treated as a mere commodity, to be bought and sold, built and done away with as profit dictates' (p. 172). More modestly, perhaps, some common themes emerge: promoting quality of life as a primary objective rather than economic growth and consumption; a broader conception of governance, in which community and environmental interests are accorded a stake in decision-making alongside those of shareholders and property-owners; and greater decentralisation and democratisation, permitting local mobilisation around environmental issues. These are, of course, irredeemably political issues: as Harvey (1993, 25) reminds us, 'all ecological projects are simultaneously political–economic projects and vice versa'.

14

A PLACE IN THE WORLD II
MILLENNIAL PROSPECTS, MILLENARIAN VISIONS

This book began with a brief glance backwards at previous accounts of the UK's human geography. Looking forward, beyond the millennium, what are the key issues which will shape future human geographies, and what implications do these have for the character of academic accounts of those geographies? The pace of recent change has been so rapid that this chapter must necessarily be selective and speculative. I will therefore concentrate on four key topics: the rise, character and consequences of 'informational capitalism'; the nature of risks to which people and societies are exposed; the question of whether 'New Labour' will be associated with 'new geographies'; and the question of national identity.

14.1 INFORMATIONAL CAPITALISM

For Castells (1996) the key to contemporary capitalism is that it is 'informational': that is, economies rise and fall according to their success in information processing. Castells implies that production of and transactions in physical commodities have become much less important than hitherto. Instead, there are 'spaces of flows' in which vast quantities of information circulate around the world and in which enormous speculative financial transactions occur. One might think that this would lead to an opening-up of developmental possibilities since economies would no longer, as in the past, be constrained by

the physical resource base. However, it is not at all clear that informational capitalism is leading to a deconcentration of economic activity; if anything, the opposite is the case. This is entirely consistent with suggestions that the world economy is increasingly organised around global cities, primarily because, despite the potential of IT, face-to-face contacts remain essential in the service sector. The tendencies discussed in chapter 5 indicate the continuing and likely future economic dominance of London and the South East over the rest of the UK. These trends have several consequences.

Castells (1997, 161) postulates that informational capitalism creates a 'sharp divide between valuable and non-valuable people and locales'. His illustrations are somewhat apocalyptic, but it is not too fanciful to see tendencies towards the extreme socio-spatial divisions he describes. Globalisation, he contends, 'proceeds selectively, including and excluding segments of economies and societies in and out of ... networks of information, wealth and power' (p. 162). One immediate consequence of informational capitalism is the spread of precarious employment, exposing people to greater levels of uncertainty and risk (see below). A second is the constitution of the 'black holes of informational capitalism'. He illustrates this with a discussion of the inner city underclass in the USA, much of which has resonances with current debates in Britain. In particular, the multiple and overlapping 'infliction of additional injuries to those who are already excluded' echoes the recognition of multiple dimensions of social exclusion. Moreover, he emphasises that the 'black holes' are socially and culturally 'out of communication with mainstream society' and goes on to argue that the 'territorial confinement of systematically worthless populations is a major characteristic of the spatial logic of the network society' (p. 164). Here he is referring to the marginalisation of sub-Saharan Africa, but it is not difficult to discern parallels. An issue neglected by Castells is that patterns of socio-spatial segregation in Britain have, I would argue, been exacerbated by mechanisms initially designed to counter social inequality. I refer to the introduction of market mechanisms in the education and social housing sectors which, when combined with the collapse of job opportunities, particularly for unskilled men, has produced communities excluded in every sense of the word (chapters 8 and 9). It is not accidental that the Commission on Social Justice (1994) characterised many peripheral housing estates as the 'Brazils of Britain' because of the fiscal haemorrhage they were experiencing. While we may regard as unlikely the development in the UK of Castells' 'wild zones' of extreme social exclusion and lawlessness, there is enough evidence that perceptions of landscapes of fear, 'no-go areas' and the like *do* shape people's behaviour.

Countering these trends towards exclusion poses formidable challenges. Current policies seem likely to operate with the grain of the market, seeking to boost sectors in which the UK has a competitive advantage. In spatial terms this means trying to develop 'clusters' of some sort, though the probability is that clusters will develop in a limited number of favoured areas (chapters 5 and 12). This will reinforce existing spatial imbalances, and pose problems of sustainability. These imbalances are likely to be exacerbated by aspects of European integration, including the Single European Market (SEM), the enlargement of the EU, and European Monetary Union (EMU). Pro-market advocates suggest that the experience of the USA is that economic integration promotes interregional convergence; they therefore argue

for as little intervention as possible with market forces. However, it is clear that EU enlargement and the SEM will pile pressures on the UK's weaker regions and that a long-drawn-out process of adjustment will take place. EMU will impose additional burdens, because nation-states will not have the option of currency devaluation. The Maastricht convergence criteria relating to the size of the public sector deficit will affect governments' ability to use public expenditure to create employment. For these and other reasons (see Armstrong, 1998, 206–9), European integration will not be spatially neutral in its effects, and will have substantial adverse impacts on the UK's peripheral regions. Moreover, optimistic comparisons with the USA's experience of economic integration often ignore the much greater flexibility in wages, and high levels of migration, which facilitate adjustment. These conditions do not apply in the UK (Martin and Tyler, 1992; chapter 6), and the likely result is higher unemployment in regions not well placed to compete under EMU. The most plausible scenario is one in which the core regions of the economy pull away from the rest. This could effectively recreate Lee's (1986) characterisation of the nineteenth-century British economy as being divided into two – a service-based, internationally oriented economy in the South East, counterposed to the rest, only this time without the manufacturing.

Elsewhere the process of adjustment to a post-industrial future will continue to be a slow and painful one. If education and human capital are the key to the future division of the 'work of nations' (Reich, 1992), then it will be a long haul indeed. The lowest levels of educational attainment in the UK are generally found in those localities where industrial employment has collapsed most dramatically. This is obviously a problem of poverty (chapter 9) but it is also one of motivation: education clearly cannot guarantee anyone a job under conditions of deficient demand. Higher levels of investment in human capital might, over the long run, make a difference (Szreter, 1997) but here we run up against a familiar problem. The fragmented social landscape of informational capitalism weakens the prospects for obtaining support for comprehensive investments in education, training (and other elements of the welfare state).

14.2 RISK

The term 'risk society' has achieved such widespread currency that it may be in danger of losing its meaning. However, the most productive uses of the term relate to situations where we lack past experience of coping with risk (e.g. in debates about the causes of climatic change), where the scale and character of risks are qualitatively different to historical precedents (e.g. insecurities in the labour market), or where there is something genuinely new and unforeseen (e.g. AIDS) (Giddens, 1998).

Risk is commonly used to describe the position now faced by many workers who formerly held comparatively secure jobs. Historically, of course, patterns of discontinuous and fragmentary participation in paid employment were the rule rather than the exception. It was only in the three decades after 1945 that generalised full employment was the norm in the advanced capitalist nations. What has succeeded that regime has provided flexibility from the point of view of capital, but this involves exposing employees to greater risks and, often, forcing them to take responsibility for providing for their own future. The creation of a welfare system capable of coping with the risk society is a major challenge. It would need to be a more active welfare state than the current passive social security system, with its benefit traps which prevent people accepting jobs. It would need to accept – and provide the resources for – periods of retraining for those wishing to change occupations or return to the labour force, without penalising individuals in those positions. Current policies seem to be pushing in two perhaps contradictory directions. The welfare-to-work programme heavily prioritises work as the sole route out of poverty and therefore emphasises reintegrating as many people as possible into the labour force. The comparatively low level of the national minimum wage shows clearly

that Labour accepts much of the neoliberal agenda of its predecessors. At the same time there are sincere efforts to make work pay (through tax credits, childcare subsidies, and so on). These need to be complemented by reforms which will facilitate moves into and out of paid work, and which will enable individuals to acquire or update skills – an 'intelligent' welfare state, or 'social investment state' (Commission on Social Justice, 1994). Again, however, the fragmented social landscape which is an *outcome* of labour-market insecurity (chapters 6 and 8) limits the prospects for *responding* to such insecurity.

Somewhat different challenges are posed by environmental risks. The issues here relate to situations where 'nature is no longer nature' (Giddens, 1998, 59), such as genetic modification or the BSE crisis, or to circumstances in which we cannot predict the long-term effects of apparently innocuous interventions (e.g. farm pesticides). These situations are likely to necessitate action at some stage but the uncertainties surrounding their long-term consequences have arguably provided a shield or excuse for government inaction. Britain has not adopted the 'precautionary principle' in which restraint is urged precisely because of scientific uncertainty. Yet the future development of informational capitalism will almost certainly involve an intensification of processes likely to have adverse environmental impacts. Current and future development trajectories may not cause the UK to run up against absolute limits to growth. However, there is enough evidence, from the neoliberal experiments of the 1980s, to urge caution and careful control of patterns of uneven development. There are also occasional hints that constraints on development in particular regions may be reached (chapter 13); water supply is the most obvious example. In one sense these issues return us to the start of the book, which drew attention to the environmentalist influence on several important texts on the UK's human geography. The physical resource base may no longer determine patterns of economic activity but current development pressures are encountering other physical constraints. More broadly, patterns of production and consumption all seem to be generating new risks. The management of these risks must involve new forms of economic intervention, regulation, taxation and public expenditure at all spatial scales (McLaren *et al.*, 1998, 308).

This raises a further important, though neglected, aspect of the 'risk society'. Many of the contemporary risks now attracting attention have become so because of government action to promote deregulation or privatisation and to minimise democratic constraints on the freedom of action of capital. Thus, in employment terms the public sector has been an important source of stable, remunerative jobs, but over the past two decades, spurred by central government, public (or formerly public) organisations have dramatically reshaped work practices. This process has gone much further in Britain than in other states (chapters 4 and 6). Again, the vigorous 'hollowing-out' of the state is not just an autonomous process, it reflects conscious political decisions (chapter 11). Likewise, although certain risks are difficult to foresee (the BSE crisis is a case in point), it is arguable that more could have been done to minimise them. In a much broader context, finally, the re-emergence of older killer diseases, such as TB, reminds us that with political will such risks could have been eliminated globally. In the case of TB this has not happened, and although at present its resurgence in Britain is confined largely to the most marginalised social groups and areas, given contemporary levels of mobility it is clearly capable of spreading. Indeed, if an illustration was required of the ways risks bear most heavily on the most marginalised sections of society, it is provided by widening health inequalities and, in some communities, absolute increases in death rates (chapter 8; Phillimore *et al.*, 1994). The risks arising from other human impacts on the environment (genetic modification, global warming) are unlikely to be so selective in their impacts.

14.3 New Labour, New Geographies?

Most of this book was completed by early 1999. Had a similar exercise been attempted early in 1981, one conclusion might have been that it was too early to chart in detail the emerging geographies of Thatcherism. To give one example, such an exercise would have taken place before the massively deflationary 1981 Budget. We can, however, point to general tendencies.

First, most commentators point to the caution with which Labour is governing, and the extent to which it has accepted basic parameters of neoliberalism. This, of course, reflects the extent to which the party has shifted towards the centre of British politics, and the difficulties attendant on greater intervention (e.g. the constraints of EU competition legislation). This helps explain the absence of large-scale redistribution – indeed, budgets have so far been presented as 'win–win' scenarios in which everybody gains. This also reflects a belief that what government can actually achieve, in a globalised economy, is strictly limited, and a pragmatism of a non-ideological 'what counts is what works' kind.

The welfare state provides a good example. There is a tendency to present social problems as small-scale and amenable to tightly focused interventions. This is certainly a rationale behind policies on welfare-to-work (chapter 6) and social exclusion (chapter 8). It also applies to education and health care (chapter 9) where various spatially targeted policies are in place to deal with areas suffering particular disadvantage. The careful targeting of redistributive measures on families with children is another illustration (a particularly welcome one). Presenting policies in this way has obvious advantages. It implies that problems are of a marginal kind, do not require major long-term resource commitments, and that (to a degree) solutions can be found from within the resources of disadvantaged communities. Outside these various zones – delimitation of which has the helpful advantage of signalling just where any New Labour supporter would choose *not* to live – life goes on as usual. This means that the processes of competition introduced elsewhere in the welfare state can proceed more-or-less unchecked. In summary, the new geography of welfare is one in which efforts are made to bring people to the starting-gate on roughly equal terms, after which market forces take over. If, in this environment, individual entities (e.g. schools) are seen to 'fail', they face heavy penalties for doing so; there are indications that Labour will push this competitive logic further than the Conservatives. Standing behind all this is New Labour's target audience – middle England – which has a strong vested interest in educational inequality (chapter 9) and low taxation; this circumscribes what can be done in the welfare state.

The more far-reaching effects of New Labour will be seen as emerging constitutional changes work through. The specific measures here (RDAs, Assemblies, reform of the Lords, city mayors, PR) are probably less important than the impact on the character of politics. Arguments here usually postulate that once the devolution genie is out of the bottle, it will not (cannot) be put back. Despite its strong instincts for control, New Labour is unlikely to be able to contain demands for further changes. For example, some suggest that the Scottish and Welsh Assemblies will witness challenges to legislation from Westminster, and that such challenges may be the first step in a move towards independence. Similarly, the asymmetric regionalism which has developed in a pragmatic way (chapter 11) may not satisfy the aspirations of some English regions. More generally, though, optimists see in constitutional changes the possibility of more cooperative and progressive forms of politics, replacing the adversarial conflicts of recent decades. Against this, pessimists suggest that Labour's preference for 'lean', 'strategic' bodies provides a convenient cloak for organisational reforms which pay little attention to democracy and community participation.

Finally, despite devolutionist rhetoric and the creation of RDAs, there seems little prospect of substantial changes to the institutional geographies which do so much to shape – and constrain – uneven development (see chapters 2 and 7). Even if given democratic legitimacy and greater powers it is hard to see the RDAs as an effective counterweight to these influences, while (to date) there have been no proposals to regulate the operation of the financial system to make it more sensitive to regional needs This highlights a central dilemma facing New Labour. It is all very well favouring more progressive social policies and being against inequality, but the causes of poverty and labour-market insecurity lie deep in the character of British capitalism (Hutton, 1995). Despite the government's progressive measures there is little sign of more fundamental economic reform to complement the rolling programme of political reform.

14.4 NATIONAL IDENTITY, EUROPEAN INTEGRATION AND THE NATION-STATE

The report of the inquiry into the death of Stephen Lawrence, a black teenager murdered by white youths in South East London, has once again provided a sharp reminder of the problems of constructing a single and homogeneous vision of national identity. After 50 years of Commonwealth immigration, non-white groups are still some distance off parity with their white counterparts and they manifestly do not receive equal treatment from all public institutions. The continued occurrence of racism of all kinds is further testimony to social division. British national identity has historically been defined in terms of a white, middle-class, southern Englishness (chapter 2), but this has been challenged in many ways – the retreat from Empire and economic decline being perhaps the most important. Growing social fragmentation and the challenges posed by European integration have contributed to an unravelling of the idea of a unitary British identity.

This is recognised in Labour's attempts to repackage or 're-brand' Britain and Britishness, although the suspicion lingers that the primary purpose is one of marketing sectors of the economy (in this case, media, communications and the creative professions) in which the UK retains international leadership. Thus, Blair's vision of a 'young country' – 'cool Britannia' – arguably speaks to young design professionals in London. Whether it has much resonance throughout the UK is more debatable. Yet if the nation-state's legitimacy and powers are eroded from above by European institutions and globalisation, and from below by devolution and regionalism, where does that leave national identity? There are competing views of how these tendencies might work out.

The more optimistic accounts argue that the growing significance of regional economies offers considerable scope to subnational governments. For Castells (1997, 358) 'the era of globalisation of the economy is also the era of localisation of polity'. Protagonists of the rise of regional economies (e.g. Cooke, 1995) share such views. There are some (e.g. Haseler, 1996, 185–6) who contend that this process, combined with adjustment to broader and looser European political institutions, will sweep away the 'constrictions imposed by the straitjacket of UKanian nationality'. In time the British people will develop a new view of their relationship with Europe. This will open up new possibilities and frontiers, and, perhaps ironically, Britain will be rescued from its post-imperial trauma and economic decline by a continent with which it has always had an ambivalent relationship.

Against this, Anderson (1996) agrees that we are seeing the partial and selective unbundling of territory, but suggests that 'states and territorialities are being qualitatively transformed' (p. 150) rather than nation-states declining in importance. Thus, instead of nested hierarchies of supranational, national and local states, political institutions are operating at multiple and overlapping levels. Furthermore, Heffernan (1998, 240) argues that European integration will not 'automatically produce a more tolerant, inclusive and cosmopolitan society', because the 'idea of Europe has always been about the politics of exclusion and division'. He therefore calls on European citizens to 'develop complex, multiple and international allegiances within and beyond their place of residence' (p. 241) rather than narrowly based nationalisms.

The perceived difficulties posed by greater European integration are not just evident in the travails of Eurosceptics in the Conservative Party. In many respects New Labour looks to the USA rather than Europe for many of its key political ideas (Marquand, 1998). On the other hand, although it has not fully embraced the cause of European integration, urging caution on monetary union, it undoubtedly displays much more enthusiasm for Europe than do the Conservatives. How this tension over Britain's place in the world is resolved will have crucial impacts on the future development of the human geographies reviewed in this book.

Finally, there are implications for academic accounts of these human geographies. We have moved a long way from the environmentalist legacy of human geography, although physical constraints have re-emerged in different ways. We have also, I hope, moved on from cartographies of distress or one-dimensional mappings of political strategies. What I have tried to show here is how key institutional structures – the character of British capitalism, the nature of British identity, the centralisation of political and social power, the character of the British state – combine to produce particular geographies. Yet we still know relatively little about how these structures interact with socio-economic processes outside a limited range of places. There is still a tendency to work with a 'South East versus old industrial regions' model of the space economy. Some of this is driven by laudable theoretical aims, of course: if geography is to contribute towards the regulationist school's analysis of contemporary capitalism then it is at least plausible that the South East is the obvious starting-point. But it leaves much of the rest of the UK untouched. If regions are 'a product of a particular combination and articulation of social relations stretched over space' (Allen *et al.*, 1998, 143), then we need an understanding of these combinations and articulations. This is a familiar call for a revival of regional geography, but in a new guise. Understanding these combinations requires an openness to a range of theoretical positions, an awareness of a large range of processes (economic, social, political, cultural) and an appreciation of how they combine and interact. We also need an understanding of how they interact in a range of places, particularly at the present time. It is not just the imminence of the millennium that provokes futurological speculation. Several key processes – globalisation, European integration, transitions in the organisation of production, the alleged 'Third Way' – are reshaping human geographies. We need to know how these processes have an impact upon the entire country. It is in Shields' (1991) 'places on the margin', as much as in Allen *et al.*'s (1998) paradigmatic neoliberal region, that the effects of these emerging geographies must be assessed, as these impacts will determine whether the kingdom will remain united.

BIBLIOGRAPHY

Abercrombie, N. and Warde, A. 1988: *Contemporary British society*. Cambridge: Polity.

Adams, W. 1996: *Future nature: a vision for conservation*. London: Earthscan.

Agyeman, J. and Tuxworth, B. 1996: The changing face of environmental policy and practice in Britain. In Buckingham-Hatfield, S. and Evans, B. (eds), *Environmental planning and sustainability*. Chichester: Wiley, 105–24.

Allanson, P. and Whitby, M. 1996: Rural policy and the British countryside. In Allanson, P. and Whitby, M. (eds), *The rural economy and the British countryside*. London: Routledge, 1–16.

Allen, J. 1988: Fragmented firms, disorganised labour? In Allen, J. and Massey, D. (eds), *The economy in question*. London: Sage, 91–135.

Allen, J. 1992: Services and the UK space economy: regionalization and economic dislocation. *Transactions, Institute of British Geographers* **NS17**, 292–305.

Allen, J. and Henry, N. 1996: Fragments of industry and employment: contract service work and the shift towards precarious employment. In Crompton, R., Gallie, D. and Purcell, K. (eds), *Changing forms of employment*. London: Routledge, 65–82.

Allen, J. and Henry, N. 1997: Ulrich Beck's *Risk Society* at work: labour and employment in the contract service industries. *Transactions, Institute of British Geographers* **22**, 180–96.

Allen, J., Massey, D. and Cochrane, A. 1998: *Rethinking the region*. London: Routledge.

Allen, J. and Thompson, G. 1997: Think global, then think again – economic globalisation in context. *Area* **29**, 213–27.

Ambrose, P. 1994: *Urban process and power*. London: Routledge.

Amin, A. 1994: Post-Fordism: models, fantasies and phantoms of transition. In Amin, A. (ed.), *Post-Fordism: a reader*. Oxford: Blackwell, 1–39.

Amin, A., Cameron, A. and Hudson, R. 1999: Welfare to work or welfare as work? Combating social exclusion in the UK. *Environment and Planning A*, forthcoming.

Amin, A. and Thrift, N. (eds) 1994: *Globalisation, institutions and regional development*. Oxford: Oxford University Press.

Amin, A. and Tomaney, J. 1991: Creating an enterprise culture in the North East. *Regional Studies* **25**, 479–88.

Anderson, B. 1983: *Imagined communities: reflections on the origins and spread of nationalism*. London: Verso.

Anderson, J. 1989: Nationalisms in a divided kingdom. In Mohan, J. (ed.), *The political geography of contemporary Britain*. Basingstoke: Macmillan, 35–50.

Anderson, J. 1996: The shifting stage of politics: new medieval and postmodern territorialities? *Environment and Planning D* **14**, 133–55.

Anderson, J. 1997: Territorial sovereignty and political identity: national problems, transnational solutions? In Graham, B.

(ed.), *In search of Ireland: a cultural geography*. London: Routledge, 215–36.

Armstrong, H. 1998: What future for regional policy in the UK? *Political Quarterly* **69**, 200–14.

Ascher, K. 1987: *The politics of privatisation: contracting-out public services*. Basingstoke: Macmillan.

Atkinson, R. 1998a: Urban crisis: new policies for the next century. In Allmendinger, P. and Chapman, M., *Planning in the millennium*. Chichester: Wiley.

Atkinson, R. 1998b: *Contemporary English urban policy and its implications for the development of an 'urban policy' in the European Union*, mimeo, University of Portsmouth.

Atkinson, R. and Moon, G. 1994: *Urban policy in Britain*. Basingstoke: Macmillan.

Bacon, R. and Eltis, W. 1976: *Britain's economic problem: too few producers*. Basingstoke: Macmillan.

Bagguley, P. 1995: Middle-class radicalism revisited. In Butler, T. and Savage, M. (eds), *Social change and the middle classes*. London: UCL Press, 293–309.

Balchin, P. 1990: *Regional policy in Britain: the North–South divide*. London: Paul Chapman.

Baldwin, P. 1990: *The politics of social solidarity*. Cambridge: Cambridge University Press.

Ball, M., Gray, F. and McDowell, L. 1989: *The transformation of Britain: contemporary social and economic change*. London: Collins.

Ball, S., Bowe, R. and Gewirtz, S. 1995: Circuits of schooling. *Sociological Review* **43**, 52–78.

Barlow, J. and Savage, M. 1986: The politics of growth: conflict and cleavage in a Tory heartland. *Capital and Class* **31**, 156–82.

Barlow, Sir M. (Chairman) 1940: *Report of the Royal Commission on the distribution of the industrial population*. London: HMSO.

Barnett, C. 1986: *The audit of war: the illusion and reality of Britain as a great nation*. Basingstoke: Macmillan.

Barnett, C. 1995: *The lost victory: British dreams, British realities, 1945–50*. Basingstoke: Macmillan.

Barratt Brown, M. 1995: The new orthodoxy. In Barratt Brown, M. and Radice, H. (eds), *Democracy versus capitalism: a response to Will Hutton and some new questions for old*

Labour. European Labour Forum Pamphlet 4. Nottingham: Spokesman, 26–41.

Barrell, R. and Pain, N. 1997: EU: an attractive investment. *New Economy* **4**, 50–4.

Bartlett, W. and Le Grand, J. (eds) 1993: *Quasi-markets and social policy*. Basingstoke: Macmillan.

Beatson, M. 1995: *Labour market flexibility*. Research Series 48, Sheffield: Employment Department.

Beatty, C. and Fothergill, S. 1996: Labour market adjustment in areas of chronic industrial decline. *Regional Studies* **30**, 627–40.

Beck, U. 1992: *Risk society: towards a new modernity*. London: Sage.

Begg, I. 1993: The service sector in regional development. *Regional Studies* **27**, 817–25.

Bell, D. 1973: *The coming of post-industrial society*. London: Heinemann.

Bennett, R. J. 1989: Whither geography and modelling in a post-welfarist world? In Macmillan, B. (ed.) *Remodelling geography*. Oxford: Blackwell, 273–89.

Bennett, R., Wicks, P. and McCoshan, A. 1994: *Local empowerment and business services*. London: UCL Press.

Beynon, H. (ed.) 1985: *Digging deeper: issues in the miners' strike*. London: Verso.

Beynon, H., Hudson, R. and Sadler, D. 1991: *A tale of two industries: the contraction of coal and steel in NE England*. Buckingham: Open University Press.

Bibby, P. and Shepherd, J. 1997: Projecting rates of urbanisation in England, 1991–2016. *Town Planning Review* **68**, 93–124.

Blackman, T. 1995: *Urban policy in practice*. London: Routledge.

Blackman, T. 1998: Facing up to underfunding: equity and retrenchment in community care. *Social Policy and Administration* **32**, 182–95.

Blowers, A. 1997: Society and sustainability. In Blowers, A. and Evans, B. (eds), *Town planning into the 21st century*. London: Routledge, 153–68.

Boddy, M. and Lovering, J. 1988: The geography of military industry in Britain. *Area* **20**, 41–51.

Boddy, M., Lovering, J. and Bassett, K. 1986: *Sunbelt city? A study of economic change in*

Britain's M4 growth corridor. Oxford: Clarendon.

Bogdanor, V. 1996: Devolution. In Halpern, D. *et al.* (eds), *Options for Britain: a strategic policy review.* Aldershot: Dartmouth, 295–314.

Bonefeld, W., Brown, A. and Burnham, P. 1996: *A Major crisis? The politics of economic policy in Britain in the 1990s.* Aldershot: Dartmouth.

Borooah, V., McGregor, P., McKee, P. and Mulholland, G. 1996: Cost-of-living differences between the regions of the United Kingdom. In Hills, J. (ed.), *New inequalities.* Cambridge: Cambridge University Press, 103–32.

Bowler, I. 1991: The agricultural pattern. In Johnston, R. J. and Gardiner, V. (eds), *The changing geography of the United Kingdom.* London: Routledge, 83–114.

Boyer, R. and Drache, D. (eds) 1996: *States against markets.* London: Routledge.

Boyne, G. and Powell, M. 1991: Territorial justice: a review of theory and evidence. *Political Geography* **10**, 262–81.

Boyne, G. and Powell, M. 1993: Territorial justice and Thatcherism. *Environment and Planning C* **11**, 35–53.

Bradbeer, J. 1999: UK environmental policy under Blair. In Savage, S. and Robins, L. (eds) *Public policy under Blair.* Basingstoke: Macmillan. (forthcoming.)

Bradford, M. 1995: Diversification and division in the English education system. *Environment and Planning A* **27**, 1595–612.

Bradford, M. and Burdett, F. 1989a: Privatisation, education and the North–South divide. In Lewis, J. R. and Townsend, A. R. (eds), *The North–South divide.* London: Paul Chapman, 192–212.

Bradford, M. and Burdett, F. 1989b: Spatial polarisation of private education in England. *Area* **21**, 47–57.

Breheny, M. 1992: Emerging constraints in the South East. In Townroe, P. and Martin, R. (eds), *Regional development in the 1990s.* London: Jessica Kingsley, 309–13.

Breheny, M. 1995: The compact city and transport energy consumption. *Transactions, Institute of British Geographers* **20**, 81–101.

Breheny, M. and Congdon, P. (eds) 1989:

Growth and change in a core region. London: Pion.

Brown, A., McCrone, D. and Paterson, L. 1998: *Politics and society in Scotland.* London: Macmillan.

Brownill, S. 1990: *Developing London's Docklands: another great planning disaster?* London: Paul Chapman.

Buck, N. 1992: Labour market inactivity and polarisation. In Smith, D. (ed.), *Understanding the underclass.* London: PSI (Policy Studies Institute), 9–31.

Bull, P. 1991: The changing geography of manufacturing activity. In Johnston, R. J. and Gardiner, V. (eds), *The changing geography of the United Kingdom.* London: Routledge, 198–232.

Burchardt, T., Hills, J. and Propper, C. 1998: *Private welfare and public policy.* York: Joseph Rowntree Foundation.

Burrows, R. 1999: The contemporary dynamics of residualism: an analysis of residential mobility, social exclusion and social housing in England. *Journal of Social Policy* **28**, 27–52.

Burrows, R. and Loader, B. (eds) 1994: *Towards a postfordist welfare state?* London: Routledge.

Burrows, R., Pleace, N. and Quilgars, D. (eds) 1997: *Homelessness and social policy.* London: Routledge.

Burrows, R. and Rhodes, D. 1998: *Unpopular places? Area disadvantage and the geography of misery in England.* Bristol: Policy Press.

Busfield, J. 1990: Social divisions in consumption: the case of medical care. *Sociology* **24**, 77–96.

Butler, D., Adonis, A. and Travers, T. 1994: *Failure in British government.* Oxford: Oxford University Press.

Butler, T. 1995: Gentrification and the urban middle classes. In Butler, T. and Savage, M. (eds), *Social change and the middle classes.* London: UCL Press, 188–204.

Byrne, D. 1992: The city. In Cloke, P. (ed.), *Policy and change in Thatcher's Britain.* London: Paul Chapman, 247–68.

Byrne, D. 1994: Planning for and against the divided city. In Burrows, R. and Loader, B. (eds), *Towards a postfordist welfare state?* London: Routledge, 136–53.

Byrne, D. 1995a: Radical geography as mere

political economy: the local politics of space. *Capital and Class* **56**, 117–38.

Byrne, D. 1995b: Deindustrialisation and dispossession. *Sociology* **29**, 95–115.

Byrne, D. and Rogers, T. 1996: Divided spaces – divided schools: an exploration of the spatial relations of social division. *Sociological Research Online* **1**(2), 1–17.

Byrne, P. 1997: *Social movements in Britain.* London: Routledge.

Cain, P. and Hopkins, A. 1993: *British imperialism: crisis and reconstruction, 1914–90.* Harlow: Longman.

Campbell, B. 1993: *Goliath: Britain's dangerous places.* London: Methuen.

Cannadine, D. 1997: Apocalypse when? British politicians and British 'decline' in the twentieth century. In Clarke, P. and Trebilcock, C. (eds), *Understanding decline.* Cambridge: Cambridge University Press, 260–83.

Carter, N. 1997: Prospects: the parties and the environment in the UK. In Jacobs, M. (ed.), *The new politics of the environment.* Oxford: Blackwell, 192–206.

Castells, M. 1996: *The rise of the network society.* Oxford: Blackwell.

Castells, M. 1997: *End of millennium.* Oxford: Blackwell.

Castells, M. and Hall, P. 1994: *Technopoles of the world.* London: Routledge.

CCBI 1997: *Unemployment and the future of work.* London: Council of Churches for Britain and Ireland.

Chambers, I. 1993: Narratives of nationalism: being 'British'. In Carter, E., Donald, J. and Squares, J. (eds), *Space and place: theories of identity and location.* London: Routledge, 145–64.

Champion, A. 1997: The facts about the urban exodus. *Town and Country Planning* **66**(3), 77–9.

Champion, A. and Townsend, A. 1990: *Contemporary Britain.* London: Edward Arnold.

Champion, A., Wong, C., Rooke, A., Dorling, D., Coombes, M. and Brunsdon, C. 1996: *The population of Britain in the 1990s: a social and economic atlas.* Oxford: Clarendon Press.

Chisholm, M. 1995: Some lessons from the review of local government in England. *Regional Studies* **29**, 573–9.

Chisholm, M. and Manners, G. (eds) 1971: *Spatial policy problems of the British economy.* Cambridge: Cambridge University Press.

Clark, G. 1992: 'Real' regulation: the administrative state. *Environment and Planning A* **24**, 615–27.

Cloke, P. (ed.) 1992: *Policy and change in Thatcher's Britain.* Oxford: Pergamon.

Cloke, P., Milbourne, P. and Thomas, C. 1995a: Poverty in the countryside: out of sight and out of mind. In Philo, C. (ed.), *Off the map: the social geography of poverty in the United Kingdom.* London: Child Poverty Action Group, 83–102.

Cloke, P., Phillips, M. and Thrift, N. 1995b: The new middle classes and the social constructs of rural living. In Butler, T. and Savage M. (eds), *Social change and the middle classes.* London: UCL Press, 220–38.

Coates, B. and Rawstron, E. 1971: *Regional variations in Britain.* London: Batsford.

Coates, D. 1994: *The question of UK decline: economy, state and society.* London: Edward Arnold.

Cochrane, A. 1992: Is there a future for local government? *Critical Social Policy* **35**, 4–19.

Cochrane, A., Peck, J. and Tickell, A. 1996: Manchester plays games: exploring the local politics of globalisation. *Urban Studies* **33**, 1319–36.

Coe, N. 1998: Exploring uneven development in producer service sectors. *Environment and Planning A* **30**, 2041–68.

Coleman, A. 1985: *Utopia on trial: vision and reality in planned housing.* London: Hilary Shipman.

Colenutt, B. and Ellis, G. 1993: Boosting the Tories. *New Statesman and Society* 30 July 1993, p. 20.

Collis, C. and Noon, D. 1996: FDI and regional development. In Collis, C. and Peck, F. (eds), *Industrial restructuring: FDI and regional development.* London: Regional Studies Association, 1–11.

Colls, R. 1992: Born-again Geordies. In Colls, R. and Lancaster, B. (eds), *Geordies: roots of regionalism.* Edinburgh: Edinburgh University Press, 1–34.

Cooke, P. (ed.) 1989: *Localities: the changing face of urban Britain.* London: Unwin Hyman.

Cooke, P. 1995: Keeping to the high road:

learning, reflexivity and associative governance in regional economic development. In Cooke, P. (ed.), *The rise of the rustbelt*. London: UCL Press, 231–45.

Cooke, P. and Morgan, K. 1993: The network paradigm: new departures in corporate and regional development. *Environment and Planning D* **11**, 543–64.

Cossey, R. 1996: Measuring income in Manchester. *BURISA* **123**, 5–9.

Coutts, K. and Rowthorn, B. 1995: Employment in the UK: trends and prospects. *Political Quarterly*, **66**, 61–78.

Cowell, R. and Owens, S. 1997: Sustainability: the new challenge. In Blowers, A. and Evans, B. (eds), *Town planning into the 21st century*. London: Routledge, 15–32.

Cowell, R. and Owens, S. 1998: Suitable locations: equity and sustainability in the minerals planning process. *Regional Studies* **32**, 797–811.

Crang, P. and Martin, R. 1991: The other side of the 'Cambridge phenomenon'. *Environment and Planning D: Society and Space* **9**, 91–116.

Crilley, D. 1990: The disorder of Short's 'new urban order'. *Transactions, Institute of British Geographers* **15**, 232–8.

Crilley, D., Bryce, C., Hall, R. *et al.* 1992: *New migrants in London's docklands*. Occasional Paper, Geography Department, London: Queen Mary and Westfield College.

Crompton, R. 1993: *Class and stratification: an introduction to current debates*. Cambridge: Polity.

Crouch, C. 1997: The terms of the neoliberal consensus. *Political Quarterly* **68**(4), 352–60.

Crouch, C. and Streeck, W. (eds) 1996: *The political economy of the world's capitalisms*. London: Sage.

CSJ (Commission on Social Justice) 1994: *Social justice: strategies for national renewal*. London: Vintage.

CSO (Central Statistical Office) various dates: *Regional trends*, London: HMSO.

Cummins, S. and MacIntyre, S. 1997: Should policies for improving diet focus on people or places? Submitted to *British Food Journal*.

Cummins, S. and MacIntyre, S. 1999: The location of food stores in urban areas: a case study in Glasgow. *British Food Journal* **101**(7), 545–53.

Currie, D. 1996: Prospects and strategies for UK economic growth. In Halpern, D. *et al.* (eds), *Options for Britain: a strategic policy review*. Aldershot: Dartmouth, 29–58.

Curtice, J. 1996: The North–South divide. In Jowell, R. *et al.* (eds), *British social attitudes: the 13th report*. Aldershot: Dartmouth, 71–88.

Curtis, S. and Rees Jones, I. 1998: Is there a place for geography in the analysis of health inequality? *Sociology of Health and Illness* **20**, 645–72.

Cutler, T. and Waine, B. 1997: The politics of quasi-markets. *Critical Social Policy* **17**, 3–26.

Davis, M. 1992: *City of quartz*. London: Verso.

Dawson, A. 1993: The reform of local government in Scotland: should it be 15 or 51, or does it matter at all? *Scottish Geographical Magazine* **109**, 111–16.

Deakin, N. and Edwards, J. 1993: *The enterprise culture and the inner city*. London: Routledge.

Denver, D. 1997: The 1997 General Election in Scotland: an analysis of the results. *Scottish Affairs* **20**, 17–33.

DETR (Department of the Environment, Transport and the Regions) 1997: *Building partnerships for prosperity*. London: HMSO, Cm. 3814.

Dex, S. and McCulloch, A. 1997: *Flexible employment: the future of Britain's jobs*. Basingstoke: Macmillan.

Dicken, P., Peck, J. and Tickell, A. 1997: Unpacking the global. In Lee, R. and Wills, J. (eds), *Geographies of economies*. London: Edward Arnold, 158–66.

Dickens, P. 1988: *One nation? Social change and the politics of locality*. London: Pluto.

Dodd, P. 1995: *Battle over Britain*. London: Demos.

Dorling, D. 1995: *A new social atlas of Britain*. Chichester: Wiley.

Dorling, D. 1997: *Death in Britain: how local mortality rates have changed, 1950s-1990s*. York: Joseph Rowntree Foundation.

Dorling, D. and Cornford, J. 1995: Who has negative equity? *Housing Studies* **10**, 151–78.

Dorling, D. and Tomaney, J. 1995: Poverty in the old industrial regions: a comparative view. In Philo, C. (ed.), *Off the map: the social geography of poverty in the UK*. London: Child Poverty Action Group, 103–22.

Dorling, D. and Woodward, R. 1995: *Social polarisation in Britain: a micro-geographical approach.* Seminar Paper **65**, Geography Department, Newcastle: University of Newcastle upon Tyne.

Dorling, D. and Woodward, R. 1996: Social polarisation 1971–91: a micro-geographical analysis of Britain. *Progress in Planning* **45**(2), 1–66.

Dorling, D., Pattie, C., Rossiter, D. and Johnston, R. 1996. Missing voters in Britain 1992–96: where and with what impact. In Farrell, D. M. *et al.* (eds), *British elections and parties yearbook, 1996.* London: Frank Cass, 37–49.

Drakeford, M. 1997: The poverty of privatization. *Critical Social Policy* **17**, 115–32.

Driver, F. 1991: Political geography and state formation: disputed territory. *Progress in Human Geography* **15**, 268–80.

DSS (Department of Social Security) 1994: *Households below average income, 1980–1991/2.* London: HMSO.

DTI (Department of Trade and Industry) 1998: *Our competitive future.* London: HMSO.

Duncan, S. 1989: What is locality? In Peet, R. and Thrift, N. J. (eds), *New models in geography* (Vol. II). London: Unwin Hyman, 221–52.

Duncan, S. 1991: The geography of gender divisions in Britain. *Transactions, Institute of British Geographers* **16**, 420–39.

Duncan, S. and Goodwin, M. 1988: *The local state and uneven development.* Cambridge: Polity.

Duncan, S. and Goodwin, M. 1989: The crisis of local government. In Mohan, J. (ed.), *The political geography of contemporary Britain.* Basingstoke: Macmillan, 69–86.

Dunford, M. 1997: Divergence, instability and exclusion: regional dynamics in Great Britain. In Lee, R. and Wills, J. (eds), *Geographies of economies.* London: Arnold, 259–77.

Dunford, M. and Perrons, D. 1994: Regional inequality, regimes of accumulation and economic development in contemporary Europe. *Transactions, Institute of British Geographers* **NS19**, 163–82.

Dunn, R., Forrest, R. and Murie, A. 1987: The geography of council house sales in England, 1979–85. *Urban Studies* **24**, 47–59.

Dury, G. H. 1968: *The British Isles: a systematic and regional geography.* London: Heinemann.

Edgell, S. and Duke, V. 1991: *A measure of Thatcherism.* London: HarperCollins.

Edwards, J. 1997: Urban policy: the victory of form over substance? *Urban Studies* **34**, 825–43.

Eisenschitz, A. and Gough, J. 1993: *The politics of local economic policy.* London: Macmillan.

Ekinsmyth, C. forthcoming: Professional workers in a risk society. *Transactions, Institute of British Geographers.*

Enthoven, A. 1985: *Reflections on the management of the NHS.* London: Nuffield Provincial Hospitals Trust.

Esping-Andersen, G. 1990: *The three worlds of welfare capitalism.* Cambridge: Polity.

Etzioni, A. 1993: *The spirit of community.* New York: Simon and Schuster.

Evans, B. 1997: From town planning to environmental planning. In Blowers, A. and Evans, B. (eds), *Town planning into the 21st century.* London: Routledge, 1–14.

Evans, R. and Harding, A. 1997: Regionalisation, regional institutions and economic development. *Policy and Politics* **25**, 19–30.

Feigenbaum, H., Henig, J. and Hamnett, C. 1998: *Shrinking the state: the political underpinnings of privatisation.* Cambridge: Cambridge University Press.

Fieldhouse, E. 1995: Thatcherism and the changing geography of political attitudes, 1964–87. *Political Geography* **14**, 3–30.

Fielding, A. 1992: Migration and social mobility: South East England as an escalator region. *Regional Studies* **26**, 1–15.

Fielding, A. 1995: Migration and middle-class formation in England and Wales, 1981–91. In Butler, T. and Savage, M. (eds), *Social change and the middle classes.* London: UCL Press, 169–87.

Flynn, N. and Taylor, A. 1986: Inside the rust-belt: an analysis of the decline of the West Midlands economy. *Environment and Planning A* **18**, 865–900.

Forrest, R. and Murie, A. 1988: *Selling the welfare state: the privatization of public housing.* London: Routledge.

Fothergill, S. and Gudgin, G. 1982: *Unequal*

growth: urban and regional employment change in the UK. London: Heinemann.

Fothergill, S. and Guy, N. 1990: *Retreat from the regions: corporate change and the closure of factories.* London: Jessica Kingsley.

Fyfe, N. 1997: Crime. In Pacione, M. (ed.), *Britain's cities: geographies of division in urban Britain.* London: Routledge, 244–61.

Fyfe, N. and Bannister, J. 1998: 'The eyes upon the street': closed-circuit television surveillance and the city. In Fyfe, N. (ed.), *Images of the street.* London: Routledge, 254–67.

Galbraith, J. K. 1992: *The culture of contentment.* London: Penguin.

Gamble, A. 1988: *The free economy and the strong state.* Basingstoke: Macmillan.

Gamble, A. 1989: Thatcherism and the new politics. In Mohan, J. (ed.), *The political geography of contemporary Britain.* Basingstoke: Macmillan, 1–17.

Gamble, A. 1994: *Britain in decline: economic policy, political strategy and the British state* (4th edition). Basingstoke: Macmillan.

Gamble, A. 1996: The crisis of Conservatism. *New Left Review* **214**, 3–25.

Gamble, A. and Kelly, G. 1996: Stakeholder capitalism and one nation socialism. *Renewal* **4**(1), 23–32.

Gamble, A., Kelly, G., Dietrich, M. and Germain, R. 1995: Banking on the regions. *New Economy* **2**(4), 257–61.

Garrahan, P. and Stewart, P. 1992: *The Nissan enigma: flexibility at work in a local economy.* London: Mansell.

Geddes, M. 1994: Public services and local economic regeneration in a post-Fordist economy. In Burrows, R. and Loader, B. (eds), *Towards a postfordist welfare state?* London: Routledge, 154–74.

Gibson, A. and Asthana, S. 1998: School performance, school effectiveness and the 1997 White Paper. *Oxford Review of Education* **24**, 195–210.

Giddens, A. 1994a: Brave new world: the new context of politics. In Miliband, D. (ed.), *Reinventing the left.* Cambridge: Polity, 21–38.

Giddens, A. 1994b: *Beyond left and right: the future of radical politics.* Cambridge: Polity.

Giddens, A. 1998: *The third way.* Cambridge: Polity.

Giggs, J. and Pattie, C. 1992: Croeso y Cymru – Welcome to Wales: but welcome to whose Wales? *Area* **24**, 268–82.

Gilbert, D. 1996: The geography of strikes, 1940–90. In Charlesworth, A., Gilbert, D., Randall, A., Southall, H. and Wrigley, C. *An atlas of industrial protest in Britain, 1750–1990.* Basingstoke: Macmillan, 181–90.

Gilroy, P. 1987: *There ain't no Black in the Union Jack.* London: Hutchinson.

Ginn, J., Arber, S., Brannen, J., *et al.* 1996: Feminist fallacies: a reply to Hakim on women's employment. *British Journal of Sociology* **47**, 167–74.

Ginsburg, N. 1992: *Divisions of welfare.* London: Sage.

Ginsburg, N. 1997: Housing. In Walker, A. and Walker, C. (eds), *Britain divided: the growth of social exclusion in the 1990s.* London: Child Poverty Action Group, 140–52.

Glyn, A. 1992: The 'productivity miracle', profits and investment. In Michie, J. (ed.), *The economic legacy, 1979–92.* Cambridge: Cambridge University Press, 77–90.

Goldthorpe, J. 1995: The service class revisited. In Butler, T. and Savage, M. (eds) *Social change and the middle classes.* London: UCL Press, 313–29.

Goldthorpe, J. and Marshall, G. 1992: The promising future of class analysis: a response to recent critiques. *Sociology* **26**, 381–400.

Goodwin, M. and Painter, J. 1996: Local governance, the crises of Fordism and the changing geographies of regulation. *Transactions, Institute of British Geographers* **NS21**, 635–48.

Gough, J. and Eisenschitz, A. 1996: The modernisation of Britain and local economic policy: promise and contradictions. *Environment and Planning D* **14**, 203–19.

Grabher, G. and Stark, D. 1997: Organizing diversity: evolutionary theory, network analysis and postsocialism. *Regional Studies* **31**, 533–44.

Graham, A. 1996: The UK 1979–95: myths and realities of Conservative capitalism. In Crouch, C. and Streeck, W. (eds), *The political economy of the world's capitalisms.* London: Sage, 117–32.

Graham, B. 1997: The imagining of place:

representation and identity in contemporary Ireland. In Graham, B. (ed.), *In search of Ireland: a cultural geography*. London: Routledge, 192–212.

Graham, B. and Shirlow, P. 1998: An elusive agenda: the development of a middle ground in Northern Ireland. *Area* **30**, 245–54.

Graham, D. and Spence, N. 1995: Contemporary deindustrialisation and tertiarisation in the London economy. *Urban Studies* **32**, 885–911.

Graham, S. and Marvin, S. 1994: Cherry-picking and social dumping: utilities in the 1990s. *Utilities Policy* **4**, 113–19.

Gray, J. 1993: *Beyond the new right*. London: Routledge.

Gray, J. 1996: *After social democracy*. London: Demos.

Green, A. 1994: *The geography of poverty and wealth*. Coventry: Institute for Employment Research.

Green, A., Gregg, P. and Wadsworth, J. 1998: Regional unemployment changes in Britain. In Lawless, P., Martin, R. L. and Hardy, S. (eds), *Unemployment and social exclusion*. London: Jessica Kingsley, 69–94.

Green, A., Owen, D. W. and Winnett, C. M. 1994: The changing geography of recession: analyses of local unemployment time series. *Transactions, Institute of British Geographers* **19**, 142–62.

Green, H., Deacon, K., Iles, N. and Downs, D. 1997: *Housing in England, 1995–96*. London: HMSO.

Gregg, P. and Wadsworth, J. 1995: A short history of labour turnover, job tenure and job security 1975–93. *Oxford Review of Economic Policy* **11**(1), 73–90.

Griffiths, M. and Johnston, R. 1991: What's in a place? *Antipode* **23**, 185–213.

Grove-White, R. 1997: Environment, risk and democracy. In Jacobs, M. (ed.), *Greening the millennium: the new politics of the environment*. Oxford: Blackwell, 109–22.

Gudgin, G. 1995: Regional problems and policy in the UK. *Oxford Review of Economic Policy* **11**, 18–63.

Hakim, C. 1995: Five feminist myths about women's employment. *British Journal of Sociology* **46**, 429–55.

Hall, P. 1985: The geography of the fifth Kondratiev. In Hall, P. and Markusen, A. (eds), *Silicon landscapes*. London: Allen & Unwin, 1–19.

Hall, P. 1991: Structural transformation in the regions of the UK. In Rodwin, H. and Sazanami, H. (eds), *Industrial change and regional economic transformation*. London: HarperCollins, 39–69.

Hall, P. 1997: The view from London Centre: twenty-five years of planning at the DOE. In Blowers, A. and Evans, B. (eds), *Town planning into the 21st century*. London: Routledge, 119–36.

Hall, P., Breheny, M., McQuaid, R. and Hart, D. 1987: *Western sunrise: the genesis and growth of Britain's major high-tech corridor*. London: Allen & Unwin.

Hall, S. 1985: Authoritarian populism: a response to Jessop *et al*. *New Left Review* **151**, 115–27.

Hallsworth, A. 1996: Short-termism and economic restructuring in Britain. *Economic Geography* **72**, 23–37.

Hamnett, C. 1989: Consumption and class in contemporary Britain. In Hamnett, C., McDowell, L. and Sarre, P. (eds), *The changing social structure*. London: Sage, 199–243.

Hamnett, C. 1994: Restructuring housing finance and the housing market. In Corbridge, S., Martin, R. L. and Thrift, N. J. (eds), *Money, power and space*. Basingstoke: Macmillan, 281–308.

Hamnett, C. 1997: A stroke of the Chancellor's pen: the social and regional impact of the Conservatives' 1988 higher rate tax cuts. *Environment and Planning A* **29**, 129–48.

Handy, C. 1995: *The empty raincoat*. London: Hutchinson.

Harding, A. 1991: The rise of urban growth coalitions, UK-style? *Environment and Planning C* **9**, 295–317.

Harloe, M., Pickvance, C. and Urry, J. (eds) 1990: *Place, policy and politics: do localities matter?* London: Unwin Hyman.

Harman, C. 1996: Globalisation: a critique of a new orthodoxy. *International Socialism* **73**, 5–33.

Harris, J. 1977: *William Beveridge: a biography*. Oxford: Oxford University Press.

Harrison, R. and Mason, C. 1996: Developments in the promotion of informal venture capital in the UK. *International Journal of Entrepreneurial Behaviour and Research* **2**, 6–33.

Harvey, D. 1989a: *The condition of postmodernity*. Oxford: Blackwell.

Harvey, D. 1989b: From managerialism to entrepreneurialism: the transformation of urban governance in late capitalism. *Geografisker Annaler* **71B**, 3–17.

Harvey, D. 1993: The nature of environment: the dialectics of social and environmental change. In Panitch, L. and Miliband, R. (eds), *Socialist register 1993*. London: Merlin, 1–51.

Harvey, D. 1996: *Justice, nature and the geography of difference*. Oxford: Blackwell.

Harvey, G. 1997: *The killing of the countryside*. London: Jonathan Cape.

Harvie, C. 1994: *Scotland and nationalism* (2nd edition). London: Routledge.

Haseler, S. 1996: *The English tribe*. Basingstoke: Macmillan.

Hay, C. 1996a: A state of disarray. *Renewal* **4**, 40–50.

Hay, C. 1996b: *Re-stating social and political change*. Buckingham: Open University Press.

Hay, C. 1997: Blaijorism: towards a one-vision polity? *Political Quarterly* **68**, 372–8.

Hay, C. 1998: Globalisation, welfare retrenchment and 'the logic of no alternative'. *Journal of Social Policy* **27**, 525–32.

Hayek, F. 1944: *The road to serfdom*. London: Routledge.

Heald, D. and Gaughan, N. 1996: Financing a Scottish Parliament. In Tindale, S. (ed.), *The state and the nations*. London: IPPR, 167–83.

Heath, A., Curtice, J., Jowell, R., Evans, G., Field, J. and Witherspoon, S. 1991: *Understanding political change: the British voter, 1964–87*. Oxford: Pergamon.

Heath, A., Jowell, R. and Curtice, J. 1985: *How Britain votes*. Oxford: Pergamon.

Heath, A. and Savage, M. 1995: Political alignments within the middle classes, 1972–89. In Butler, T. and Savage, M. (eds), *Social change and the middle classes*. London: UCL Press, 275–92.

Hebbert, M. 1993: 1992: myth and aftermath. *Regional Studies* **27**, 709–18.

Hechter, M. 1975: *Internal colonialism: the Celtic fringe in British national development*. London: Routledge & Kegan Paul.

Heffernan, M. 1998: *The meaning of Europe*. London: Arnold.

Helm, D. 1998: Environmental policy: objectives, instruments and institutions. *Oxford Review of Economic Policy* **14**, 1–19.

Henderson, J. 1987: Semiconductors, Scotland and the International Division of Labour. *Urban Studies* **24**, 5, 389–408.

Henry, N. 1992: New industrial spaces: locational logic of a new production era? *International Journal of Urban and Regional Research* **16**, 3, 375–96.

Henry, N., Pinch, S. and Russell, S. 1996: In pole position? Untraded interdependencies, new industrial spaces and the British motor sport industry, *Area* **28**, 1, 25–36.

Hepple, L. 1989: Destroying local leviathans and designing landscapes of liberty: public choice theory and the poll tax. *Transactions, Institute of British Geographers* **NS14**, 387–99.

Hewitson, R. 1987: *The heritage industry: Britain in a climate of decline*. London: Methuen.

Higgs, G., Senior, M. and Williams, H. 1998: Spatial and temporal variations of mortality and deprivation 1: widening health inequalities. *Environment and Planning A* **30**, 1661–82.

Hill, S. and Munday, M. 1992: The UK regional distribution of foreign direct investment: analysis and determinants. *Regional Studies* **26**, 535–44.

Hills, J. (ed.) 1996: *New inequalities*. Cambridge: Cambridge University Press.

Hindess, B. 1987: *Freedom, equality and the market*. London: Tavistock.

Hirst, P. 1994: *Associative democracy: new forms of economic and social governance*. Cambridge: Polity.

Hirst, P. and Thompson, G. 1996: *Globalisation in question*. Cambridge: Polity.

Hirst, P. and Zeitlin, J. (eds) 1989: *Reversing industrial decline? Industrial policy in Britain and her competitors*. Oxford: Berg.

Hobsbawm, E. and Ranger, T. (eds) 1983: *The invention of tradition*. Cambridge: Cambridge University Press.

Hoggett, P. 1987: Decentralisation as an

emerging private and public sector paradigm. In Hoggett, P. and Hambleton, R. (eds), *Decentralisation and democracy: localising public services*. Bristol: SAUS (School for Advanced Urban Studies).

Hoggett, P. (ed.) 1997: *Contested communities: experiences, struggles, policies*. Bristol: Policy Press.

House, J. W. (ed.) 1973: *The UK space: resources, environment and the future*. London: Weidenfeld & Nicolson.

House of Commons 1972: *Expenditure Committee, second report: regional development incentives*. London: HMSO.

House of Commons Treasury Committee 1997: *Second report, session 1997–98: the Barnett formula*. London: HMSO, Cm. 341.

House of Lords 1985: *Report of the Select Committee on Overseas Trade*, HC-238. London: HMSO.

Hudson, R. 1989a: Rewriting history and reshaping geography: the nationalised industries and the political economy of Thatcherism. In Mohan, J. F. (ed.), *The political geography of contemporary Britain*. Basingstoke: Macmillan, 113–29.

Hudson, R. 1989b: *Wrecking a region*. London: Pion.

Hudson, R. 1989c: Labour market changes and new forms of work in 'old' industrial regions. *Environment and Planning D: Society and Space* **7**, 1–28.

Hudson, R. 1994: New production concepts, new production geographies? Reflections on changes in the automobile industry. *Transactions, Institute of British Geographers* **19**, 331–45.

Hudson, R. 1997a: The end of mass production and of the mass collective worker? In Lee, R. and Wills, J. (eds), *Geographies of economies*. London: Arnold, 302–10.

Hudson, R. 1997b: Regional futures: industrial restructuring, new high volume production concepts and spatial development strategies in the new Europe. *Regional Studies* **31**, 467–78.

Hudson, R. and Sadler, D. 1992: *Manufacturing success? Reindustrialisation policies in Derwentside in the 1980s*. OP-25, Geography Department, Durham: Durham University.

Hudson, R. and Williams, A. 1995 (second edition): *Divided Britain*. Chichester: Wiley.

Hutton, W. 1995: *The state we're in*. London: Jonathan Cape.

Hutton, W. 1997: *The state to come*. London: Vintage.

Hutton, W. 1998: The state we should be in. *Marxism Today* **Nov./Dec.**, 34–7.

Imrie, R. and Thomas, H. 1993: *British urban policy and the urban development corporations*. London: Paul Chapman.

Ingham, G. 1984: *Capitalism divided? The city and industry in British social development*. Basingstoke: Macmillan.

Ingham, G. 1995: British capitalism: empire, merchants and decline. *Social History* **XX**, 339–54.

Inland Revenue 1997: *Inland Revenue statistics*. London: HMSO.

Jackson, P. 1987: *Race and racism: essays in social geography*. London: Unwin Hyman.

Jackson, P. 1991: Mapping meanings: a cultural critique of locality studies. *Environment and Planning A* **23**, 215–28.

Jacobs, M. 1997: The new politics of the environment. In Jacobs, M. (ed.), *Greening the millennium: the new politics of the environment*. Oxford: Blackwell, 1–17.

James, W., Nelson, M., Ralph, A. and Leather, S. 1997: The contribution of nutrition to inequalities in health. *British Medical Journal* **314**, 1545–9.

Jefferson, C. and Trainor, M. 1993: Public sector relocation and regional development. *Urban Studies* **33**, 37–48.

Jenkins, S. 1995: *Accountable to none: the Tory nationalisation of Britain*. London: Hamish Hamilton.

Jessop, B. 1990: *State theory: putting capitalist states in their place*. Cambridge: Polity.

Jessop, B. 1992: From social democracy to Thatcherism: twenty-five years of British politics. In Abercrombie, N. and Warde, A. (eds), *Social change in contemporary Britain*. Cambridge: Polity, 14–39.

Jessop, B. 1993: Towards a Schumpeterian workfare state? Preliminary remarks on post-Fordist political economy. *Studies in Political Economy* **40**, 7–39.

Jessop, B. 1995: Towards a Schumpeterian workfare regime in Britain? Reflections on regulation, governance and the welfare state. *Environment and Planning A* **27**, 1613–26.

Jessop, B., Bonnett, K. and Bromley, S. 1990: Farewell to Thatcherism? Neoliberalism and new times. *New Left Review* **179**, 81–102.

Jessop, B., Bonnett, K., Bromley, S. and Ling, T. 1984: Authoritarian populism, two nations, and Thatcherism. *New Left Review* **147**, 32–60.

Jessop, B., Bonnett, K., Bromley, S. and Ling, T. 1988: *Thatcherism: a tale of two nations*. Cambridge: Polity.

Jewell, H. 1994: *The North–South divide: the origins of northern consciousness in England*. Manchester: Manchester University Press.

John, P. and Whitehead, A. 1997: The renaissance of English regionalism in the 1990s. *Policy and Politics* **25**, 7–17.

Johnston, R. and Pattie, C. 1992a: Is the seesaw tipping back? The end of Thatcherism and changing voting patterns in Great Britain, 1979–92. *Environment and Planning A* **24**, 1491–505.

Johnston, R. and Pattie, C. 1992b: Unemployment, the poll tax and the British general election of 1992. *Environment and Planning C* **10**, 467–83.

Johnston, R. and Pattie, C. 1996: Local government in local governance: the 1994–95 restructuring of local government in England. *International Journal of Urban and Regional Research* **20**, 671–96.

Johnston, R., Pattie, C. and Russell, A. 1993: Dealignment, spatial polarisation and economic voting. *European Journal of Political Research* **23**, 67–90.

Jones, B. 1996: The social constitution of labour markets. In Crompton, R., Gallie, D. and Purcell, K. (eds), *Changing forms of employment*. London: Routledge, 109–132.

Jones, M. 1997: The degradation of labour market programmes. *Critical Social Policy* **52**, 91–104.

Karn, V., Coleman, D., Salt, J., Ratcliffe, P. and Peach, C. 1996: *Ethnicity in the 1991 census*. London: HMSO.

Kavanagh, D. 1997: *The reordering of British politics*. Oxford: Oxford University Press.

Kawachi, I. and Kennedy, B. 1997: Health and social cohesion: why care about income inequality? *British Medical Journal* **314**, 1037–40.

Kearns, A. 1992: Active citizenship and urban governance. *Transactions, Institute of British Geographers* **17**, 20–34.

Keeble, D. 1989: High technology industry and regional development in Britain. *Environment and Planning C* **7**, 153–72.

Keeble, D. 1991: Deindustrialisation and new industrialisation processes and regional restructuring in the EC. In Wilding, T. and Jones, P. (eds), *Deindustrialisation and new industrialisation in Britain and Germany*. London: Anglo German Foundation, 40–65.

Keeble, D. 1992: High technology industry and the restructuring of the UK space economy. In Townroe, P. and Martin, R. (eds), *Regional development in the 1990s*. London: Jessica Kingsley, 172–82.

Keeble, D. and Bryson, J. 1996: Small-firm creation and growth, regional development and the North–South divide in Britain. *Environment and Planning A* **28**, 909–34.

Keeble, D., Bryson, J. and Wood, P. 1991: Small firms, business services growth and regional development in the UK: some empirical findings. *Regional Studies* **25**, 439–57.

Keeble, D. and Walker, S. 1994: New firms, small firms and dead firms: spatial patterns and determinants in the UK. *Regional Studies* **28**, 411–27.

Keep, E. and Mayhew, K. 1994: The changing structure of training provision. In Buxton, T., Chapman, P. and Temple, P. (eds), *Britain's economic performance*. London: Routledge, 308–41.

Keith, M. 1987: 'Something happened': explanations of the 1981 riots in London. In Jackson, P. (ed.), *Race and racism*. London: Allen & Unwin, 275–303.

Keith, M. 1993: *Race, riots and policing: lore and disorder in a multi-racist society*. London: UCL Press.

Keith, M. and Rogers, A. (eds) 1991: *Rhetoric and reality in the inner city*. London: Mansell.

Kendall, J. and Knapp, M. 1996: *The voluntary sector in the UK*. Manchester: Manchester University Press.

Kennedy, P. 1993: *Preparing for the 21st century*. London: HarperCollins.

King, D. 1995: *Actively seeking work?* Oxford: Clarendon.

Kitson, M. and Michie, J. 1996: Manufacturing capacity, investment and employment. In Michie, J. and Grieve Smith, J. (eds), *Towards full employment.* Oxford: Oxford University Press, 24–51.

Klein, R. and Millar, J. 1995: Do-it-yourself social policy. *Social Policy and Administration* **29**, 303–16.

Kozol, J. 1991: *Savage inequalities: children in America's schools.* New York: Random House.

Kozul-Wright, R. 1996: Transnational corporations and the nation state. In Michie, J. and Grieve Smith, J. (eds), *Managing the global economy.* Cambridge: Cambridge University Press, 135–71.

Krieger, J. 1986: *Reagan, Thatcher and the politics of decline.* Cambridge: Polity.

Krugman, P. 1996: *The self-organising economy.* Oxford: Blackwell.

Lash, S. and Urry, J. 1987: *The end of organised capitalism.* Cambridge: Polity.

Lattimer, M. and Holly, K. 1992: *Charity and NHS Reform.* London: Directory of Social Change.

Lee, C. H. 1986: *The British economy since 1700: a macroeconomic perspective.* Cambridge: Cambridge University Press.

Lee, R. 1996: Moral money? LETS and the creation of local economic geographies in Southeast England. *Environment and Planning A* **28**, 1377–94.

Levitas, R. 1996: The concept of social exclusion and the new Durkheimian hegemony. *Critical Social Policy* **16**, 5–20.

Lewis, J. 1984: Regional policy and planning: convergence and contradiction. In Bornstein, S., Held, D. and Krieger, J. (eds), *The state in capitalist Europe.* London: Allen & Unwin, 138–55.

Lewis, J. and Townsend, A. (eds) 1989: *The North–South divide.* London: Paul Chapman.

Leys, C. 1995: Still a question of hegemony. *New Left Review* **181**, 119–28.

Leyshon, A. 1995: Missing words: what happened to the geography of poverty? *Environment and Planning A* **27**, 1021–8.

Leyshon, A. and Thrift, N. 1989: South goes north? The rise of the provincial financial centre. In Lewis, J. and Townsend, A. (eds), *The North–South divide.* London: Paul Chapman, 114–56.

Leyshon, A. and Thrift, N. 1995: Geographies of financial exclusion: financial abandonment in Britain and the United States. *Transactions, Institute of British Geographers* **NS20**, 312–41.

Leyshon, A. and Thrift, N. 1997: *Money/space: geographies of monetary transformation.* London: Routledge.

Lipietz, A. 1992: *Towards a new economic order: postfordism, ecology and democracy.* Cambridge: Polity.

Lister, R. 1998: From equality to social inclusion: New Labour and the welfare state. *Critical Social Policy* **18**, 215–26.

Lloyd, P. and Shutt, J. 1985: Recession and restructuring in the North-West region, 1974–82. In Massey, D. and Meegan, R. (eds), *Politics and method: contrasting studies in industrial geography.* London: Methuen, 16–60.

Lovering, J. 1990a: The Labour party and the peace dividend. *Capital and Class* **41**, 7–14.

Lovering, J. 1990b: Fordism's unknown successors. *International Journal of Urban and Regional Research* **14**, 159–74.

Lovering, J. 1995a: Opportunity or crisis? The remaking of the British arms industry. In Turner, R. (ed.), *The British economy in transition.* London: Routledge, 88–122.

Lovering, J. 1995b: Creating discourses rather than jobs: the crisis in the cities and the transition fantasies of intellectuals and policy-makers. In Healey, P., Cameron, S., Davoudi, S. *et al.* (eds), *Managing cities: the new urban context.* Chichester: Wiley, 109–26.

Lovering, J. 1996: New myths of the Welsh economy. *Planet* **116**, 6–16.

Lovering, J. 1998: Labour and the defence industry: allies in 'globalisation'. *Capital and Class* **65**, 9–20.

Lovering, J. 1999: Theory led by policy: the inadequacies of the 'new regionalism'. *International Journal of Urban and Regional Research* **23**, 379–95.

Lowe, P. 1996: Blueprint for a rural economy. In Allanson, P. and Whitby, M. (eds), *The rural economy and the British countryside.* London: Routledge, 187–202.

Lowe, P. and Flynn, A. 1989: Environmental politics and policy in the 1980s. In Mohan, J. (ed.), *The political geography of contemporary Britain*. London: Macmillan, 255–79.

Lowe, P., Murdoch, J. and Cox, G. 1995: A civilised retreat? Anti-urbanism, rurality and the making of an Anglo-centric culture. In Healey, P., Cameron, S., Davoudi, S. *et al.* (eds), *Managing cities: the new urban context*. Chichester: Wiley, 63–82.

Lowe, P., Clark, J., Seymour, S. and Ward, N. 1997: *Moralizing the environment: countryside change, farming and pollution*. London: UCL Press.

McAllister, I. 1997: Regional voting. *Parliamentary Affairs* **50**, 641–57.

McAllister, L. 1998: The Welsh devolution referendum: definitely, maybe? *Parliamentary Affairs* **51**, 149–65.

McCormick, J. and Alexander, R. 1996: Firm foundations: securing the Scottish Parliament. In Tindale, S. (ed.) *The State and the Nations*. London: IPPR, 99–166.

McCrone, D. 1991: 'Excessive and unreasonable': the politics of the poll tax in Scotland. *International Journal of Urban and Regional Research* **15**, 443–52.

McCrone, D. 1992: *Understanding Scotland: the sociology of a stateless nation*. London: Routledge.

McDowell, L. 1991: Life after father and Ford: the new gender order of postfordism. *Transactions, Institute of British Geographers* **16**, 400–19.

McDowell, L. and Court, G. 1994: Gender divisions of labour in the postfordist economy. *Environment and Planning A* **26**, 1397–418.

MacKay, R. 1994: Automatic stabilisers, European union and national unity. *Cambridge Journal of Economics* **18**, 571–85.

MacKay, R. 1995: Non-market forces, the nation-state and the European Union. *Papers in Regional Science* **74**, 209–31.

MacKintosh, M. and Wainwright, H. (eds) 1987: *A taste of power: the politics of local economics*. London: Verso.

McLaren, D., Bullock, S. and Yousuf, N. 1998: *Tomorrow's world: Britain's share in a sustainable future*. London: Earthscan.

McLoone, P. and Boddy, F. A. 1994: Deprivation and mortality in Scotland, 1981 and 1991. *British Medical Journal* **309**, 1465–70.

Malmberg, A. 1996: Industrial geography: agglomeration and local milieu. *Progress in Human Geography* **20**, 392–403.

Marquand, D. 1997: After euphoria: the dilemmas of New Labour. *Political Quarterly* **68**(4), 335–43.

Marquand, D. 1998: The Blair paradox. *Prospect*, May, 19–24.

Marr, A. 1996: *Ruling Britannia: the failure and future of British democracy*. London: Penguin.

Marsden, T., Murdoch, J., Lowe, P. , Munton, R. and Flynn, A. 1993: *Constructing the countryside*. London: UCL Press.

Marsh, D. and Rhodes, R. (eds) 1992: *Implementing Thatcherite policies*. Buckingham: Open University Press.

Marshall, G. 1988: Classes in Britain: Marxist and official. *European Sociological Review* **4**, 141–54.

Marshall, G. 1997: *Repositioning class: social inequality in industrial societies*. London: Sage.

Marshall, G., Roberts, S. and Burgoyne, C. 1996: Social class and underclass in Britain and the USA. *British Journal of Sociology* **47**, 22–44.

Marshall, G., Newby, H., Rose, D. and Vogler, C. 1988: *Social class in modern Britain*. London: Hutchinson.

Marshall, J. N. (ed.) 1988: *Uneven development in the service economy: understanding the location and role of producer services*. Oxford: Oxford University Press.

Marshall, J. N. 1994: Business reorganization and the development of corporate services in metropolitan areas. *Geographical Journal* **160**, 41–9.

Marshall, J. N. 1996: Civil service reorganisation and urban and regional development in Britain. *Service Industries Journal* **16**, 347–67.

Marshall, J. N., Hopkins, W. and Richardson, R. 1997: Civil service relocation and the regions. *Regional Studies* **31**, 607–13.

Marshall, J. N. and Wood, P. 1992: The role of services in urban and regional development. *Environment and Planning A* **24**, 1255–70.

Marshall, M. 1987: *Long waves of regional development*. Basingstoke: Macmillan.

Martin, R. 1986: Thatcherism and Britain's

industrial landscape. In Martin, R. and Rowthorn, R. (eds), *The geography of de-industrialisation*. Basingstoke: Macmillan, 238–90.

Martin, R. 1988a: The political economy of Britain's North–South divide. *Transactions, Institute of British Geographers* **13**, 389–418.

Martin, R. 1988b: Industrial capitalism in transition: the contemporary reorganisation of the British space economy. In Massey, D. and Allen, J. (eds), *Uneven redevelopment: cities and regions in transition*. London: Hodder, 202–31.

Martin, R. 1989a: Regional imbalance as consequence and constraint in national economic renewal. In Green, F. (ed.), *The restructuring of the UK economy*. Brighton: Harvester Wheatsheaf, 80–100.

Martin, R. 1989b: The growth and geographical anatomy of venture capital in the UK. *Regional Studies* **23**, 389–403.

Martin, R. 1989c: De-industrialisation and the state: Keynesianism, Thatcherism and the regions. In Mohan, J. (ed.), *The political geography of contemporary Britain*. Basingstoke: Macmillan, 87–112.

Martin, R. 1992: Has the British economy been transformed? In Cloke, P. (ed.), *Policy and change in Thatcher's Britain*. Oxford: Pergamon, 123–58.

Martin, R. 1993: Remapping British regional policy: the end of the north–south divide? *Regional Studies* **27**, 797–806.

Martin, R. 1994: Economic theory and human geography. In Gregory, D. J., Smith, G. and Martin, R. L. (eds), *Human geography: society, space and social science*. Basingstoke: Macmillan, 21–53.

Martin, R. 1995: Income and poverty inequalities across regional Britain: the north–south divide lingers on. In Philo, C. (ed.), *Off the map: the social geography of poverty in the United Kingdom*. London: Child Poverty Action Group, 23–44.

Martin, R. 1998: Regional dimensions of Europe's unemployment crisis. In Lawless, P., Martin, R. L. and Hardy, S. (eds), *Unemployment and social exclusion*. London: Jessica Kingsley, 11–48.

Martin, R. and Minns, R. 1995: Undermining the financial basis of regions: the spatial structure and implications of the UK pension fund system. *Regional Studies* **29**, 125–44.

Martin, R. and Rowthorn, R. (eds) 1986: *The geography of de-industrialisation*. Basingstoke: Macmillan.

Martin, R. and Sunley, P. 1997: The post-Keynesian state and the space economy. In Lee, R. and Wills, J. (eds), *Geographies of economies*. London: Arnold, 278–89.

Martin, R., Sunley, P. and Wills, J. 1996: *Union retreat and the regions*. London: Jessica Kingsley.

Martin, R. and Townroe, P. (eds) 1992: *Regional development in the 1990s: the British Isles in transition*. London: Jessica Kingsley.

Martin, R. and Tyler, P. 1992: The regional legacy. In Michie, J. (ed.), *The economic legacy 1979–92*. London: Academic Press, 80–97.

Mason, C. and Harrison, R. 1999: Financing entrepreneurship: venture capital and regional development. In Martin, R. (ed.), *Money and the space economy*. Chichester: Wiley, 157–84.

Massey, D. 1984: *Spatial divisions of labour* (1st edition). London: Macmillan.

Massey, D. 1986: The legacy lingers on: the impact of Britain's international role on its internal geography. In Martin, R. L. and Rowthorn, R. E. (eds), *The geography of de-industrialisation*. Basingstoke: Macmillan, 31–52.

Massey, D. 1988: Uneven development: social change and spatial divisions of labour. In Massey, D. and Allen, J. (eds), *Uneven redevelopment: cities and regions in transition*. London: Hodder, 250–76.

Massey, D. 1991: The political place of locality studies. *Environment and Planning A* **23**, 267–81.

Massey, D. 1995: Reflections on gender and geography. In Butler, T. and Savage, M. (eds), *Social change and the middle classes*. London: UCL Press, 330–44.

Massey, D. 1997: *Spatial divisions of labour* (2nd edition). London: Macmillan.

Massey, D. and Meegan, R. 1982: *The Anatomy of job loss*. London: Methuen.

Massey, D. and Meegan, R. (eds) 1985: *Politics and method*. London: Methuen.

Massey, D., Quintas, P. and Wield, D. 1992:

High-tech fantasies: science parks in society, science and space. London: Routledge.

Matless, D. 1994: Doing the English village, 1945–90: an essay in imaginative geography. In Cloke, P., Doel, M., Matless, D., Phillips, M. and Thrift, N. (eds), *Writing the rural: five cultural geographies.* London: Paul Chapman, 7–88.

Maud Commission 1969: *Report of the Royal Commission on local government in England* (Chairman: Lord Redcliffe-Maud). London: HMSO.

Mawson, J. and Spencer, K. 1997: The government offices for the English regions: towards regional governance? *Policy and Politics* **25**, 71–84.

Meegan, R. 1989: Paradise postponed. In Cooke, P. (ed.), *Localities: the changing face of urban Britain.* London: Unwin Hyman, 198–234.

Melling, J. 1991: Industrial capitalism and the welfare of the state. *Sociology* **25**, 219–39.

Middleton, R. 1996: *Government versus the market.* Cheltenham: Edward Elgar.

Miller, W. 1998: The periphery and its paradoxes. *West European Politics* **21**, 167–196.

Minford, P. and Storey, P. 1991: Regional policy and market forces. In Bowen, A. and Mayhew, K. (eds), *Reducing Regional Inequalities.* London: Kogan Page.

Ministry of Defence 1998: *UK defence statistics, 1998.* London: HMSO.

Minns, R. 1996: The social ownership of capital. *New Left Review* **219**, 42–61.

Minns, R. and Tomaney, J. 1995: Regional government and local economic development: the realities of economic power in the UK. *Regional Studies* **29**, 202–8.

Mitchell, J. 1996: *Strategies for self-government: the campaigns for a Scottish parliament.* Edinburgh: Edinburgh University Press.

Modood, T. (ed.) 1997: *Ethnic minorities in Britain: diversity and disadvantage.* London: Policy Studies Institute.

Mohan, J. (ed.) 1989: *The political geography of contemporary Britain.* Basingstoke: Macmillan.

Mohan, J. 1995a: *A National Health Service? The restructuring of health care in Britain since 1979.* Basingstoke: Macmillan.

Mohan, J. 1995b: Missing the boat: unemployment, poverty and debt in South East England. In Philo, C. (ed.), *Off the map: the social geography of poverty in the United Kingdom.* London: Child Poverty Action Group, 133–52.

Mohan, J. 1996: Review essay. *Environment and Planning A* **28**, 1519–24.

Mohan, J. 1998: Uneven development, territorial politics and the NHS reforms. *Political Studies* **LXVII**, 308–329.

Mohan, J. and Lee, R. 1989: Unbalanced growth? Public services and labour shortages in a European core region. In Breheny, M. and Congdon, P. (eds), *Growth and change in a core region.* London: Pion, 33–54.

Moore, B., Rhodes, J. and Tyler, P. 1986: *The effects of government regional economic policy.* London: HMSO.

Morgan, K. 1997: The learning region: institutions, innovation and regional renewal. *Regional Studies* **31**, 491–503.

Morgan, K. and Sayer, A. 1988: *Microcircuits of capital: 'sunrise' industry and uneven development.* Cambridge: Polity.

Morris, J. 1995: McJobbing a region? Industrial restructuring and the widening socio-economic divide in Wales. In Turner, R. (ed.), *The British economy in transition.* London: Routledge, 46–66.

Morris, L. 1993: Is there a British underclass? *International Journal of Urban and Regional Research* **17**, 404–13.

Morris, L. 1994: *Dangerous classes.* London: Routledge.

Munday, M., Morris, J. and Wikinson, B. 1995: Factories or warehouses? A Welsh perspective on Japanese transplant manufacturing. *Regional Studies* **29**, 1–17.

Munton, R. 1997: Engaging sustainable development: some observations on progress in the UK. *Progress in Human Geography* **21**, 147–63.

Murdoch, J. 1997: The shifting territory of government. *Area* **29**, 109–18.

Murie, A. 1997: The social rented sector, housing and the welfare state in the UK. *Housing Studies* **12**, 437–61.

Murphy, E. 1991: *After the asylums.* London: Penguin.

Murphy, P. and Caborn, R. 1996: Regional government: an economic imperative. In Tindale, S. (ed.), *The state and the nations.* London: IPPR, 184–221.

Myrdal, G. 1957: *Economic theory and under-developed regions*. London: Duckworth.

Nairn, T. 1977: *The break-up of Britain: crisis and neo-nationalism*. London: Verso.

Nairn, T. 1988: *The enchanted glass: Britain and its monarchy*. London: Radius.

Nairn, T. 1997: *Faces of nationalism: Janus revisited*. London: Verso.

Nairn, T. 1998: Breaking up is hard to do. *Marxism Today* **Nov./Dec.**, 40–3.

Newton, S. and Porter, D. 1988: *Modernization frustrated: the politics of industrial decline in Britain since 1900*. London: Unwin Hyman.

Noble, M. and Smith, G. 1996: Two nations? Changing patterns of income and wealth in two contrasting areas. In Hills, J. (ed.), *New inequalities*. Cambridge: Cambridge University Press, 292–318.

Norris, P. 1997: Are we all Green now? Public opinion on environmentalism in Britain. *Government and Opposition* **32**, 320–39.

North, D. 1998: Rural industrialisation. In Ilbery, B. (ed.), *The geography of rural change*. Harlow: Longmans, 161–88.

Offe, C. 1984: *Contradictions of the welfare state*. London: Hutchinson.

Ohmae, K. 1990: *The borderless world: power and strategy in the interlinked economy*. New York: Harper Business.

O'Neill, M. 1997: *Green parties and political change in contemporary Europe*. Aldershot: Ashgate.

ONS (Office for National Statistics) 1996: *Family spending: a report on the Family Expenditure Survey*. London: HMSO.

ONS (Office for National Statistics) 1997: *Labour force survey: historical supplement*. London: HMSO.

ONS (Office for National Statistics) 1998: *Manufacturing, production and construction statistics: Summary Volume (PA 1002)*. London: HMSO.

OPCS, various dates: *General household survey*. London: HMSO.

O'Reilly, K. and Rose, D. 1998: Changing employment relations: plus ça change, plus c'est la même chose? *Work, Employment and Society* **12**, 713–33.

Osborne, D. and Gaebler, T. 1992: *Reinventing government*. Reading, MA: Addison-Wesley.

Osmond, J. 1985: Coping with a dual identity. In Osmond, J. (ed.), *The national question again: Welsh political identity in the 1980s*. Llandysul: Gomer, xix–xlvi.

Osmond, J. 1988: *The divided kingdom*. Llandysul: Gomer.

Owen, D. 1996: Labour force participation rates, self-employment and unemployment. In Karn, V. (ed.), *Employment, education and housing among the ethnic minority populations of Britain*. London: HMSO 29–66.

Owens, S. 1995: A response to Michael Breheny. *Transactions, Institute of British Geographers* **20**, 381–4.

Owens, S. 1997: 'Giants in the path': planning, sustainability and environmental values. *Town Planning Review* **68**, 293–304.

Pahl, R. 1984: *Divisions of labour*. Oxford: Blackwell.

Pahl, R. 1988: Some remarks on informal work, social polarisation and the social structure. *International Journal of Urban and Regional Research* **12**, 247–67.

Pahl, R. 1989: Is the emperor naked? Some comments on the adequacy of sociological theorising in urban and regional research. *International Journal of Urban and Regional Research* **13**, 4, 709–20.

Painter, J. 1995: *Politics, geography and 'political geography'*. London: Arnold.

Pakulski, J. and Waters, M. 1996: *The death of class*. London: Sage.

Paton, C. 1992: *Competition and planning in the NHS*. London: Chapman & Hall.

Pattie, C. and Johnston, R. 1995: 'It's not like that round here': region, economic evaluations and voting at the 1992 British General Election. *European Journal of Political Research* **28**, 1–32.

Pattie, C., Russell, A. and Johnston, R. 1991: Going Green in Britain? Votes for the Green Party and attitudes to Green issues in the late 1980s. *Journal of Rural Studies* **7**, 285–97.

Pattie, C., Denver, D., Mitchell, J. and Bochel, H. 1998: The 1997 Scottish referendum: an analysis of the results. *Scottish Affairs* **22**, 1–15.

Pattie, C., Johnston, R., Dorling, D., Rossiter, D., Tunstall, H. and MacAllister, I. 1997: New Labour, new geography? The electoral geography of the 1997 British General Election. *Area* **29**, 253–59.

Peach, C. 1996: Does Britain have ghettos? *Transactions, Institute of British Geographers* **21**, 216–35.

Peck, J. 1991: Letting the market decide (with public money): TECs and the future of labour market programmes. *Critical Social Policy* **31**, 4–17.

Peck, J. 1995: Moving and shaking: business elites, state localism and urban privatism. *Progress in Human Geography* **19**, 16–46.

Peck, J. 1996: *Work-place: the social regulation of labour markets.* New York: Guilford.

Peck, J. 1999: New Labour, new labourers? Making a new deal for the workless class. *Environment and Planning C* **17**, 345–72.

Peck, J. and Dicken, P. 1994: Tootal: internationalisation, corporate restructuring and 'hollowing out'. In Nilsson, J.-E., Dicken, P. and Peck, J. (eds), *The internationalisation process.* London: Paul Chapman, 109–29.

Peck, J. and Jones, M. 1995: Training and enterprise councils: Schumpeterian workfare state or what? *Environment and Planning A* **27**, 1361–96.

Peck, J. and Tickell, A. 1994: Searching for a new institutional fix: the *after*-Fordist crisis and global-local disorder. In Amin, A. (ed.), *Post-Fordism: a reader.* Oxford: Blackwell, 280–316.

Peck, J. and Tickell, A. 1995a: The social regulation of uneven development: 'regulatory deficit', England's South East, and the collapse of Thatcherism. *Environment and Planning A* **27**, 15–40.

Peck, J. and Tickell, A. 1995b: Business goes local: dissecting the business agenda in Manchester. *International Journal of Urban and Regional Research* **19**, 55–78.

Penrose, J. 1993: Reification in the name of change: the impact of nationalism on social constructions of nation, people and place in Scotland and the UK. In Jackson, P. and Penrose, J. (eds), *Constructions of race, place and nation.* London: UCL Press, 27–49.

Phelps, N., Lovering, J. and Morgan, K. 1998: Tying the firm to the region or tying the region to the firm? Early observations on the case of LG in South Wales. *European Urban and Regional Studies* **5**, 119–37.

Phillimore, P., Beattie, A. and Townsend, P.

1994: Widening inequality of health in northern England, 1981–91. *British Medical Journal* **308**, 1125–8.

Phillips, D. 1998: Black minority ethnic concentration, segregation and dispersal in Britain. *Urban Studies* **35**, 1681–702.

Philo, C. (ed.) 1995: *Off the map: the social geography of poverty in the United Kingdom.* London: Child Poverty Action Group.

Piachaud, D. and Webb, J. 1996: *The price of food: missing out on mass consumption.* LSE: STICERD.

Pierson, C. 1998: Contemporary challenges to welfare state development. *Political Studies* **XLVI**, 777–94.

Pike, A. 1998: Making performance plants from branch plants? In situ restructuring in the automobile industry in the UK. *Environment and Planning A* **30**, 881–900.

Pimlott, B. 1989: Is the post-war consensus a myth? *Contemporary Record* **2**(6), 12–14.

Pinch, S. 1993: Social polarisation: a comparison of evidence from Britain and the USA. *Environment and Planning A* **26**, 774–95.

Pinch, S. 1994: Labour flexibility and the changing welfare state: is there a post-Fordist model? In Burrows, R. and Loader, B. (eds), *Towards a postfordist welfare state.* London: Routledge, 203–22.

Pinch, S. 1997: *Worlds of welfare.* London: Routledge.

Pinch, S. and Henry, N. 1997: Discursive aspects of technological innovation: the case of the British motor sport industry. *Environment and Planning A.*

Pinch. S. and Henry, N. 1999: Paul Krugman's geographical economics, industrial clustering and the British motor sport industry. *Regional Studies*, forthcoming.

Pinch, S., Henry, N. and Turner, D. 1997: *In pole position: explaining the supremacy of Britain's 'Motor Sport Valley'.* Working Paper 62, School of Geography, Birmingham: University of Birmingham.

Pinch, S., Mason, C. and Witt, S. 1989: Labour flexibility and industrial restructuring in the UK 'sunbelt'. *Transactions, Institute of British Geographers* **14**, 418–34.

Pinch, S., Mason, C. and Witt, S. 1991: Flexible employment strategies in British industry: evidence from the UK 'sunbelt'. *Regional Studies* **25**(3), 207–18.

Plender, J. 1997: *The stakeholder solution*. London: Cape.

Pocock, D. C. D. 1979: The novelists' image of the North. *Transactions, Institute of British Geographers* **NS4**, 62–76.

Pollert, A. 1988: The 'flexible firm': fixation or fact? *Work, Employment and Society* **2**, 281–316.

Pollert, A. (ed.) 1991: *Farewell to flexibility?* Oxford: Blackwell.

Popay, J., Williams, G., Thomas, C. and Gatrell, A. 1998: Theorising inequalities in health: the place of lay knowledge. *Sociology of Health and Illness* **20**, 619–44.

Potter, J. 1995: Branch plant economies and flexible specialisation: evidence from Devon and Cornwall. *Tijdschrift voor Economische en Social Geografie* **86**, 162–76.

Power, A. and Tunstall, R. 1997: *Dangerous disorder: riots and violent disturbances in thirteen areas of Britain*. York: Joseph Rowntree Foundation.

Pryke, M. 1994: Looking back on the space of a boom: (re)developing spatial matrices in the City of London. *Environment and Planning A* **26**, 235–64.

Putnam, R. 1993: *Making democracy work: civic traditions in modern Italy*. Princeton, NJ: Princeton University Press.

RCEP (Royal Commission on Environmental Pollution) 1994: *Eighteenth report: transport and the environment*. London: HMSO, Cm. 2674.

RCEP 1997: *Twentieth report: transport and the environment – developments since 1994*. London: HMSO, Cm. 3752.

Redwood, J. 1988: *Popular capitalism*. London: Routledge.

Reich, R. 1992: *The work of nations*. New York: Vintage.

Rhodes, R. 1997: *Understanding governance*. Buckingham: Open University Press.

Rhys, G. 1995: The transformation of the motor industry in the UK. In Turner, R. (ed.), *The British economy in transition*. London: Routledge, 142–68.

Richardson, R. and Marshall, J. N. 1996: The growth of telephone call centres in peripheral areas of Britain. *Area* **28**, 308–17.

Robinson, F. and Storey, D. 1981: Employment change in manufacturing industry in Cleveland, 1965–76. *Regional Studies* **15**, 161–72.

Robinson, F. and Shaw, K. 1994: *Who runs the North?* Newcastle-upon-Tyne: UNISON Northern Region.

Robson, B. 1988: *Those inner cities: reconciling the social and economic aims of urban policy*. Oxford: Oxford University Press.

Robson, B. 1994: No city, no civilisation. *Transactions, Institute of British Geographers* **19**, 131–41.

Robson, B. 1998: Lowering the flag. *Town and Country Planning* **67**(4), 100–1.

Robson, B., Parkinson, M., Bradford, M. *et al.* 1994: *Assessing the impact of urban policy*. London: HMSO.

Rogerson, R. 1997: *Quality of life in Britain*. Glasgow: University of Strathclyde.

Rootes, C. 1991: The greening of British politics? *International Journal of Urban and Regional Research* **15**, 287–97.

Rootes, C. 1992: The new politics and the new social movements. *European Journal of Political Research* **22**, 171–91.

Rose, G. 1989: Locality studies and waged labour: an historical critique. *Transactions, Institute of British Geographers* **NS14**, 317–28.

Rowntree Commission 1995: *Report of the inquiry into income and wealth*. York: Joseph Rowntree Foundation.

Rowthorn, R. 1986: De-industrialisation in Britain. In Martin, R. and Rowthorn, R. (eds), *The geography of de-industrialisation*. Basingstoke: Macmillan, 1–30.

RSA 1998: *Redefining work*. London: Royal Society of Arts.

Rubery, J. 1989: Labour market flexibility in Britain. In Green, F. (ed.), *The restructuring of the UK economy*. Hemel Hempstead: Harvester, 155–76.

Runciman, W.G. 1990: How many classes are there in contemporary British society? *Sociology* **24**, 377–96.

Sadler, D. 1992: *The global region: production, state policies and uneven development*. Oxford: Pergamon.

Sadler, D. 1993: Place-marketing, competitive places and the construction of hegemony in Britain in the 1980s. In Philo, C. and Kearns, G. (eds), *Selling places: the city as cultural capital*. Oxford: Pergamon, 175–92.

Sadler, D. 1997: The role of supply chain management strategies in the 'Europeanization' of the automobile production system. In Lee, R. and Wills, J. (eds), *Geographies of economies*. London: Arnold, 311–20.

Samuel, R. 1989: Introduction: exciting to be English. In Samuel, R. (ed.), *Patriotism: the making and unmaking of British national identity*. London: Routledge, x–lxv.

Samuel, R. 1998: *Island stories: unravelling Britain*. London: Verso.

Sancton, A. 1976: British socialist theory of the division of power by area. *Political Studies* **XXIV**, 158–70.

Sarlvik, B. and Crewe, I. 1983: *Decade of dealignment*. Cambridge: Cambridge University Press.

Saunders, P. 1990: *A nation of homeowners*. London: Unwin Hyman.

Savage, M. 1987: Understanding political alignments in Britain: do localities matter? *Political Geography Quarterly* **6**, 53–76.

Savage, M., Barlow, J., Dickens, P. and Fielding, A. 1992: *Property, bureaucracy and culture: middle-class formation in contemporary Britain*. London: Routledge.

Savage, M., Dickens, P. and Fielding, A. 1988: Some social and political implications of the contemporary fragmentation of the 'service class' in Britain. *International Journal of Urban and Regional Research* **12**, 455–75.

Sayer, R. A. 1995: *Radical political economy: a critique*. Oxford: Blackwell.

Schmitt, J. and Wadsworth, J. 1994: The rise in economic inactivity. In Glyn, A. and Miliband, D. (eds), *Paying for inequality: the economic cost of social injustice*. London: Rivers Oram Press, 114–29.

Scott, A. 1988a: Flexible production systems and regional development: the rise of new industrial spaces in North America and Western Europe. *International Journal of Urban and Regional Research* **12**, 171–86.

Scott, A. 1988b: *New industrial spaces*. London: Pion.

Scott, J. 1991: *Who rules Britain?* Cambridge: Polity.

SEU (Social Exclusion Unit) 1998: *Bringing Britain together: a national strategy for neighbourhood renewal*. London: HMSO.

Shaw, K. 1993: The political economy of urban regeneration in the north east: the rise of the growth coalition or local corporatism revisited? *Regional Studies* **27**, 251–9.

Shaw, K. and Robinson, F. 1998: Learning from experience? Reflections on two decades of British urban policy. *Town Planning Review* **69**, 49–63.

Shields, R. 1991: *Places on the margin: alternative geographies of modernity*. London: Routledge.

Short, J. 1981: *Public expenditure and taxation in the UK regions*. Farnborough: Gower.

Short, J., Fleming, S. and Witt, S. 1986: *Housebuilding, planning and community action*. London: Routledge & Kegan Paul.

Sibley, D. 1995: *Geographies of exclusion*. London: Routledge.

Silver, H. 1994: Social exclusion and social solidarity: three paradigms. *International Labour Review* **133**, 531–78.

Slater, S., Marvin, S. and Newson, M. 1994: Land use planning and the water sector. *Town Planning Review* **65**, 375–97.

Smith, G., Smith, T. and Wright, G. 1997: Poverty and schooling: choice, diversity or division? In Walker, A. and Walker, C. (eds), *Britain divided: the growth of social exclusion in the 1990s*. London: Child Poverty Action Group, 123–39.

Smith, N. 1987: Dangers of the empirical turn: comments on the CURS initiative. *Antipode* **19**, 59–68.

Smith, S. 1989a: The politics of 'race' and a new segregationism. In Mohan, J. (ed.), *The political geography of contemporary Britain*. Basingstoke: Macmillan, 151–71.

Smith, S. 1989b: Society, space and citizenship: a human geography for the 'new times'. *Transactions, Institute of British Geographers* **NS14**, 144–56.

Smith, S. 1993: Immigration and nation-building in Canada and the UK. In Jackson, P. and Penrose, J. (eds), *Constructions of race, place and nation*. London: UCL Press, 50–77.

Southall, H. 1988: The origins of the depressed areas: unemployment, growth and regional structure in Britain before 1914. *Economic History Review* **XLI**, 236–58.

Southall, H., Dorling, D., Ell, P. and Gregory, I. 1998: Mapping and analysing 200 years of the Census. *Statistical News* **121**, 14–19.

Stamp, L. D. and Beaver, S. 1971: *The British*

Isles: a geographic and economic survey. London: Longman.

Steed, M. 1986: The core–periphery dimension in British politics. *Political Geography Quarterly* **5**, S91–S103.

Stoker, G. 1989: Creating a local government for a post-Fordist society: the Thatcherite project. In Stoker, G. and Stewart, J. (eds), *The future of local government*. London: Macmillan.

Stoker, G. 1991: *The politics of local government*. Basingstoke: Macmillan.

Stone, I. 1995: Symbolism and substance in the modernisation of a traditional industrial economy: the case of Wearside. In Turner, R. (ed.), *The British economy in transition*. London: Routledge, 169–98.

Stone, I., and Peck, F. 1996: The foreign-owned manufacturing sector in UK peripheral regions 1978–93: restructuring and comparative performance. *Regional Studies* **30**, 55–68.

Storper, M. 1995: The resurgence of regional economies. *European Urban and Regional Studies* **2**, 191–221.

Storper, M. and Scott, A. J. 1989: The geographical foundations and social regulation of territorial production complexes. In Dear, M. and Wolch, J. (eds), *The power of geography*. London: Allen & Unwin, 21–40.

Sunley, P. 1990: Striking parallels: a comparison of the geographies of the 1926 and 1984–85 coalmining disputes. *Environment and Planning D: Society and Space* **8**, 35–52.

Supple, B. 1994: Fear of failing: economic history and the decline of Britain. *Economic History Review* **XLVII**, 441–58.

Szreter, S. 1997: British economic decline and human resources. In Clarke, P. and Trebilcock, C. (eds), *Understanding decline*. Cambridge: Cambridge University Press, 73–102.

Taylor, B. 1997: Green in word … In Jowell, R. *et al.* (eds), *British social attitudes: the end of Conservative values. 14th report*. Aldershot: Ashgate, 113–36.

Taylor, I. 1996: Fear of crime, urban fortunes and suburban social movements. *Sociology* **30**, 317–37.

Taylor, I., Evans, K. and Fraser, P. 1996: *A tale of two cities: global change, local feeling and everyday life in the North of England*. London: Routledge.

Taylor, J. and Wren, C. 1997: UK regional policy: an evaluation. *Regional Studies* **31**, 835–48.

Taylor, M. 1999: The small firm as a temporary coalition. *Entrepreneurship and regional development* **11**, 1–19.

Taylor, M. and Thrift, N. 1983: The role of finance in the evolution and functioning of industrial systems. In Hamilton, F. and Linge, G. (eds), *Spatial analysis, industry and the industrial environment, Vol. III. Regional economies and industrial systems*. Chichester: Wiley, 359–85.

Taylor, P. 1989: Britain's changing role in the world-economy. In Mohan, J. (ed.), *The political geography of contemporary Britain*. Basingstoke: Macmillan, 8–14.

Taylor, P. 1991a: The English and their Englishness: 'a curiously mysterious, elusive and little understood people'. *Scottish Geographical Magazine* **107**, 146–61.

Taylor, P. 1991b: The changing political geography. In Johnston, R. and Gardiner, V. (eds), *The changing geography of the United Kingdom*. London: Routledge.

Taylor, P. 1993: The meaning of the North: England's foreign country within? *Political Geography Quarterly* **12**, 136–55.

Taylor, S. and Taylor, B. 1996: Transport and the environment. In Halpern, D., Woods, S., White, S. and Cameron, G. (eds), *Options for Britain*. Aldershot: Dartmouth, 235–72.

Taylor-Gooby, P. 1997: In defence of second-best theory: state, class and capital in social policy. *Journal of Social Policy* **26**, 171–92.

Thornley, A. 1993: *Urban planning under Thatcherism*. London: Routledge.

Tickell, A. 1998: A tax on success? Privatization, employment and the 'windfall tax'. *Area* **30**, 83–90.

Timmins, N. 1995: *The five giants: a biography of the welfare state*. London: HarperCollins.

Tomaney, J. 1999: New Labour and the English question. *Political Quarterly* **70**, 75–82.

Tomlinson, B. 1992: *Report of the inquiry into London's health service, medical education and research*. London: HMSO.

Townsend, A. 1983: *The impact of recession*. Beckenham: Croom Helm.

Townsend, A. 1993: The urban–rural growth

cycle in the Thatcher years. *Transactions, Institute of British Geographers* **18**, 207–21.

Townsend, A. 1997: *Making a living in Europe.* London: Routledge.

Townsend, A. and Peck, F. 1984: Contrasting experiences of recession and spatial restructuring: British Shipbuilders, Plessey and Metal Box. *Regional Studies* **18**, 319–38.

Townsend, A. and Peck, F. 1985: The geography of mass redundancy in named corporations. In Pacione, M. (ed.), *Progress in industrial geography.* London: Croom Helm, 174–218.

Townsend, P. 1995: Persuasion and conformity: an assessment of the Borrie Report on Social Justice. *New Left Review* **213**, 137–50.

Trade and Industry Select Committee 1995: *Fourth Report, session 1994–95.* London: HMSO.

Turner, R. 1995a: After coal. In Turner, R. (ed.), *The British economy in transition.* London: Routledge, 23–45.

Turner, R. (ed.) 1995b: *The British economy in transition: from the old to the new.* London: Routledge.

Turok, I. 1992: Property-led urban regeneration: panacea or placebo? *Environment and Planning A* **24**, 361–79.

Turok, I. 1993: Inward investment and local linkages: how deeply embedded is 'Silicon Glen'? *Regional Studies* **27**, 401–17.

Turok, I. 1999: Urban labour markets: the causes and consequences of change. *Urban Studies.*

Turok, I. and Webster, D. 1998: The new deal: jeopardised by the geography of unemployment? *Local Economy* **February**, 309–28.

Twigg, J. 1992: *Carers: research and practice.* London: HMSO.

Urry, J. 1987: Some social and spatial aspects of services. *Environment and Planning D* **5**, 5–26.

Urry, J. 1995: A middle-class countryside? In Butler, T. and Savage, M. (eds), *Social change and the middle classes.* London: UCL Press, 205–19.

Walker, R. 1992: The brave new world of the service economy. In Sayer, A. and Walker, R. (eds), *The new social economy: reworking the division of labour.* Oxford: Blackwell, 56–107.

Walker, R. and Huby, M. 1989: Social security spending in the UK: bridging the North–South economic divide. *Environment and Planning C* **7**, 321–40.

Ward, N. 1996: Pesticides, pollution and sustainability. In Allanson, P. and Whitby, M. (eds), *The rural economy and the British countryside.* London: Routledge, 40–61.

Warde, A. 1985: Spatial change, politics and the division of labour. In Gregory, D. and Urry, J. (eds), *Social relations and spatial structures.* London: Macmillan, 190–212.

Warde, A. 1990: Production, consumption and social change: reservations concerning Peter Saunders' sociology of consumption. *International Journal of Urban and Regional Research* **14**(2), 228–48.

Weakliem, D. and Heath, A. 1995: Regional differences in class alignment: a comment on Johnston and Pattie. *Political Geography* **14**, 643–51.

Webster, C. 1990: Conflict and consensus: explaining the British health service. *Twentieth Century British History* **1**, 115–51.

Westergaard, J. 1995: *Who gets what? The hardening of class inequality in the late twentieth century.* Cambridge: Polity.

White, S. 1998: Interpreting the 'Third Way': not one route, but many. *http: //www.net-nexus.org/library/papers/white22.htm.*

Whitelegg, J. 1997: *Critical mass: transport, environment and society in the twenty-first century.* London: Pluto.

Whyley, E., McCormick, J. and Kempson, E. 1998: *Paying for peace of mind: access to home contents insurance for low-income households.* Bristol: Policy Press.

Wiener, M. 1981: *English culture and the decline of the industrial spirit, 1850–1980.* Cambridge: Cambridge University Press.

Wilcox, S. 1996: *Housing Review 1996/7.* York: Joseph Rowntree Foundation.

Wilkinson, R. 1996: *Unhealthy societies: the afflictions of inequality.* London: Routledge.

Wilkinson, R. 1997: Health inequalities: relative or absolute material standards? *British Medical Journal* **314**, 591–5.

Wilks-Heeg, S. 1996: Urban Experiments Limited Revisited: Urban policy comes full circle? *Urban Studies* **33**, 1263–79.

Williams, C. 1997: *Consumer services and economic development*. London: Routledge.

Williams, C. and Windebank, J. 1995: Social polarisation of households in contemporary Britain: a 'whole economy' perspective. *Regional Studies* **29**, 8, 723–28.

Williams, R. 1983: *Towards 2000: resources for a journey of hope*. London: Penguin.

Winckler, V. 1990: Restructuring the Civil Service: reorganisation and relocation. *International Journal of Urban and Regional Research* **14**, 135–57.

Winter, M. 1996: *Rural politics: policies for agriculture, forestry and the environment*. London: Routledge.

Winter, M. and Gaskell, P. 1998: The *Agenda 2000* debate and CAP reform in Great Britain: is the environment being sidelined? *Land Use Policy* **15**, 217–31.

Wolch, J. 1990: *The shadow state*. New York: Foundation Center.

Wood, P. 1996: Business services, the management of change and regional development in the UK. *Transactions, Institute of British Geographers* **21**, 649–65.

Wood, P. A. 1991: Flexible accumulation and the rise of business services. *Transactions, Institute of British Geographers* **16**, 160–72.

Woodward, R. 1995: Approaches towards the study of social polarisation in the UK. *Progress in Human Geography* **19**(1), 75–89.

Wright, A. 1993: *Citizens and subjects*. London: Routledge.

Wright, E. 1985: *Classes*. London: Verso.

Wright, P. 1985: *On living in an old country: the national past in contemporary Britain*. London: Verso.

Young, I. M. 1990: *Justice and the politics of difference*. Princeton, NJ: Princeton University Press.

Young, S. 1998: The United Kingdom: a mirage beyond the participation hurdle? In Lafferty, W. and Eckerberg, K. (eds), *From the Earth Summit to Local Agenda 21*. London: Earthscan, 179–203.

INDEX

post-Fordism, 16, 63, 149, 170, 179, 189, 193
'post-materialist' politics, 166
postfordism, 15, 170, 179
postmodern social theory, 119
Potter, J., 57
poverty, 1, 131, 217
  distribution of child, 4
  ethnic minorities, 129
  regional variations, 102
  rural, 127
  spatial concentration, 53, 127
Powell, Enoch, 31–2
private education, 138, 139
Private Finance Initiative (PFI), 39, 109
private health care, 138, 139
private health insurance, 138, 139, 145
privatisation, 18, 36, 37, 38, 39, 69, 86, 214, 219
  coal industry, 53
  housing, 138, 139–40
  pensions, 26
  public space, 133
  public transport, 208, 209
  Trustee Savings Bank, 114
  utilities, 43–4, 54–5
producer services, 73, 74–7
production
  agricultural productivism, 210
  changes in the organisation of, 14–17, 61, 85, 86, 108, 170, 189, 221–2
  Conservative restructuring, 37
  deindustrialisation and reindustrialisation, 46–63
  globalisation of, 26–7
  rationalisation, 185
  social relations of, 10–12
  sunbelts and new industrial spaces, 64–83
  see also industry
productivity, 37, 61, 62
property
  dependence of City Challenge on, 191–2
  local taxation based on values, 171, 172
  meshing of markets with the economic system, 111
  sector growth, 62, 70
public and private
  boundaries redrawn in industry, 52–5
  mix for welfare, 138–42
  partnerships in local governance, 176
'public choice' theory, 169
public disorder
  Notting Hill riots, 31
  'symbolic locations', 133
  urban 'riots', 133, 153
public expenditure
  cutbacks and restrictions, 36, 37, 80, 185
  hidden regional policies, 108–10
  inequitable distribution, 105–6

  need for 'regional audit', 113
public sector
  boundaries with private sector redrawn, 52–5
  commercialisation, 175
  contribution to employment growth in Wales, 197
  de facto incomes policy, 85–6
  financing of Urban Development Corporations, 188
  geographies of, 78–80
  re-shaped work practices, 90–91, 219
public transport, 165, 206, 207, 208–9
Putnam, R., 42–3, 134, 150, 195

'quality of life', 4–6, 214, 215
quangos, 39–40, 44, 175

race
  and class, 115, 120
  and national identity, 31–2
  and public disorder, 133
  and social segregation, 128–30
  and the underclass, 122
racism, 128, 129, 220
rail privatisation, 208–9
rates, 171
Ravenscraig steel plant, 52
Rawstron, E., 101
RCEP (Royal Commission on Environmental Pollution), 209
recession
  1970s, 2, 129, 185
  1980s, 50, 52, 53, 105, 109, 129
  1990s, 7, 39, 83, 98, 142
regimes (modes) of accumulation, 15, 16, 34, 61, 92, 170
regional aggregates working parties (RAWPs), 204
'regional associationalism', 197
'regional competitiveness', 197–8
regional development
  contribution of services, 74
  conventional accounts challenged by 70s recession, 2
  implications of high-technology industry, 68
  implications of regional policies, 184–5
  influence of financial system, 110–11
  limits to strategy of public ownership, 52–3
  new industrial spaces, 80–83
  regional government and, 77
  role of the public sector, 79
  see also uneven development
Regional Development Agencies (RDAs), 178, 191, 193–5, 220
regional development banks (RDBs), 114
regional economies, 14, 57, 61
  resurgence, 81, 197, 221

Milton Keynes UK
Ingram Content Group UK Ltd.
UKHW051849071024
449327UK00025B/1896